ELECTRONIC STRUCTURE METHODS FOR COMPLEX MATERIALS

Electronic Structure Methods for Complex Materials

The orthogonalized linear combination
of atomic orbitals

WAI-YIM CHING and PAUL RULIS
University of Missouri-Kansas City, USA

OXFORD
UNIVERSITY PRESS

UNIVERSITY PRESS

Great Clarendon Street, Oxford, OX2 6DP,
United Kingdom

Oxford University Press is a department of the University of Oxford.
If furthers the University's objective of excellence in research, scholarship,
and education by publishing worldwide. Oxford is a registered trade mark of
Oxford University Press in the UK and in certain other countries

First Edition published in 2012
Impression: 1

British Library Cataloguing in Publication Data
Data available
Library of Congress Cataloging in Publication Data
Data available

ISBN 978–0–19–957580–0

Printed and bound by
CPI Group (UK) Ltd, Croydon, CR0 4YY

Wai-Yim Ching expresses his deep appreciation to his wife Mon Yin and daughter Kunyu for their encouragement and understanding. This book is dedicated to them.
This book is also dedicated to Etsuko. Her own determination has been a sustaining inspiration.

Preface

The aim of this book is to broadly introduce a robust method for electronic structure calculation that is particularly well-suited for application to large and complex systems. This method, called the orthogonalized linear combination of atomic orbitals (OLCAO) method, is an all-electron density functional theory (DFT)-based method that uses local atomic orbitals in the basis expansion. We endeavor to illustrate its utility by presenting a large number of examples of how it has been applied to many complex systems and different types of materials. These results are based on the authors' work, culled from research conducted over the last 35 years up to and including current projects. We have specifically avoided delving into underlying issues associated with DFT, the merits of its different implementations, or the subtleties of the various types of exchange-correlation functionals that are often used, as there are already many excellent resources for these topics. Further, discussions about other DFT-based electronic structure methods are minimized, except in cases where they are related to the OLCAO method or are used in conjunction with it. Our motivation for writing this book is to formally present a detailed account of the OLCAO method that brings attention to what it has been used for, what it is presently capable of, and what research and development could be done with it in the future. It is expected that researchers in condensed matter physics, materials theory, nanotechnology, engineering, and computational biology will be able make use of this method in their own work. This book is ideal for graduate students in physics, chemistry, materials science, engineering, or any other field where the application of DFT based electronic structure methods is seen as a useful tool for theoretical and simulation studies. This book will also be a useful reference to those researchers that are interested in fundamental materials properties because of the extensive list of materials systems and the many areas of application that were all obtained through one approach.

The content of the book can be roughly divided onto three parts as seen from a study of the listed chapters in the Table of Contents. Chapters 1–4 comprise the first part. Chapters 1 and 2 provide context through a brief historical account and description of the background of the method while Chapters 3 and 4 detail the technical aspects of the method with the focus being on the special advantages that make it ideal for complex systems. Chapters 5–12 comprise the second part of the book and contain the bulk of the research material. Results of the application of the OLCAO method to a variety of materials systems ranging from simple semiconductors to large complex biomolecules are given in Chapters 5 through 10. The classification of these materials systems is loosely defined and some topics that appear in one chapter could be argued to belong in another. However, care has been taken to make the presentation of the diverse collection of materials systems as clear and coherent as possible. Chapter 11 focuses on the application of the OLCAO method to the calculation of core level spectra from a variety of different materials systems. The last chapter, Chapter 12, discusses potential further extensions of the OLCAO method with some of the developments already in progress. The four Appendices comprise the third part of the book. These pages deal with practical aspects of performing the calculations and also discuss the computational and programming aspects of the OLCAO method. In particular, Appendix C can be regarded as an introductory users guide to the program. It is the authors' intention to release the OLCAO code to the general public under an open source license after the publication of this book.

Many of the references that are cited are, understandably, from the authors' own work, with some of them being unpublished results. We may have inadvertently and unintentionally omitted the work of other colleagues who were directly or indirectly involved in the development of this method or have been working on similar methods. We sincerely apologize for such omissions.

Without contributions at various levels from the authors' past and present collaborators, the publication of this book would not be possible. They have contributed immensely toward the publication of many of the results that were obtained using the OLCAO method. WYC is particularly indebted to his Ph. D. advisor, the late Professor Joseph Callaway (Louisiana State University), and his mentors, Professors Chun C. Lin and David L. Huber (University of Wisconsin-Madison), for introducing him to this exciting and rewarding field of research in the early stage of his career. Additionally, WYC would like to recognize the efforts of many of his past and present collaborators especially Drs. Yong-Nian Xu, Zhong-Quan Gu, Ming-Zhu Huang, Y. P. Li, Guan Lin Zhao, Xue-Fu Zhong, D. J. Lam, Kai W. Wong, Y. C. Jean, Jun Chen, A. N. Caruso, R. F. Rajter, and more. Several past and present students also made significant contributions including: Lee Wei Song, A. R. Murray, F. Zandiehnadem, Dong Li, F. Gan, Hongyi Yao, Shang-Di Mo, Brian K. Brikeen, Lizhi Ouyang, Paul Rulis, Yu Chen, Hongzhi Yao, Lei Liang, Altaf Hussain, Yuxiang Mo, Sitaram Aryal, Liaoyuan Wang, Jay Eifler, and C. Dharmawardhana. PR would especially like to thank his Ph. D. advisor, mentor, friend, and co-author in this endeavor, WYC, for his infectious passion.

We are particularly thankful to the many collaborators that we have had the pleasure of working with including: Professors Bruce N. Harmon, S. S. Jaswal, Isao Tanaka, Y. Ikuhara, T. Mizoguchi, T. Sekine, M. Kohyama, M. Matsunaga, M. Yoshiya, F. Oba, K. Ogasawara, L. Randaccio, W. Wong-Ng, E. Z. Kurmaev, A. Moewes, Yet-Ming Chiang, Ralf Riedel, C. Barry Carter, S. J. Pennycook, Sashi Satpathy, Anil Misra, Roger H. French, Adrian Parsegian, and Rudolf Podgornik. WYC expresses his gratitude to Professor Manfred Rühle (Max Planck Institute for Metal Research) who inspired him to work on ceramic microstructures when he spent a year in his laboratory. We have profoundly benefited from their advice, wisdom, stimulating interaction, and friendship.

Preface

The authors express their gratitude for special technical help provided by Lei Liang and Jay Eifler in the preparation of this book. We also acknowledge the vital contributions of Professor Lizhi Ouyang in the development and application of the OLCAO method to biomolecular systems. Finally, we would like to recognize the exceptionally constructive role that the National Energy Research Scientific Computing Center played in much of the recent research work presented in the book.

Wai-Yim Ching and Paul Rulis
Kansas City, Missouri, 2012

Contents

8 Application to Non-Crystalline Solids and Liquids

9 Application to Impurities, Defects, and Surfaces

Contents

Electronic Structure Methods in Materials Theory

1

1.1 Introduction

Within the last two decades, there has been tremendously deep and broad progress in the fundamental study of materials. Many of the key advances were either directed or strongly influenced by computational studies of electronic structure. The ability of computational studies to play this role is primarily due to the rise of various powerful computational methods based on sophisticated algorithms and increasingly accurate theoretical models. It is because these methods have been fueled with ever increasing computational power that they have come to dominate the landscape in many areas of materials research. These methods originated from modern condensed matter physics and quantum chemistry, and presently the heart of these methods is based on density functional theory (Hohenberg and Kohn, 1964, Kohn and Sham, 1965). The density functional approach has been successfully applied to very diverse areas of solid state physics, chemistry, Earth science, and biology, all of which are closely connected to materials science and engineering. More sophisticated theories and methodologies have enabled researchers to deal with materials of increasing complexity and to explore new phenomenon and functionalities in different types of materials ranging from polymers and nano-particles to bio-ceramics and DNA strands. Computational studies can predict their behaviors in extreme thermodynamic conditions and in many cases have actually led to the discovery of new materials with new physical properties (Oganov, 2010).

Concurrently with the theoretical and methodological developments, advances in computer science and hardware, software, and networking technologies have played a pivotal role. Examples of such advances include the increase in operating frequency for CPUs and memory modules, and the burgeoning amount of data storage available on hard disks. Other critical advances center on the communication hierarchy between components (e.g. CPUs, cache, memory modules, network nodes, etc.) such that latency is reduced while throughput is increased. Finally, post-processing and visualization capabilities have enabled data mining and exploration to levels of detail that were previously unthinkable. Thus, computational projects that, two decades ago, would have required years of computer time can now be accomplished in a few hours or minutes on a laptop computer. There are now many massively parallel petascale supercomputers installed at supercomputer centers all over the world conducting computational research in materials science, weather prediction, climatology, geosciences, cosmology and astrophysics, fluid dynamics, energy science, fundamentals of life science, and applications in medicine. With exascale computational facilities on horizon (Geist, 2010), this is indeed an exciting time for computational research.

1.2 One electron methods

The electronic structures of materials are obtained from the quantum mechanical calculation of systems containing nuclei and electrons. In essence, one must find the solution of the Schrödinger equation for a collection of interacting atoms. With the exception of small molecules or simple crystals, the electronic structure studies of most materials must rely on approximations. These approximations started from the Thomas Fermi model, and later included the Hartree–Fock and Born–Oppenheimer approximations. Density functional theory in its local density approximation (LDA) (Kohn and Sham, 1965) shifted the focus from the wave function to the charge density and put one electron methods on a firm theoretical ground. Although extremely useful, DFT left the exchange-correlation functional that embodies the interacting characteristics of the electrons undefined. Development of an exchange-correlation functional that is rigorously accurate (or at least accurate enough within computational feasibility) represents the main point of business for many present day theoreticians. Thus, in recent years DFT has advanced to include different levels of sophistication in the exchange-correlation functional that can be pictured as rungs on a ladder. At the lowest point is the LDA followed by the generalized gradient approximation (GGA), and then meta-GGA, hybrid functionals, and so on (Perdew and Schmidt, 2001).

The various implementations of DFT (computational methods) can be divided into groups according to a few rough criteria. For crystals and infinite solids, periodic boundary conditions are necessary whereas for molecules and clusters they are not. The basis expansion of the Bloch functions for periodic systems can be either numerical, localized orbital functions, or plane wave functions. There are also many successful methods that use a mixture of different types of basis functions. A number of approaches exist and are in common use for constructing the potentials in the Hamiltonian with a broad division between all electron based methods that include core states and pseudopotential based methods with no cores states. Broad material differences, such as metallic or insulating, magnetic or non-magnetic, openly spaced or compact structures, high symmetry or totally disordered, and so on can affect the choice of which particular method to use, the manner in which the solution is obtained, and the way the results are presented (either in real space or in reciprocal space). For example, calculations of large molecules are always done in real space with solutions in the form of energy eigenvalues, while the traditional way of representing the electronic structures of semiconductors and metals is to present their band structures along symmetry axes of the Brillouin zone in reciprocal space. Often, within each method different expressions for the exchange-correlation functional are available. This can act as a further refinement to match a method to a system under study. With all of this complexity, it is not surprising that no single method can treat all problems equally well in terms of functionality, accuracy, efficiency, and ease of use. Each method overlaps with other methods in some areas while maintaining a distinct advantage (or disadvantage) in another area. Hence the boundaries between different types of methods are usually fuzzy and the choice of which particular method and package to use depends on the materials and the problem at hand.

1.3 Quantum chemical approaches and solid state methods

It is relatively easy to divide the computational methods according to whether they originate from either a quantum chemical approach or from a solid state physics approach. They have different historical perspectives and tend to focus their applicability to different systems. Quantum chemical methods usually use localized orbitals and are applied to molecules and clusters with open surfaces. The most popular quantum chemical packages include: GAUSSIAN-09 (Frisch et al., 2009), GAMESS (Schmidt et al., 1993), ADF (Fonseca Guerra et al., 1998, Te Velde et al., 2001), DMol^3 (Delley et al., 1995), DV-Xα (Adachi et al., 2006, Ellis and Painter, 1970), NW-Chem (Valiev et al., 2010), Multiple Scattering (Rehr et al., 2009), and others. It is not unusual though to use quantum chemical methods to study the properties of solids.

The solid state methods are mostly preferred by condensed matter physicists and are applied to crystals and models of materials systems with periodic boundary conditions. Reciprocal space is the preferred arena for results presentation and the plane wave based methods are the dominant choice for the basis expansion. The most popular packages include: LAPW (Blaha et al., 1990, Schwarz and Blaha, 2003) (Wien2k, etc), plane wave based pseudo potential methods (VASP (Kresse and Furthmüller, 1996b, Kresse and Furthmüller, 1996a, Kresse and Hafner, 1993, Kresse and Hafner, 1994), CASTEP (Clark et al., 2005, Segall et al., 2002), PWSCF (Paolo et al., 2009), etc), LMTO (Methfessel, 1988, Methfessel et al., 1989, Methfessel et al., 2000, Skriver, 1984, Wills et al., 2000), Crystal (Dovesi et al., 2005), LCAO (Eschrig, 1989), Layer-KKR (Maclaren et al., 1990), tight-binding (Horsfield and Bratkovsky, 2000), Siesta (Artacho et al., 2008, Soler et al., 2002), LCPAO (Han et al., 2006, Ozaki and Terakura, 2001) Quantum Simulator (Ishibashi et al., 2007), and many others. It should be stressed that the distinctions between the above two classes are not always clear and some of these methods and packages can be applied to both areas.

1.4 The OLCAO method

The main objective of this book is to describe the orthogonalized linear combination of the atomic orbitals (OLCAO) method which is an extension of the LCAO method. Although it is classified as a solid state method, it is more appropriate to regard it as a hybrid between quantum chemical methods and solid state methods. It shares many of the strengths and limitations of both types of methods. It is an all electron method using local orbitals (atomic orbitals) for the expansion of the Bloch wave function but it can be used for either periodic solids or isolated clusters of atoms (large molecules). The interaction integrals are evaluated in real space but the results are more likely to be presented via reciprocal space. It is particularly suitable for complex materials systems that may contain different types of elements, large numbers of atoms, low or no symmetry, and so on. In the limit of very large number of atoms (N), OLCAO is essentially an order N method or O(N) method. This and other details will be explained and demonstrated in the rest of the book.

References

Adachi, H., Mukoyama, T., & Kawai, J. (2006), *Hatree-Fock-Slater Method for Materials Science: The Dv-Xα Method for Design and Characterization of Materials* (Berlin: Springer-Verlag).

Artacho, E., Anglada, E., Dieguez, O., et al. (2008), *Journal of Physics: Condensed Matter*, 20, 064208.

Blaha, P., Schwarz, K., Sorantin, P., & Trickey, S. B. (1990), *Computer Physics Communications*, 59, 399–415.

Clark, S. J., Segall, M. D., Pickard, C. J., et al. (2005), *Zeitschrift fur Kristallographie*, 220, 567–70.

Delley, B., Seminario, J. M., & Politzer, P. (1995), *Theoretical and Computational Chemistry*, Volume 2, 221–54.

Dovesi, R., Orlando, R., Civalleri, B., et al., *Zeitschrift fur Kristallographie*, 220, 571–73.

Ellis, D. E. & Painter, G. S. (1970), *Physical Review B*, 2, 2887.

Eschrig, H. (1989), *Optimized Lcao Method and the Electronic Structure of Extended Systems* (Berlin: Springer-Verlag).

Fonseca Guerra, C., Snijders, J. G., Te Velde, G., & Baerends, E. J. (1998), *Theoretical Chemistry Accounts: Theory, Computation, and Modeling (Theoretica Chimica Acta)*, 99, 391–403.

Frisch, M. J., Trucks, G. W., Schlegel, H. B., et al. (2009), Gaussian 09. Wallingford, CT: Gaussian, Inc.

Geist, A. (2010), *SciDAC Review*, 16, 52–9.

Han, M. J., Ozaki, T. & Yu, J. (2006), *Physical Review B*, 73, 045110.

Hohenberg, P. & Kohn, W. (1964), *Physical Review*, 136, B864.

Horsfield, A. P. & Bratkovsky, A. M. (2000), *Journal of Physics: Condensed Matter*, 12, R1.

Ishibashi, S., Tamura, T., Tanaka, S., Kohyama, M., & Terakura, K. (2007), *Physical Review B*, 76, 153310.

Kohn, W. & Sham, L. J. (1965), *Physical Review*, 140, A1133.

Kresse, G. & Hafner, J. (1993), *Physical Review B*, 47, 558.

Kresse, G. & Hafner, J. (1994), *Physical Review B*, 49, 14251.

Kresse, G. & Furthmüller, J. (1996a), *Physical Review B*, 54, 11169.

Kresse, G. & Furthmüller, J. (1996b), *Computational Materials Science*, 6, 15–50.

Maclaren, J. M., Crampin, S., Vvedensky, D. D., Albers, R. C., & Pendry, J. B. (1990), *Computer Physics Communications*, 60, 365–89.

Methfessel, M. (1988), *Physical Review B*, 38, 1537.

Methfessel, M., Rodriguez, C. O., & Andersen, O. K. (1989), *Physical Review B*, 40, 2009.

Methfessel, M., Van Schilfgaarde, M. & Casali, R. A. (2000), *Electroinc Structure and Physical Properties of Solids: The Uses of the Lmto Method* (Berlin: Springer-Verlag).

Oganov, A. R. (2010), *Modern Methods of Crystal Structure Prediction* (Weinheim: Wiley-VCH Verlag GmbH & Co. KGaA).

Ozaki, T. & Terakura, K. (2001), *Physical Review B*, 64, 195126.

Paolo, G. Et al. (2009), *Journal of Physics: Condensed Matter*, 21, 395502.

Perdew, J. P. & Schmidt, K. (2001), Jacob's ladder of density functional approximations for the exchange-correlation energy. *In:* Van Doren, V. E., Van Alsenoy, K. & Geerlings, P. (eds) *Density Functional Theory and Its Applications to Materials* (Melville, NY: American Institute of Physics).

Rehr, J. J., Kas, J. J., Prange, M. P., Sorini, A. P., Takimoto, Y. & Vila, F. (2009), *Comptes Rendus Physique*, 10, 548–59.

Schmidt, M. W., Baldridge, K. K., Boatz, J. A., et al. (1993), *Journal of Computational Chemistry*, 14, 1347–63.

Schwarz, K. & Blaha, P. (2003), *Computational Materials Science*, 28, 259–73.

Segall, M. D. & Et Al. (2002), *Journal of Physics: Condensed Matter*, 14, 2717.

Skriver, H. L. (1984), *The Lmto Method: Muffin-Tin Orbitals and Electronic Structure* (Berlin, New York: Springer-Verlag).

Soler, J. M., Artacho, E., Gale, J. D., Garcia, A. J., Junquera, J., Ordejon, P. & Portal, D. S. (2002), *J. Phys.: Condens. Matter.*, 14, 2745.

Te Velde, G., Bickelhaupt, F. M., Baerends, E. J., Fonseca Guerra, C., Van Gisbergen, S. J. A., Snijders, J. G. & Ziegler, T. (2001), *J. Comput. Chem.*, 22, 931–67.

Valiev, M., Bylaska, E. J., Govind, N., Kowalski, K., Straatsma, T. P., Van Dam, H. J. J., Wang, D., Nieplocha, J., Apra, E., Windus, T. L. & De Jong, W. A. (2010), *Computer Physics Communications*, 181, 1477–89.

Wills, J. M., Eriksson, O., Alouani, M. & Price, D. L. (2000), *Electronic Structure and Physical Properties of Solids: The Uses of the Lmto Method* (Berlin: Springer-Verlag).

2

Historical Account of the LCAO Method

2.1 Early days of the band theory of solids

It is fitting to review the development of methods for the electronic structure calculation of simple molecules and crystals from the early stages up until the present day. The most significant event in the development of modern physics was Schrödinger's publication of his wave equation method for quantum mechanics in 1926 (Schrodinger, 1926). Shortly thereafter, Pauling and others worked to derive expressions for the solutions of the wave equation in the potential fields of atoms, molecules, and solids (Pauling, 1927a, Pauling, 1927b). The precepts of molecular orbital theory were suggested by Hund and Mulliken (Hund, 1926, Hund, 1927a, Hund, 1927b, Hund, 1930, Mulliken, 1928), Felix Bloch's theorem (Bloch, 1929) for the periodic lattice was established and the concept of the Brillouin zone was introduced (Bouckaert et al., 1936). In the mid 1930s, work by Slater (Slater, 1934), Fröhlich (Fröhlich, 1932), Seitz (Ewing and Seitz, 1936), Mott and Jones (Mott and Jones, 1936), Herring (Herring, 1937) and many others addressed the problem of free electrons in metals and the band structure concept for solids was firmly established.

The next phase of progress in electronic structure theory resulted in practical calculations for crystals of various types. The concept of a linear combination of atomic orbitals (LCAO) was a natural one during the development of different types of methods for solving the Schrödinger equation of a many electron system because it attempts to define a solid structure directly in terms of its constituent components (Callaway, 1964, Fletcher, 1971) Hence, many methods at that time were intimately related to the LCAO concept. Examples include the method of tight binding, the cellular method, the Green's function method (also called the KKR method), the empirical pseudopotential method, and the orthogonalized plane wave (OPW) method. Most modern day electronic structure methods such as the first-principles pseudopotential method and the augmented plane wave method (APW) can be traced to these early ideas. An interesting point is that the pioneers of these early developments were all academically related to some degree and yet their interests diverged to different methods and types of crystal systems.

2.2 Origin of the LCAO method

The LCAO concept started at the very beginning of the quantum theory of solids when the free electron model of metals was formulated. Later on, other related methods were developed primarily to solve or avoid certain practical difficulties associated with implementing LCAO. A typical example is the Hückel method, which was introduced in the 1930s and which was important for molecular quantum chemistry (Hückel, 1931). The work of Conyers Herring to develop the OPW method is particularly noteworthy because it was considered to be the first method capable of a practical calculation of a crystal band structure (Coulson, 1947, Herring, 1940, Jones, 1934, Morita, 1949, Shockley, 1937, Wallace, 1947). It is then fair to say that the early ideas of LCAO led to the OPW method and that the OPW method enabled the first substantially realistic solid state calculations. This in turn led to the development of different and improved approaches for LCAO calculations that finally led to the present day OLCAO method.

All band structure methods in the early days were related to each other and were based on the LCAO concept. This was natural because crystals are composed of atoms or cohesive molecules and atoms and molecules are described by atomic orbitals on the basis of fundamental quantum mechanics. What the different methods tried to do was balance certain technical difficulties with the desire to achieve improved accuracy and a greater range of applicability beyond monatomic crystals. This was true then and is still true today. In these early endeavors, the pioneering researchers in band theory and quantum chemistry seldom used a single method. They used multiple methods, applied them to different crystals or molecules, assessed their pros and cons and then modified the method or invented a new one to overcome any obstacles encountered.

In using the LCAO concept for different types of calculations, several approaches were used. The most common one to take root in chemistry circles was the Hückel method (Hückel, 1931) and later the Extended Hückel method (EHM). Roald Hoffmann received the 1981 Nobel Prize in Chemistry for contributions made to the understanding of chemical reactions based on EHM. In the Hückel method, an orthogonal basis based on π electron interactions is used, and the bonding and anti-bonding were simple and the orbital interactions were fitted to experimentally known quantities. In the same spirit as the Hückel method, a simple implementation of LCAO in solid state physics is the so-called tight binding (TB) method (Fletcher and Wohlfarth, 1951), and later the extended TB (ETB) method. In fact, the tight binding method is synonymous with the LCAO method and the terms are often used without much distinction. The approach was pioneered by Slater and Koster (Slater and Koster, 1954), and in it the matrix elements for energy were tabulated for different crystal structures so that the full band structure of the crystal in the Brillouin zone could be easily obtained. In later years, the Slater–Koster parameters were obtained by fitting them to other crystal band structures obtained by more accurate methods (Papaconstantopoulos, 1986). In this way, it was not necessary to calculate the interaction integrals explicitly and the TB method could be directly applied to more complex crystals. It was also possible to obtain properties other than the band structure. Even now with the availability of many accurate and sophisticated computational packages, the TB method still has certain appealing characteristics and is routinely applied to materials such as carbon nanotubes (Dresselhaus et al., 2005).

Before DFT or other similar one-electron methods became popular, the Hartree–Fock (HF) method was regarded as the gold standard method of considerable rigor especially by quantum chemists. Although it ignored the effects of correlation (Cramer, 2002, Hartree and Hartree, 1936), the premise of the Hartree–Fock method was that it treated exchange exactly and that the correlation effect could always be added later either through configurational interaction or some other means. The HF method is still robust and actively used nowadays but tends to be limited to calculations of relatively small molecules or very simple crystals because of the computational demand associated with solving the Hartree–Fock equation. Another important method that uses the LCAO approach is the discrete variational method (DV-Xα) with exchange-correlation parameter α which adopted a numerical basis (Adachi et al., 2006, Ellis and Painter, 1970). Because of its conceptual simplicity and the non-empirical nature of the calculation, this method has a large number of followers especially among researchers who need a simple and quick method to provide meaningful results to interpret experimental data.

Another method in the same class is the multiple scattering method of Johnson and Slater (Johnson, 1966) which is actually rooted from the Green's Function methods developed in the 1950s and early 1960s (Johnson et al., 1973, Kohn and Rostoker, 1954, Korringa, 1947, Morse, 1956). It later led to a much more popular and accurate linear muffin tin orbital (LMTO) method that is particularly effective for compact crystals (Skriver, 1984). The LMTO method has been developed into a very popular method in electronic structure theory (Tank and Arcangeli, 2005).

The above discussion is certainly not comprehensive, nor is it intended to be. Some other methods that are either motivated by or closely related to the LCAO concept have not been explicitly mentioned. The critical observation is that many modern practical methods for performing such calculations originated from the same idea and that they are all intimately related to each other. It is difficult to classify them into very distinct categories and each can be said to have special merits for application to different materials and for understanding different properties. There are several excellent books discussing the origin and practice of the LCAO method as well as its relation to other quantum chemical or solid state methods (Harrison, 1980, Levin, 1977, Phillips, 1973, Trinddle, 2008). Interested readers should consult these references.

2.3 Use of Gaussian orbitals in LCAO calculations

One of the most important mathematical facilitations in the calculation of molecular and crystalline solid properties is the use of Gaussian type orbitals

(GTOs) to describe the underlying molecular or solid state wave function. This approach began with the Hartree–Fock calculations of atomic wave functions in atomic physics (Clementi and Roetti, 1974). GTOs have some unique and advantageous mathematical properties such as the fact that the product of two GTOs can be transformed into another GTO and that integration and differentiation can be represented with simple analytic expressions. These properties make the computation of orbital interaction integrals rather straightforward (Boys, 1950, Huzinaga, 1965, Shavitt, 1963). This is in contrast to the use of Slater type orbitals (STO) for the atomic wave functions. Although the form of STOs are more natural and accurate for describing atomic wave functions, they are difficult to express in analytical integrals and numerical techniques must be used instead (Fonseca Guerra et al., 1998, Te Velde et al., 2001). Extensive tables and listings of atomic wave functions expressed in the form of GTOs obtained from Hartree–Fock calculations have been published as reference material (Huzinaga, 1965) and they were used in early calculations of the band structures of simple crystals. These GTO based atomic wave functions are typically limited to a description of the occupied atomic states and even in the present only a few listings have wave functions for the excited states.

One fact about the use of GTOs and the related Gaussian transformation technique is that as the atomic number Z increases, the maximum principle quantum number n and orbital quantum number ℓ (s, p, d, f, ...) also increases Hence, the analytic expressions for the interaction integrals can quickly become very complicated to derive and cumbersome to implement. An important and relatively recent, contribution from people like Obara and Saika (Obara and Saika, 1986) has been the development of recursive algorithms for deriving the necessary formulas. This is extremely valuable and has made it much easier to treat heavier atoms and obtain the matrix elements of other physical operators.

It is worth mentioning that GTOs were developed mostly by atomic physicists and quantum chemists and were only later adopted by solid sate physicists. This is best exemplified by the prevalence of the successive versions of the Gaussian package (Frisch et al., 2009) initiated by J.A. Pople. Indeed, he and Walter Kohn, the inventor of DFT, shared the Nobel Prize in Chemistry in 1998. In the last two decades, the implementation of DFT in the Gaussian package has greatly reshaped the course of quantum chemical calculations in almost all areas of chemistry, pharmacology, and biochemistry.

The use of GTOs in LCAO based band structure calculations started in the late 1960s and 1970s lead by the group of C.C. Lin and his collaborators (Chaney et al., 1971a, Chaney et al., 1971b, Lafon and Lin, 1966). This work was later followed by the group of J. Fry and J. Callaway (Ching and Callaway, 1973, Laurent et al., 1981, Rath et al., 1973, Wang and Callaway, 1974, Wang and Callaway, 1977). During this period, accurate LCAO methods were applied to a large number of crystals including transition metals and the foundation was laid for the OLCAO method. Many of these calculations focused on elemental metals and simple crystals. The accurate treatment of spin-polarized transition metals with $3d$ electrons paved the way for attacking problems on magnetism and magnetic materials. These were the boom years for the LCAO method. It was soon realized that the expansion of the Bloch functions need not always use atomic orbitals (expressed as GTOs), and that the use of single GTOs for the expansion could be equally effective and offered greater variational flexibility. It should also be mentioned that accurate calculations using the LCAO method are not limited to the use of the GTOs. Other groups have used STOs or other less common orbital definitions for the atomic basis functions. These methods will not be discussed in this book.

2.4   Beginning of the OLCAO method

The OLCAO method is the direct extension of the LCAO method. The name was coined by Ching and Lin and was inspired by the OPW method (Ching and Lin, 1975a, Ching et al., 1977). It was realized that for calculations of more complex systems with no crystal symmetry, the secular equation to be solved would necessarily be large. Because the core states usually do not play any significant role they could be eliminated by an orthogonalization process, in the same way as was done in the OPW method. This is particularly valuable for systems with high Z atoms because they have many more core states. The OLCAO method was carefully tested on crystalline Si-III (Ching and Lin, 1975a, Ching et al., 1977) and then applied to models of amorphous Si (a-Si) (Ching and Lin, 1975b) and hydrogenated a-Si (H-a-Si) (Ching et al., 1979). It was then used to study amorphous SiO_2 (a-SiO_2) glass (Ching, 1981) and metallic glasses (Jaswal and Ching, 1982, Zhao and Ching, 1989), thus formally introducing a more rigorous calculation for disordered or non-crystalline systems. These early calculations were all non-self-consistent, relatively small in size, and of limited accuracy. Over the intervening years, improvements to the OLCAO method removed such limitations.

The accuracy of the OLCAO method was significantly improved in the 1980s by implementing a fully self-consistent field (SCF) calculation following the work of Sambe and Felton (Sambe and Felton, 1975) and of Harmon et al. (Harmon et al., 1982). This was an important step in the maturing of the OLCAO method. Over the next 20 years the OLCAO method was applied to a vast collection of condensed matter materials systems primarily by the electronic structure group (ESG) at the University of Missouri-Kansas City (UMKC). These applications covered both crystalline and non-crystalline materials with increasingly complicated structures. Early steps include calculations on metallic and insulating glasses, the first realistic calculations of the $Nd_2Fe_{17}B$ permanent magnet (Gu and Ching, 1987); early calculations on the electronic structure and optical properties of the $YBa_2Cu_3O_7$ superconductor and later the charge transfer salt organic superconductors (Laurent et al., 1978). Other notable contributions include the calculation of C_{60} and alkali-doped C_{60} crystals, non-linear optical crystals (Ching and Huang, 1993, Huang and Ching, 1993), laser host crystals such as yttrium alumina garnet (YAG) and related Y-Al-O crystals (Xu and Ching, 1999, Xu et al., 1995), the systematic investigation of a large number of spinel nitrides (Mo et al., 1999), complex biomolecules such as the vitamin B_{12} cobalamins with all its side chains

(Ouyang et al., 2003), grain boundary structures (Ching et al., 1995), and core level spectroscopy (Mo et al., 1996). These results will be presented and discussed in this book starting from Chapter 5.

2.5 Current status and future trends of the OLCAO method

In recent years, the OLCAO method has been applied to many more different types of complex crystals and has been made far more efficient and versatile. Advances include the use of data compression techniques, the creation of a systematic data base of basis functions, and the construction of efficient analysis tools for studying the computed results. The experience accumulated from these studies had made it possible to perform fully self-consistent calculations on very large systems such as non-crystalline structures and materials containing defects and microstructures.

The strength of the OLCAO method is in its ability to deal with the electronic structures of very large systems at an appreciably high level of accuracy. The method can and has benefited substantially by being paired with plane-wave based *ab initio* methods such as the VASP package. The plane-wave method is primarily used to obtain accurate structural models that are then fed to the OLCAO method for specific calculations such as core level spectroscopy. This will be described in Chapter 11.

Unfortunately, at the time of publication of this book, the application of the OLCAO method in materials research is limited to only a few people in a handful of groups, mostly those with close past or current collaborations with the authors. An area of urgent need is the creation of a complete distributable package so that the method can become more widely known and used. This work is currently under way and is anticipated to be complete soon after the publication of this book. The creation of this package includes primarily technical developments such as a user friendly installation scheme, a parallelized code base that can take advantage of the cluster or supercomputer resources that are often available to researchers now, and a user manual and usage tutorial. We believe that with further development and implementation of theory and when used in combination with many other excellent computational packages, the OLCAO method will be a very popular method for studying fundamental interactions at the atomistic level through quantum mechanical calculations, especially for large systems with nano-scale features and those relevant to complex biomolecular systems include liquid phases.

References

Adachi, H., Mukoyama, T., & Kawai, J. (2006), *Hatree-Fock-Slater Method for Materials Science: The Dv-Xα Method for Design and Characterization of Materials* (Berlin: Springer-Verlag).

Bloch, F. (1929), *Zeitschrift für Physik A Hadrons and Nuclei*, 52, 555–600.

Bouckaert, L. P., Smoluchowski, R., & Wigner, E. (1936), *Physical Review*, 50, 58.

Boys, S. F. (1950), *Proceedings of the Royal Society of London. Series A. Mathematical and Physical Sciences*, 200, 542–54.

Callaway, J. (1964), *Energy Band Theory* (New York: Academic Press).

Chaney, R. C., Lafon, E. E., & Lin, C. C. (1971a), *Physical Review B*, 4, 2734.

Chaney, R. C., Lin, C. C., & Lafon, E. E. (1971b), *Physical Review B*, 3, 459.

Ching, W. Y. & Callaway, J. (1973), *Phys. Rev. Lett.*, 30, 441–3.

Ching, W. Y. & Lin, C. C. (1975a), *Physical Review B*, 12, 5536.

Ching, W. Y. & Lin, C. C. (1975b), *Phys. Rev. Lett.*, 34, 1223–6.

Ching, W. Y. & Lin, C. C. (1977), *Phys. Rev. B*, 16, 2989.

Ching, W. Y., Lam, D. J., & Lin, C. C. (1979), *Phys. Rev. Lett.*, 42, 805–8.

Ching, W. Y. (1981), *Phys. Rev. Lett.*, 46, 607–10.

Ching, W. Y. & Huang, M. Z. (1993), *Phys. Rev. B*, 47, 9479–91.

Ching, W. Y., Gan, F., & Huang, M.-Z. (1995), *Phys. Rev. B*, 52, 1596–611.

Clementi, E. & Roetti, C. (1974), *Atomic Data and Nuclear Data Tables*, 14, 177–478.

Coulson, C. A. (1947), *Nature*, 159, 265–6.

Cramer, C. J. (2002), *Essentials of Computational Chemistry* (Chichester: John Wiley & Sons Ltd.).

Dresselhaus, M. S., Dresselhaus, G., Saito, R., & Jorio, A. (2005), *Physics Reports*, 409, 47–99.

Ellis, D. E. & Painter, G. S. (1970), *Physical Review B*, 2, 2887.

Ewing, D. H. & Seitz, F. (1936), *Physical Review*, 50, 760.

Fletcher, G. C. & Wohlfarth, E. P. (1951), *Philosophical Magazine Series 7*, 42, 106–9.

Fletcher, G. C. (1971), *The Electron Band Theory of Solids*(Amsterdam: Noord-Hollandsche U.M.).

Fonseca Guerra, C., Snijders, J. G., Te Velde, G., & Baerends, E. J. (1998), *Theoretical Chemistry Accounts: Theory, Computation, and Modeling (Theoretica Chimica Acta)*, 99, 391–403.

Frisch, M. J., Trucks, G. W., Schlegel, H. B., et al. (2009), Gaussian 09. Wallingford, CT: Gaussian, Inc.

Fröhlich, H. (1932), *Annalen der Physik*, 405, 229–48.

Gu, Z. & Ching, W. Y. (1987), *Phys. Rev. B*, 36, 8530–46.

Harmon, B. N., Weber, W. & Hamann, D. R. (1982), *Physical Review B*, 25, 1109.

Harrison, W. A. (1980), *Electronic Structure and Properties of Solids* (San Francisco: W. H. Freeman & Co.).

Hartree, D. R. & Hartree, W. (1936), *Proceedings of the Royal Society of London. Series A—Mathematical and Physical Sciences*, 154, 588–607.

Herring, C. (1937), *Physical Review*, 52, 361.

Herring, C. (1940), *Physical Review*, 57, 1169.

Huang, M. Z. & Ching, W. Y. (1993), *Phys. Rev. B*, 47, 9464–78.

Hückel, E. (1931), *Zeitschrift für Physik A Hadrons and Nuclei*, 70, 204–86.

Hund, F. (1926), *Z. Phys.*, 36, 657–74.

Hund, F. (1927a), *Z. Phys.*, 40, 742–64.

Hund, F. (1927b), *Z. Phys.*, 43, 788–804.

Hund, F. (1930), *Z. Phys.*, 63, 719–51.

Huzinaga, S. (1965), *The Journal of Chemical Physics*, 42, 1293–302.

Jaswal, S. S. & Ching, W. Y. (1982), *Phys. Rev. B*, 26, 1064–6.

Johnson, K. H. (1966), *The Journal of Chemical Physics*, 45, 3085–95.

Johnson, K. H., Norman, J. G. J., & Connolly, J. W. D. (1973), *Computational Methods for Large Molecules and Localized States in Solids* (New York: Plenum Press).

Jones, H. (1934), *Proceedings of the Royal Society of London. Series A*, 144, 225–34.

Kohn, W. & Rostoker, N. (1954), *Physical Review*, 94, 1111.

Korringa, J. (1947), *Physica*, 13, 392–400.

Lafon, E. E. & Lin, C. C. (1966), *Physical Review*, 152, 579.

Laurent, D. G., Wang, C. S., & Callaway, J. (1978), *Physical Review B*, 17, 455.

Laurent, D. G., Callaway, J., Fry, J. L., & Brener, N. E. (1981), *Physical Review B*, 23, 4977.

Levin, A. A. (1977), *Solid State Quantum Chemistry* (New York: McGraw-Hill).

Mo, S.-D., Ching, W. Y., & French, R. H. (1996), *J. Am. Ceram. Soc.*, 79, 627–33.

Mo, S.-D., Ouyang, L., Ching, W. Y., et al. (1999), *Phys. Rev. Lett.*, 83, 5046–9.

Morita, A. (1949), *Science Repts. Tohoku Univ.*, 33, 92–8.

Morse, P. M. (1956), *Proceedings of the National Academy of Sciences*, 42, 276–86.

Mott, N. F. & Jones, H. (1936), *The Theory of the Properties of Metals and Alloys* (Oxford, UK: Oxford Press).

Mulliken, R. S. (1928), *Physical Review*, 32, 186.

Obara, S. & Saika, A. (1986), *The Journal of Chemical Physics*, 84, 3963–74.

Ouyang, L., Randaccio, L., Rulis, P., et al. (2003), *Journal of Molecular Structure: THEOCHEM*, 622, 221–7.

Papaconstantopoulos, D. A. (1986), *Handbook of the Band Structure of Elemental Solids* (New York: Plenum Press).

Pauling, L. (1927a), *J. Am. Chem. Soc.*, 49, 765–92.

Pauling, L. (1927b), *Proc. R. Soc. London, Ser. A*, 114, 181–211.

Phillips, J. C. (1973), *Bonds and Bands in Semiconductors* (New York: Academic Press).

Rath, J. & Callaway, J. (1973), *Physical Review B*, 8, 5398.

Sambe, H. & Felton, R. H. (1975), *The Journal of Chemical Physics*, 62, 1122–26.

Schrodinger, E. (1926), *Physical Review*, 28, 1049.

Shavitt, I. (1963), *Methods in Computational Physics* (New York: Academic Press).

Shockley, W. (1937), *Physical Review*, 51, 129.

Skriver, H. L. (1984), *The Lmto Method: Muffin-Tin Orbitals and Electronic Structure* (Berlin, New York: Springer-Verlag).

Slater, J. C. (1934), *Reviews of Modern Physics*, 6, 209.

Slater, J. C. & Koster, G. F. (1954), *Physical Review*, 94, 1498.

Tank, R. W. & Arcangeli, C. (2005), *An Introduction to the Third-Generation Lmto Method* (Berlin: Wiley-VCH Verlag GmbH & Co. KGaA).

Te Velde, G., Bickelhaupt, F. M., Baerends, E. J., et al. (2001), *Journal of Computational Chemistry*, 22, 931–67.

Trinddle, C. (2008), *Electronic Structure Modeling: Connections between Theory and Software* (Boca Raton: CRC Press).

Wallace, P. R. (1947), *Physical Review*, 71, 622.

Wang, C. S. & Callaway, J. (1974), *Physical Review B*, 9, 4897.

Wang, C. S. & Callaway, J. (1977), *Physical Review B*, 15, 298.

Xu, Y. N., Ching, W. Y., Jean, Y. C., & Lou, Y. (1995), *Phys. Rev. B*, 52, 12946–50.

Xu, Y. N. & Ching, W. Y. (1999), *Phys. Rev. B*, 59, 10530–5.

Zhao, G. L. & Ching, W. Y. (1989), *Phys. Rev. Lett.*, 62, 2511–14.

3 Basic Theory and Techniques of the OLCAO Method

The OLCAO method is derived from the traditional LCAO method with the addition of numerous modifications and extensions. Atomic orbitals are used in the basis expansion where the radial part is expanded in terms of Gaussian-type of orbitals (GTOs). The OLCAO method was originally designed for the purpose of studying amorphous solids represented by large periodic atomic models (Ching and Lin, 1975a). As long as the periodic model is sufficiently large, say much larger than the mean free path of an electron in the solid, this is a very effective way of studying the electronic structure of disordered solids and non-crystalline materials. It was later found to be quite effective when applied to complex crystals, microstructural models, and multifarious biological systems. Over the years, the method has been systematically upgraded and refined in terms of its computational efficiency, accuracy, ease of use, its range of applicability to different elements and types of systems, and the inclusion of more rigorous theory. It can now compute the electronic structure of almost any kind of solid modeled with periodic boundary conditions. Recent versions of OLCAO require substantially less intermediate storage space and CPU time than older versions on identical hardware. Methods have been devised and applied to achieve a high degree of transferability of the atomic orbital basis. A wide variety of ancillary programs have been developed to make deep analysis of complex results more straightforward. Explicit calculation of momentum matrix elements, inclusion of core-hole screening, and spin-polarization are among the many components of theory that have been implemented. A brief summary of the structure of the OLCAO package is outlined in Appendix C.

3.1 The atomic basis functions

We shall describe the OLCAO method strictly as a band structure method and we will therefore retain the wave vector k description of the electronic state. The use of the Bloch theorem circumvents the problem of free surfaces encountered when using finite cluster models and makes it possible to describe a system of atoms as an infinitely extended solid. In the case where it is necessary to consider the system as an isolated group of atoms (e.g. a molecule) the periodic boundary conditions of the Bloch theorem can still be met by placing the system in a simulation cell that is sufficiently large to render interaction with neighboring cells negligible.

Within the OLCAO description, the solid state wave function $\Psi_{n\vec{k}}(\vec{r})$ of band index n (or energy levels in the $k = 0$ description) and wave vector k is expanded in terms of Bloch sums $b_{i\gamma}(\vec{k}, \vec{r})$:

$$\Psi_{n\vec{k}}(\vec{r}) = \sum_{i,\gamma} C_{i\gamma}^n\left(\vec{k}\right) b_{i\gamma}\left(\vec{k}, \vec{r}\right) \tag{3.1}$$

The Bloch sum $b_{i\gamma}\left(\vec{k}, \vec{r}\right)$ is constructed from the linear combination of atomic orbitals u_i centered at each atomic site:

$$b_{i\gamma}\left(\vec{k}, \vec{r}\right) = \left(\frac{1}{\sqrt{N}}\right) \sum_{v} e^{i\left(\vec{k}\bullet\vec{R}_v\right)} u_i\left(\vec{r} - \vec{R}_v - \vec{t}_\gamma\right) \tag{3.2}$$

where γ labels the atoms in the cell and i represents all the collective quantum numbers of the atom. The γ represents different types of atoms as well as non-equivalent atoms of the same type within the unit cell or the simulation cell. Here, R_v is the lattice vector and t_γ is the position of the γth atom in the cell.

The atomic orbital u_i consists of a radial part and an angular part:

$$u_i(\vec{r}) = \left[\sum_{j=1}^N A_j r^l e^{(-\alpha_j r^2)}\right] \cdot \mathcal{Y}_l^m(\theta, \phi) \tag{3.3}$$

The radial part is represented by a linear combination of a suitable number of Cartesian GTOs and the angular part consists of the real spherical harmonics. The orbital quantum number i collectively represents the principal quantum number n and the angular momentum quantum numbers (l, m). The spin quantum number s can also be included as a simple extension but it will be ignored now for clarity.

Each of the GTOs in Eq. (3.3) is characterized by a decaying exponent α_j. The way the $\{\alpha_j\}$ are chosen for the u_i deserves some comment. A simple and effective way is to choose a set of N pre-determined exponentials $\{\alpha_j\}$ ranging from a minimum α_{min} to a maximum α_{max}, distributed as a geometric series. The values of N, α_{min}, and α_{max} vary from element to element but can be guided by the experience gained from many test calculations that have been done in the past. Typical values for N are from 16 to 30 while values for α_{min} and α_{max} are 0.1 to 0.15 and 10^6 to 10^9 respectively depending on the size of the atom or the atomic number Z. Smaller atoms need fewer expansion terms and a smaller value of α_{max}. In the present version of the OLCAO code, α_{min} is set at 0.12 for all atoms but this can be flexible to suit special needs. A value of $\alpha_{min} < 0.05$ can result in over-extended atomic functions that lead to numerical instability due to over completeness of the basis set. It will also create an excessive number of multi-center integrals because of the long interaction range. A value of $\alpha_{min} > 0.15$ can result in wave functions that are too localized and thus detrimental to the accuracy of the calculation.

Once the exponential set $\{\alpha_j\}$ has been determined, the expansion coefficients A_j can be obtained in several ways. They can be obtained as the normalized eigenvector coefficients of a single-atom eigenvalue problem solved self-consistently within the same density functional theory using a basis of

Gaussian functions, or they can be obtained by linearly fitting to atomic wave functions calculated by the self-consistent Hartree–Fock method or from other *ab initio* atomic calculations. (Clementi and Roetti, 1974, Huzinaga, 1984) For almost all calculations, it is desirable to use the same set $\{\alpha_i\}$ for *all* the atoms of the same element and for *all* the orbitals of different quantum numbers i. In other words, in the OLCAO method, the basis function is element specific and pre-determined. This greatly reduces the total number of analytic integrals that need to be evaluated, which would otherwise be enormous for a model structure containing several thousand atoms. Although it is possible to have slightly different basis functions for the same type of atom in more delicate systems, such as those found in the biomolecules, a fixed basis set in $\{\alpha_i\}$ is generally satisfactory since the solid state wave function adjusts itself through the coefficients $C_{i\gamma}^n$ in the self-consistent iterations.

The set of atomic orbitals u_i in Eq. (3.3) includes the core orbitals, the occupied valence orbitals, and a variable number of additional empty orbitals. For calculations of large amorphous systems, a minimal basis (MB) is usually sufficient which includes the core orbitals and the orbitals in the valence shell of the atom, occupied or unoccupied. This retains the full advantage of the Mulliken scheme (Mulliken, 1955) for charge analysis (to be discussed later). For example, the MB for Fe consists of $1s$, $2s$, $2p_x$, $2p_y$, $2p_z$, $3s$, $3p_x$, $3p_y$, $3p_z$ (core orbitals) and $4s$, $4p_x$, $4p_y$, $4p_z$, $3d_{xy}$, $3d_{yz}$, $3d_{xz}$, $3d_{x^2-y^2}$, $3d_{3z^2-r^2}$ (valence shell orbitals). If additional empty orbitals of the next unoccupied shell are added, the basis set is referred to as a full basis (FB) set. So a FB set for Fe consists of MB plus $5s$, $5p$, and $5d$ orbitals. For most purposes, a FB is more than sufficient to give accurate results. In some special cases such as in spectral calculations where unoccupied states at high energy are of interest, an additional shell of the excited atomic basis is added to the FB to form an extended basis (EB) set. Alternatively, one can also add additional single GTOs to the FB to form the EB for additional variational freedom in the basis set. For special-purpose high precision calculations, u_i can be further optimized in some way as practiced by many quantum chemists in molecular calculations. In summary, there is a great deal of flexibility in the choice of atomic basis set for a given problem with a good balance between the accuracy needed and the time it will take to execute the program in a reasonable time span.

We have maintained a large data base of the atomic basis functions u_i for almost all the elements in the periodic table with the exception of some very heavy and rare elements. These are based mostly on the results of our past calculations and other additional tests. The details of the data base are listed in Appendix A. In the Appendix, the N, α_{min}, α_{max} for each element whose basis functions are in the data base are listed together with the proper specification of MB, FB, and EB for that element. These are the default values used in the current package and can be easily changed to suit particular circumstances. It bears repeating that in the current implementation of the OLCAO method and the data base, the N, α_{min}, and α_{max} for the same element but different angular momentum states are kept the same. This strategy reduces the total number of multi-center integrals that need to be calculated for a given system with virtually no compromise in accuracy.

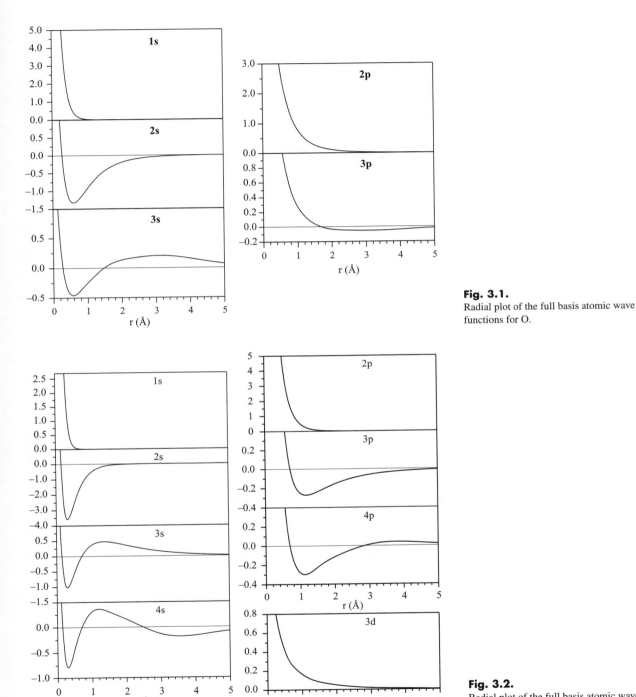

Fig. 3.1.
Radial plot of the full basis atomic wave functions for O.

Fig. 3.2.
Radial plot of the full basis atomic wave functions for Si.

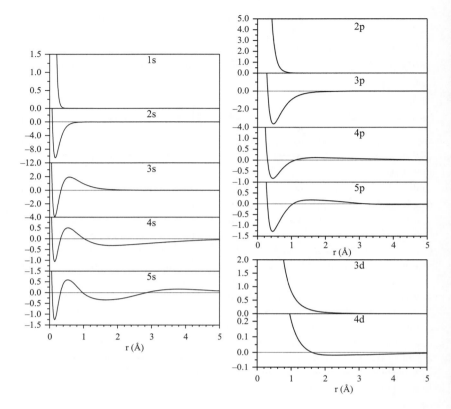

Fig. 3.3.
Radial plot of the Fe full basis atomic
orbital wave functions.

In Fig. 3.1, Fig. 3.2, and Fig. 3.3, we plot the FB atomic functions for
elements O, Si, and Fe that are in current use in the data base. Table 3.1 lists
the real spherical harmonics (in angular and Cartesian form) up to $\ell = 3$, or the
f-orbitals, which are the angular functions used in the current implementation.
In principle, extension to $\ell = 4$ or g-orbitals is possible but at the moment this
is unnecessary in the OLCAO method.

3.2   Bloch functions and the Kohn–Sham equation

Following the standard practice of density functional theory, the next step in the
OLCAO method is to iteratively solve the one-electron Kohn–Sham equation.
In atomic units, it is given by:

$$\left[-\vec{\nabla}^2 + V_{e-n}(\vec{r}) + V_{e-e}(\vec{r}) + V_{xc}\left[\rho\,(\vec{r}) \right] \right] \Psi_{n\vec{k}}(\vec{r}) = E_n\left(\vec{k}\right) \Psi_{n\vec{k}}(\vec{r}) \quad (3.4)$$

The first term in Eq. (3.4) is the kinetic energy term while $V_{e-n}(\mathbf{r})$,
$V_{e-e}(\mathbf{r})$, and $V_{xc}[\rho(\mathbf{r})]$ are the electron–nuclear, electron–electron Coulomb,
and exchange-correlation parts of the potential respectively. The exchange-
correlation potential $V_{xc}[\rho(\mathbf{r})]$ depends on the electron density $\rho(\mathbf{r})$. The

Table 3.1. Real spherical harmonics in angular and Cartesian coordinates.

s type $\left\{ Y_0^0 = \sqrt{\frac{1}{4\pi}} = \sqrt{\frac{1}{4\pi}} \right\}$

p type $\left\{ \begin{array}{l} Y_1^{-1} = \sqrt{\frac{3}{4\pi}}\sin(\theta)\sin(\phi) = \sqrt{\frac{3}{4\pi}}\frac{y}{r} \\[6pt] Y_1^0 = \sqrt{\frac{3}{4\pi}}\cos(\theta) = \sqrt{\frac{3}{4\pi}}\frac{z}{r} \\[6pt] Y_1^1 = \sqrt{\frac{3}{4\pi}}\sin(\theta)\cos(\phi) = \sqrt{\frac{3}{4\pi}}\frac{x}{r} \end{array} \right\}$

d type $\left\{ \begin{array}{l} Y_2^{-2} = \sqrt{\frac{15}{4\pi}}\sin(\theta)^2\sin(\phi)\cos(\phi) = \sqrt{\frac{15}{4\pi}}\frac{xy}{r^2} \\[6pt] Y_2^{-1} = \sqrt{\frac{15}{4\pi}}\sin(\theta)\cos(\theta)\sin(\phi) = \sqrt{\frac{15}{4\pi}}\frac{yz}{r^2} \\[6pt] Y_2^0 = \sqrt{\frac{5}{16\pi}}(3\cos(\theta)^2 - 1) = \sqrt{\frac{5}{16\pi}}\frac{2z^2-x^2-y^2}{r^2} \\[6pt] Y_2^1 = \sqrt{\frac{15}{4\pi}}\sin(\theta)\cos(\theta)\cos(\phi) = \sqrt{\frac{15}{4\pi}}\frac{xz}{r^2} \\[6pt] Y_2^2 = \sqrt{\frac{15}{16\pi}}\sin(\theta)^2(2\cos(\phi)^2 - 1) = \sqrt{\frac{15}{16\pi}}\frac{x^2-y^2}{r^2} \end{array} \right\}$

f type $\left\{ \begin{array}{l} Y_3^{-3} = \sqrt{\frac{35}{32\pi}}\sin(\theta)^3\sin(\phi)(4\cos(\phi)^2 - 1) = -\sqrt{\frac{35}{32\pi}}y\frac{y^2-3x^2}{r^3} \\[6pt] Y_3^{-2} = \sqrt{\frac{105}{4\pi}}\sin(\theta)^2\cos(\theta)\sin(\phi)\cos(\phi) = \sqrt{\frac{105}{4\pi}}\frac{xyz}{r^3} \\[6pt] Y_3^{-1} = \sqrt{\frac{21}{32\pi}}\sin(\theta)(5\cos(\theta)^2 - 1)\sin(\phi) = \sqrt{\frac{21}{32\pi}}y\frac{4z^2-x^2-y^2}{r^3} \\[6pt] Y_3^0 = \sqrt{\frac{7}{16\pi}}\cos(\theta)(5\cos(\theta)^2 - 3) = \sqrt{\frac{7}{16\pi}}z\frac{2z^2-3x^2-3y^2}{r^3} \\[6pt] Y_3^1 = \sqrt{\frac{21}{32\pi}}\sin(\theta)(5\cos(\theta)^2 - 1)\cos(\phi) = \sqrt{\frac{21}{32\pi}}x\frac{4z^2-x^2-y^2}{r^3} \\[6pt] Y_3^2 = \sqrt{\frac{105}{16\pi}}\sin(\theta)^2\cos(\theta)(2\cos(\phi)^2 - 1) = \sqrt{\frac{105}{16\pi}}z\frac{x^2-y^2}{r^3} \\[6pt] Y_3^3 = \sqrt{\frac{35}{32\pi}}\sin(\theta)^3\cos(\phi)(4\cos(\phi)^2 - 3) = \sqrt{\frac{35}{32\pi}}x\frac{x^2-3y^2}{r^3} \end{array} \right\}$

electron density of the solid is obtained from the following summation over the occupied states:

$$\rho(\vec{r}) = \sum_{occ} |\Psi_{n\vec{k}}(\vec{r})|^2 \qquad (3.5)$$

so that Eq. (3.4) can be solved self-consistently.

In the local density approximation (LDA) of density functional theory, the exchange and correlation part of the potential $V_{xc}(\mathbf{r})$ is responsible for simplifying the many electron interactions. It is derivable from an exchange-correlation energy functional ε_{xc} for the exchange-correlation energy $E_{xc}(\mathbf{r})$

$$E_{xc} = (\vec{r}) \int \rho(\vec{r})\varepsilon_{xc}[\rho(\vec{r})]d\vec{r} \qquad (3.6)$$

With the local approximation $V_{xc}(\mathbf{r})$ takes the form:

$$V_{xc}(\vec{r}) = \frac{d(\rho\varepsilon_{xc}[\rho])}{d\rho} = -\frac{3}{2}\alpha\left[\frac{3}{\pi}\rho(\vec{r})\right]^{\frac{1}{3}} \tag{3.7}$$

The simplest form for V_{xc} is the Kohn–Sham approximation with α in Eq. (3.7) equal to 2/3. Not too long ago, α was used as an adjustable parameter between 2/3, the Kohn–Sham limit, and 1 the full Slater limit (Schwarz, 1972, Slater, 1951). There are several other popular forms of V_{xc} in the LDA which were obtained with different treatments aimed at improving the exchange and correlation energy term by making use of highly accurate results for the uniform electron gas system (Barth and Hedin, 1972, Ceperley and Alder, 1980, Gunnarsson and Lundqvist, 1976, Hedin and Lundqvist, 1971, Vosko et al., 1980). The default form used in the current OLCAO code is the Wigner interpolation formula (Wigner, 1934) to account for the correlation effect in addition to the exchange effect such that:

$$V_{xc}(\vec{r}) = \rho(\vec{r})^{1/3}\left[-0.984 - \frac{0.944 + 8.90\rho(\vec{r})}{(1 + 12.57\rho(\vec{r})^{1/3})^2}\right]^{\frac{1}{3}} \tag{3.8}$$

In recent years, rapid advances have been made to further improve the exchange-correlation part of the potential beyond its local approximation for homogeneous electron systems. These potentials include the generalized gradient approximation (GGA) (Perdew et al., 1996, Perdew and Wang, 1992) and other hybrid methods aimed at certain classes of systems (Becke, 1986, Lee et al., 1988). These potentials have been implemented in many other electronic structure methods and will not be discussed in detail here.

In the OLCAO method under LDA, the total energy of the system can be evaluated from the following expression:

$$E_T = \sum_{n,\vec{k}}^{occ} E_n(\vec{k}) + \int \rho(\vec{r})\left(\varepsilon_{xc} - V_{xc} - \frac{V_{e-e}}{2}\right)d\vec{r} + \frac{1}{2}\sum_{\gamma,\delta}\frac{Z_\gamma Z_\delta}{\vec{R}_\gamma - \vec{R}_\delta}, \tag{3.9}$$

where the first term is the sum over one-electron band energies and the last term is a sum over the lattice sites. The factor of 1/2 accounts for the double counting in the nuclear Coulomb repulsion. The total energy expression E_T derived from DFT is a very important physical quantity in electronic structure theory. It is widely used for inter-atomic force calculation and stability studies as well as for structural optimization of different materials. In the OLCAO method, total energy is used as a criterion for convergence in the self-consistent potential and in the relative comparison of the stability of systems with the same number of atoms of the same elemental composition. It has not been fully utilized for force calculations due to complications associated with atomic site dependent basis functions. However, a simpler scheme of force calculation using the OLCAO method based on E_T and the finite difference method has been demonstrated (Ouyang and Ching, 2001).

3.3 The site-decomposed potential function

An important feature of the OLCAO method is the real space description of the crystal's charge density $\rho_{cry}(\vec{r})$ and the one-electron crystal potential $V_{cry}(\vec{r})$ as sums of atom-centered functions consisting of Gaussians. When carefully constructed, these site-specific atom-centered potential functions are transferable. Thus, it is possible to obtain a self-consistent potential from the calculation of a simpler system (say quartz, α-SiO$_2$) and then use it in the calculation of a larger and more complicated system (amorphous SiO$_2$ glass) without performing self-consistent calculations on the more complex system. In recent years however, the exceptional increase in computational power has rendered full self-consistent treatment of very complex systems possible (see examples in later chapters). Much of this can be attributed to the strategic choice to use atom-centered potential functions consisting of Gaussians. We write:

$$\rho_{cry}(\vec{r}) = \sum_A \rho_A \left(\vec{r} - \vec{t}_A\right), \quad \rho_A(\vec{r}) = \sum_{j=1}^{N} B_j e^{-\beta_j r^2} \qquad (3.10)$$

Similarly, each part of the potential function (the electron–nuclear and electron–electron Coulomb interaction V_{Coul} and the exchange-correlation potential V_{xc}) are also expressed as atom-centered functions as follows:

$$V_{Coul}(\vec{r}) = \sum_A V_C \left(\vec{r} - \vec{t}_A\right), \quad V_C(\vec{r}) = -\frac{Z_A}{r} e^{-\zeta r^2} - \sum_{j=1}^{N} D_j e^{-\beta_j r^2} \quad (3.11)$$

$$V_{xc}(\vec{r}) = \sum_A V_x \left(\vec{r} - \vec{t}_A\right), \quad V_x(\vec{r}) = \sum_{j=1}^{N} F_j e^{-\beta_j r^2} \qquad (3.12)$$

The first term in $V_C(\vec{r})$ accounts for the potential near the nuclei where Z_A is the atomic mass number of the atom at that site. The use of a rapidly decaying exponential factor in Eq. (3.11) facilitates the numerical evaluation of the integral using error functions (to be described later) and the exponential factor ζ in Eq. (3.11) can be flexible (usually taken as a fixed number 20 in the current implementation).

The crystal potential is then written as the sum of atom-centered potentials:

$$V_{cry}(\vec{r}) = \sum_A V_A \left(\vec{r} - \vec{t}_A\right), \quad V_A(\vec{r}) = V_C(\vec{r}) + V_x(\vec{r}) \qquad (3.13)$$

Here, we adopt a very crucial approximation which is central to the efficiency of the OLCAO method. The same exponential set $\{\beta_j\}$ is used in the Gaussian expansions in Eqs (3.10)–(3.12). This approximation reduces the number of multi-center integrals between GTOs that need to be evaluated by several orders of magnitude and enables the OLCAO method to be applied to large complex systems and still retain the *ab initio* character of the method. The exponential set $\{\beta_j\}$ is pre-determined for each atom whereas the coefficients B_j, D_j, F_j, and so on are updated in each cycle of the self-consistent iterations using a linear fitting procedure to the $\rho(\vec{r})$ data. Because the accuracy of the calculation depends largely on how accurately the true charge density can be represented by Eq. (3.10), the optimal choice of the fitting set $\{\beta_j\}$ is extremely important. This is usually achieved based on experience attained through the application of OLCAO to a large number of simple crystals with a comparison of calculated physical properties such as equilibrium lattice parameters, bulk modulus, optical properties, and so on to experimentally determined values. It can also be obtained by fitting Gaussians to more accurately calculated atomic charge densities. Like in the case of the basis functions, the $\{\beta_j\}$ set is also characterized by β_{min}, β_{max} and a number of terms N cast as a geometrical series. The accuracy of the fit is carefully monitored by comparing the integrated charge from the fitted functions to the total number of electrons in the simulation cell. The values chosen for β_{min}, β_{max}, and N depend on Z and sometimes the crystal under study, and are guided by the desire to minimize the total fitting error. As a rule of thumb, the heavier the atom, the larger are N and β_{max}. The choice of β_{min} is more subtle but it usually ranges between 0.08 to 0.15 depending on the element and the system being studied. Again, for smaller β_{min}, the potential range is longer and vice versa for larger values. We have a complete data base of $\{\beta_j\}$ sets for the most common elements in electronic structure calculations which are listed in Appendix B where the default values for β_{min}, β_{max} and N for these elements are listed.

It should be noted that although the transferable atom-centered functions $V_A(\vec{r})$ and $\rho_A(\vec{r})$ in Eqs (3.10) and (3.13) consist of spherical Gaussian functions around each atom, the superposition of them is non-spherical and can accurately represent different types of bonding in different types of structural configurations without shape approximation. This is illustrated in Fig. 3.4.

The concentric circles surrounding the atomic centers in Fig. 3.4 have diameters corresponding to the full width at half of the maximum height (FWHM) of the normalized Gaussian packet with exponential β_j. Although each Gaussian function is spherical, their superposition is clearly not and can accurately represent the charge density and potential in three dimensions. Such an analytic approach is totally different from the usual Cartesian numerical grids employed in many other computational packages where the charge density and the potential at each grid point must be evaluated. Accurate calculations will then require a fine grid with a large number of grid points. Data storage can be a real problem in this case making it difficult to apply that approach to large complex materials. In contrast, the method of Gaussian orbitals employed in the OLCAO method does not have this problem. Both the charge distribution and potential in real space are calculated analytically via the technique of Gaussian transformation (see next section) and the data management for very large systems is not prohibitive.

In solving the Kohn–Sham equation, Eq. (3.4), iteratively, the new $\rho(\vec{r})$ is calculated from the occupied state wave functions $\Psi_{n\vec{k}}(\vec{r})$. It is a common practice to blend the old $\rho(\vec{r})$ with the newly determined $\rho(\vec{r})$ for the next iteration to improve numerical stability. The coefficients B_j, D_j, F_j, in Eqs (3.10)–(3.12) are constantly updated until convergence is obtained when the

Fig. 3.4.
A sketch showing the superposition of atom-centered potential functions in a hypothetical two-atom crystal in two dimensions.

total energy changes by less than a predetermined value, typically 0.00001 or 0.0001 or other similar criterion. Various acceleration schemes for fast convergence can be used. In general, insulating systems converge quickly because of the existence of a band gap that unambiguously separates the occupied and unoccupied states. In metallic systems, the presence of a Fermi level may complicate the matter slightly, a sufficiently large number of k-space sampling points may be necessary to determine the Fermi level accurately. Also, for non-crystalline systems or crystals with many non-equivalent sites of the same element, each site can be treated as independent to increase the accuracy in the crystal potential representation based on the super-positions of the potential functions at different sites. On the other hand, for crystals with high symmetry, the $V_A(\vec{r})$ and $\rho_A(\vec{r})$ for each equivalent site in the crystal are kept the same through symmetry operations. In the intermediate case, where there are a large number of sites that are similar but not crystallographically equivalent, it is desirable to group the sites based on their degree of similarity where the best compromise can be reached between accuracy and computational efficiency.

The OLCAO method's site-decomposed charge density $\rho_A(\vec{r})$ and potential $V_A(\vec{r})$ in real space are very useful outputs for calculating many physical observables. They can also be utilized in creative simulation studies. More importantly, the expansion of ρ_A and V_A in terms of Gaussians facilitates the analytic evaluation of multi-center integrals in the Hamiltonian matrix elements via the technique of Gaussian transformation. This is now discussed in the section below.

3.4 The technique of Gaussian transformation

The Gaussian transformation technique is at the core of the OLCAO method and the technique has a long history dating back to at least to 1950. (Boys, 1950, Boys and Shavitt, 1960, Huzinaga, 1965, Shavitt and Karplus, 1965) Its effectiveness in calculating the integrals of molecular orbitals has been recognized for a long time in quantum chemistry. The popular Gaussian package (Frisch et al., 2009) is a prime example. Gaussian transformation's application to problems in condensed matter physics is comparatively smaller where plane wave basis expansions are usually preferred for periodic crystals satisfying the Bloch theorem. Here we outline the main steps in the use of the Gaussian transformation technique in the OLCAO method.

The band structure of the crystal is obtained by solving the Kohn–Sham equation, Eq. (3.4), at various k-points in the Brillouin zone (BZ), or equivalently by solving the secular equation:

$$|H_{i\gamma,j\delta}(\vec{k}) - S_{i\gamma,j\delta}(\vec{k})E(\vec{k})| = 0 \tag{3.14}$$

where $S_{i\gamma,j\delta}(\vec{k})$ and $H_{i\gamma,j\delta}(\vec{k})$ are the overlap and Hamiltonian matrix respectively:

$$S_{i\gamma,j\delta}\left(\vec{k}\right) = \left\langle b_{i\gamma}\left(\vec{k},\vec{r}\right)|b_{j\delta}\left(\vec{k},\vec{r}\right)\right\rangle$$

$$= \sum_{\mu} e^{-i\vec{k}\bullet\vec{R}_{\mu}} \int u_i\left(\vec{r}-\vec{t}_{\gamma}\right) u_j\left(\vec{r}-\vec{R}_{\mu}-\vec{t}_{\delta}\right) d\vec{r} \tag{3.15}$$

$$H_{i\gamma,j\delta}(\vec{k}) = \left\langle b_{i\gamma}(\vec{k},\vec{r})|H|b_{j\delta}(\vec{k},\vec{r})\right\rangle$$

$$= \sum_{\mu} e^{-i\vec{k}\bullet\vec{R}_{\mu}} \int u_i\left(\vec{r}-\vec{t}_{\gamma}\right)[-\nabla^2 + V_{coul}(\vec{r}) + V_{ex}(\vec{r})]u_j$$

$$(\vec{r}-\vec{R}_{\mu}-\vec{t}_{\delta})d\vec{r} \tag{3.16}$$

For calculations with large periodic cells or supercells, the lattice sums in Eqs (3.15) and (3.16) converge very rapidly and entail the summation over the neighboring cells only. Moreover, for a large cell the corresponding BZ is usually very small and a single k- point at the zone center Γ is sufficient. Thus, in spite of the large number of atoms in a supercell, the total number of interaction integrals that need to be evaluated in Eqs (3.15) and Eq. (3.16) is still kept at a manageable level.

With the one-electron LDA potential and charge density expressed as site decomposed atom-centered Gaussian functionals as shown in Eqs (3.11)–(3.13), let us now consider u_i to be an s-type function ($\ell = 0$) consisting of a simple Gaussian orbital with exponential α_1 centered at the atomic site A,

$$|s_A\rangle = e^{-\alpha_1 \vec{r}_A^2}; \quad \vec{r}_A = \vec{r} - \vec{A} \tag{3.17}$$

The merit of using GTOs in the basis expansion and in the potential function representation is that all the interaction integrals in Eq. (3.15) and Eq. (3.16) can be expressed in analytical forms. It can be shown that the integrals involved are of the following five types:

Overlap integral

$$I1 = \langle s_A | s_B \rangle = \int e^{-\alpha_1 \vec{r}_A^2} e^{-\alpha_2 \vec{r}_B^2} d\vec{r} \qquad (3.18)$$

Kinetic energy integral

$$I2 = \langle s_A | -\nabla^2 | s_B \rangle = \int e^{-\alpha_1 \vec{r}_A^2} (-\nabla^2) e^{-\alpha_2 \vec{r}_B^2} d\vec{r} \qquad (3.19)$$

Three-center integral

$$I3 = \langle s_A | e^{-\alpha \vec{r}_C^2} | s_B \rangle = \int e^{-\alpha_1 \vec{r}_A^2} e^{-\alpha_3 \vec{r}_C^2} e^{-\alpha_2 \vec{r}_B^2} d\vec{r} \qquad (3.20)$$

*Three-center over **r** integral*

$$I4 = \left\langle s_A \left| \frac{1}{r_C} e^{-\alpha \vec{r}_C^2} \right| s_B \right\rangle = \int e^{-\alpha_1 \vec{r}_A^2} \frac{1}{r_C} e^{-\alpha_3 \vec{r}_C^2} e^{-\alpha_2 \vec{r}_B^2} d\vec{r} \qquad (3.21)$$

The matrix elements of the momentum operator can be expressed in a similar fashion:

Momentum integral

$$I5 = \left\langle s_A | \vec{P} | s_B \right\rangle = -i\hbar \int e^{-\alpha_1 \vec{r}_A^2} \left(\frac{\partial}{\partial x_B}, \frac{\partial}{\partial y_B}, \frac{\partial}{\partial z_B}, \right) e^{-\alpha_2 \vec{r}_B^2} d\vec{r} \qquad (3.22)$$

$I1$, $I2$, and $I5$ are two center integrals involving a Gaussian at site A and another at site B. $I3$ and $I4$ are three-center integrals involving another Gaussian from the potential function centered at site C. Three-center integrals are more time consuming to evaluate than two-center integrals, and because there are so many more of them they account for the bulk of the time spent in computing these integrals. The two-center integral $I1$ can be considered as a special case of $I3$ with $\alpha_3 = 0$.

By applying the technique of Gaussian transformation as illustrated in Fig. 3.5, it can be shown that $I3$ is reduced to a closed form:

$$I3 = \left\langle s_A | e^{-\alpha_3 \vec{r}_A^2} | s_B \right\rangle = \left[\frac{\pi}{\alpha_T} \right]^{\frac{3}{2}} e^{\left[\alpha_T \vec{E}^2 - \alpha_1^2 - \alpha_2^2 - \alpha_3^2 \right]} \qquad (3.23)$$

where $\alpha_T = \alpha_1 + \alpha_2 + \alpha_3$ and $\vec{E} = (\alpha_1 \vec{A} + \alpha_2 \vec{B} + \alpha_3 \vec{C})/(\alpha_1 + \alpha_2 + \alpha_3)$.

The kinetic energy integral $I2$ and the momentum integral $I5$ can be obtained from the overlap integral $I1$ by direct differentiation:

$$\left\langle s_A | -\nabla^2 | s_B \right\rangle = \left[6\alpha_2 + 4\alpha_2^2 \frac{\partial}{\partial \alpha_2} \right] \langle s_A | s_B \rangle \qquad (3.24)$$

$$\langle s_A | p_x | s_B \rangle = \left\langle s_A | -i\hbar \frac{\partial}{\partial x_B} | s_B \right\rangle = -i\hbar \frac{\partial}{\partial x_B} \langle s_A | s_B \rangle \qquad (3.25)$$

*I*4 cannot be expressed in closed form, but can be expressed in terms of standard error functions.

$$I4 = \left\langle s_A \left| \frac{1}{\vec{r}_C} e^{-\alpha \vec{r}_C^2} \right| s_B \right\rangle = \left[\frac{2\pi}{\alpha_T t} \right] e^{\left(-\alpha_1 |\vec{A} - \vec{C}|^2 - \alpha_2 |\vec{B} - \vec{C}|^2 + \frac{(\alpha_1 + \alpha_2)^2 |\vec{D} - \vec{C}|^2}{\alpha_T} \right)}$$

$$\int_0^t e^{-\frac{z^2}{2}} dz \tag{3.26}$$

where $D = \frac{\alpha_1 \vec{A} + \alpha_2 \vec{B}}{\alpha_1 + \alpha_2}$, and $t = \sqrt{2}(\alpha_1 + \alpha_2)|\vec{D} - \vec{C}|^2 \cdot \frac{1}{\sqrt{\alpha_T}}$.

We now consider the two kinds of three-center integrals around three sites *A*, *B*, and *C*, that is

$$\left\langle s_A \left| e^{-\alpha_3 r_C^2} \right| s_B \right\rangle \equiv \left\langle e^{-\alpha_1 r_A^2} \left| e^{-\alpha_3 r_C^2} \right| e^{-\alpha_2 r_B^2} \right\rangle \tag{3.27}$$

$$\left\langle s_A \left| \frac{1}{r_C} e^{-\alpha_3 r_C^2} \right| s_B \right\rangle \equiv \left\langle e^{-\alpha_1 r_A^2} \left| \frac{1}{r_C} e^{-\alpha_3 r_C^2} \right| e^{-\alpha_2 r_B^2} \right\rangle \tag{3.28}$$

where $\mathbf{r}_A$ is the distance between the electron and site *A*, and so on. We define

$$\alpha_T = \alpha_1 + \alpha_2 + \alpha_3, \tag{3.29}$$

$$\vec{E} = \frac{\left(\alpha_1 \vec{A} + \alpha_2 \vec{B} + \alpha_3 \vec{C} \right)}{\alpha_T}, \tag{3.30}$$

$$\overrightarrow{AE} = |\overrightarrow{AE}| = |\vec{A} - \vec{E}|, \text{ etc.} \tag{3.31}$$

where *A* is the vector from origin to site *A*. The relation between various points and vectors is illustrated in Fig. 3.5.

It follows that

$$\vec{r}_A = \vec{r}_E + \overrightarrow{AE}, \text{ etc.} \tag{3.32}$$

$$\alpha_1 \overrightarrow{AE} + \alpha_2 \overrightarrow{BE} + \alpha_3 \overrightarrow{CE} = 0 \tag{3.33}$$

which, when combined with Eq. (3.27), gives

$$\left\langle s_A | e^{-\alpha_3 r_C^2} | s_B \right\rangle = e^{\left(-\alpha_1 \overrightarrow{AE}^2 - \alpha_2 \overrightarrow{BE}^2 - \alpha_3 \overrightarrow{CE}^2 \right)} \int e^{-\alpha_T r_E^2} d\tau$$

$$= \left(\frac{\pi}{\alpha_T} \right)^{\frac{3}{2}} e^{\left(-\alpha_1 \overrightarrow{AE}^2 - \alpha_2 \overrightarrow{BE}^2 - \alpha_3 \overrightarrow{CE}^2 \right)} \tag{3.34}$$

To evaluate Eq. (3.28) we introduce

$$\vec{D} = \frac{\alpha_1 \vec{A} + \alpha_2 \vec{B}}{\alpha_1 + \alpha_2} \tag{3.35}$$

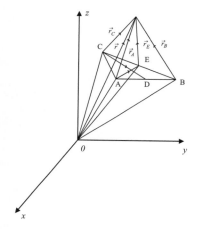

Fig. 3.5.

Vector relations between various points in the method of Gaussian transformation.

and take the z-axis along $\overrightarrow{CD}$ so that

$$\left\langle s_A \left| \frac{1}{r_C} e^{-\alpha_3 r_C^2} \right| s_B \right\rangle = e^{-\alpha_1 \overrightarrow{CA}^2 - \alpha_2 \overrightarrow{CB}^2} \int e^{-\alpha_T r_C^2 r_C^{-1}} e^{2(\alpha_1+\alpha_2)\vec{r}_C \bullet \overrightarrow{CD}} d\tau$$

$$= 2\pi e^{-\alpha_1 \overrightarrow{CA}^2 - \alpha_2 \overrightarrow{CB}^2} \int_0^\infty \int_{-1}^1 e^{-\alpha_T r_C^2 r_C^{-1}} e^{2(\alpha_1+\alpha_2)\vec{r}_C \bullet \overrightarrow{CD}\cos\theta} d(\cos\theta) dr_C$$

$$= \left(\frac{\pi}{(\alpha_1+\alpha_2)\overrightarrow{CD}} \right) e^{-\alpha_1 \overrightarrow{CA}^2 - \alpha_2 \overrightarrow{CB}^2}$$

$$\int_0^\infty e^{-\alpha_T r^2} \left(e^{2(\alpha_1+\alpha_2)\overrightarrow{CD}r_C} - e^{-2(\alpha_1+\alpha_2)\overrightarrow{CD}r_C} \right) dr \qquad (3.36)$$

Upon letting

$$\beta = \frac{(\alpha_1+\alpha_2)\overrightarrow{CD}}{\alpha_T} \qquad (3.37)$$

$$t = \frac{\sqrt{2}(\alpha_1+\alpha_2)\bar{C}\vec{D}}{\sqrt{\alpha_T}} \qquad (3.38)$$

the above reduces to

$$\left\langle s_A \left| \frac{1}{r_C} e^{-\alpha_3 r_C^2} \right| s_B \right\rangle = \left(\frac{\pi}{(\alpha_1+\alpha_2)\overrightarrow{CD}} \right) e^{\left(-\alpha_1 \overrightarrow{CA}^2 - \alpha_2 \overrightarrow{CB}^2 + \frac{(\alpha_1+\alpha_2)\overrightarrow{CD}^2}{\alpha_T} \right)}$$

$$\left(\int_{-B}^\infty e^{-\alpha_T y^2} dy - \int_B^\infty e^{-\alpha_T y^2} dy \right)$$

$$= \left(\frac{2\pi}{\alpha_T t} \right) e^{\left(-\alpha_1 \overrightarrow{CA}^2 - \alpha_2 \overrightarrow{CB}^2 + \frac{(\alpha_1+\alpha_2)\overrightarrow{CD}^2}{\alpha_T} \right)} \int_0^t e^{-z^2/2} dz \qquad (3.39)$$

The analytic expressions in Eqs (3.18)–(3.22) pertain to four types of integrals between the simplest s-type of GTO. The corresponding expressions for integrals involving p-type, d-type, and f-type of GTOs can be obtained from those of lower ℓ quantum number by successive differentiations with respect to Cartesian components of lattice coordinates. For example,

$$\langle p_A^x | s_B \rangle = \int x_A e^{-\alpha_1 \vec{r}_A^2} e^{-\alpha_2 \vec{r}_B^2} d\vec{r}$$

$$= \int \frac{1}{2\alpha_1} \frac{\partial}{\partial A_x} e^{-\alpha_1 \vec{r}_A^2} e^{-\alpha_2 \vec{r}_B^2} d\vec{r} = \frac{1}{2\alpha_1} \frac{\partial}{\partial A_x} \langle s_A | s_B \rangle \qquad (3.40)$$

$$\langle p_A^x | p_B^y \rangle = \frac{1}{2\alpha_2} \frac{\partial}{\partial B_y} \langle p_A^x | s_B \rangle \text{ etc.} \qquad (3.41)$$

Analytic formulae for integrals of type $I1$ to $I5$ up to $\ell = 3$ (f-orbitals) have been derived. Repeated differentiations quickly result in long complicated messy formulae. Many techniques using recursive computations of these integrals have been suggested (Obara and Saika, 1986). It is possible to use special computer software such as Maple or Mathematica to derive such expressions to check for accuracy and correctness.

It is also possible to derive the generalized form for the integrals involving the Cartesian GTO (Browne and Poshusta, 1962):

$$x^n y^l z^m e^{-\alpha r^2} \tag{3.42}$$

where $n + l + m = 0$, 1, 2, 3 for an s, p, d, f type of GTO. For example, the generalized overlap integrals can be written as:

$$s_{nlm,n'l'm'}(\alpha_1, \alpha_2, \vec{A}, \vec{B}) = \int x_A^n y_A^l z_A^m e^{-\alpha_1 \vec{r}_A^2} x_B^{n'} y_B^{l'} z_B^{m'} e^{-\alpha_2 \vec{r}_B^2} d\tau$$

$$= S_{n,n'}(\alpha_1, \alpha_2, A_x, B_x) S_{l,l'}(\alpha_1, \alpha_2, A_y, B_y) S_{m,m'}(\alpha_1, \alpha_2, A_z, B_z), \tag{3.43}$$

where:

$$S_{n,n'}(\alpha_1, \alpha_2, A_x, B_x) = e^{-\hbar \overline{AB}_x} \sum_S \binom{n}{s} \overline{AD}_x^{n-s} \sum_{S'} \binom{n'}{s'} \overline{BD}_x^{n'-s'} F_{s+sa'}(\beta)$$

where:

$$\beta = \alpha_1 + \alpha_2, \; h = \frac{\alpha_1 \alpha_2}{\beta}, \; D = \frac{\alpha_1 \vec{A} + \alpha_2 \vec{B}}{\beta}, \; \overline{AB}_x = B_x - A_x,$$

and

$$F_n(\beta) = N_n \beta^{-(n+1)/2}, \; N_n = \pi^{1/2} \begin{cases} 1 & \text{if } n = 0 \\ 0 & \text{if } n = \text{odd} \\ (n-1)!/2^{n/2} & \text{if } n = \text{even} \end{cases} \tag{3.44}$$

For cubic or rectangular supercells used to model amorphous solids, it was claimed (E. E. Lafon, private communication) that all the integrals can be factorized as above such that the number of multi-center integrals that need to be evaluated goes with 3N, not N^3 where N is the total number of orbitals. This approach has not been sufficiently tested but could potentially further facilitate the large scale computation associated with complex systems using GTOs. However, the efficiency of actual calculations for large complex systems depends on many factors. There is no fixed recipe for the specific execution of the OLCAO method. For most calculations, over 90% of the multi-center integrals calculated are either zero or negligibly small when the inter-atomic separation distances are large. Specific filtering techniques that have been implemented in the non-generalized form of the integral calculations may prove to be more efficient. In this respect, the OLCAO method based on GTOs for basis expansion is truly an order N method for large systems.

3.5 The technique of core orthogonalization

For large complex systems such as amorphous materials, supercell models for defects and microstructures, or complicated crystals with large Z elements, the

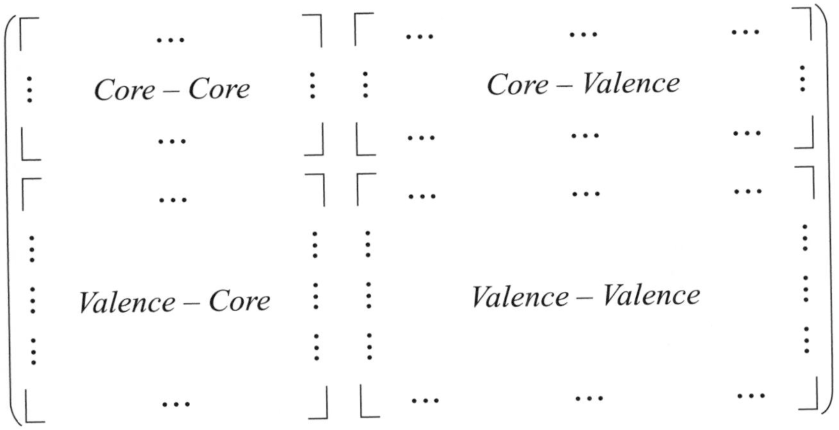

$$\left(\begin{array}{ll} \begin{bmatrix} & \cdots & \\ \vdots & Core - Core & \vdots \\ & \cdots & \end{bmatrix} & \begin{bmatrix} \cdots & \cdots & \cdots \\ Core - Valence & & \vdots \\ \cdots & \cdots & \cdots \end{bmatrix} \\ \begin{bmatrix} & \cdots & \\ \vdots & & \vdots \\ \vdots & Valence - Core & \vdots \\ \vdots & & \vdots \\ & \cdots & \end{bmatrix} & \begin{bmatrix} \cdots & \cdots & \cdots \\ \vdots & & \vdots \\ Valence - Valence & & \vdots \\ \vdots & & \vdots \\ \cdots & \cdots & \cdots \end{bmatrix} \end{array}\right)$$

Fig. 3.6.
Illustration of matrix representation for the orthogonalization scheme.

matrix equation, Eq. (3.14), is large and its solution requires a large amount of computer time and memory. However, in many cases, the core states of a solid (typically those $\sim 30\,\text{eV}$ below the highest occupied state) are of little physical interest, the calculation can be reduced to only the non-core valence electron states and any of the unoccupied states in the full basis or extended basis expansion. Core states can be eliminated from the secular equation, Eq. (3.14), to reduce its dimension by the orthogonalization process described below (Ching and Lin, 1975b).

The overlap and the Hamiltonian matrices in Eq. (3.14) are rearranged by interchanging the rows and the columns such that all core states are in the upper-left quadrant and all the non-core states are in the lower right quadrant as illustrated in Fig. 3.6 below. The matrix elements between Bloch sums in Eq. (3.14) can be divided into three groups: (1) core–core, (2) core–valence and valence–core, and (3) valence–valence. Here we use the term "valence" to include all non-core orbitals for simplicity. By assuming that the matrix elements between core Bloch sums on different sites are effectively zero, and hence negligible, and also by imposing the orthogonality condition, the dimension of the matrix equation can be effectively reduced. The orthogonalized matrix elements in the valence–valence block can be expressed as the original non-orthogonalized elements plus correction terms involving the core–valence and the valence–core matrix elements.

Let us use the superscripts v and c to denote the valence (or non-core) and the core portions of the Bloch sums and v' to denote the orthogonalized valence (or non-core) Bloch sums. Expanding the orthogonalized Bloch sum $b_{i\alpha}^{v'}(\vec{k}, \vec{r})$ in terms of the original non-orthogonalized ones, we have:

$$b_{i\alpha}^{v'}\left(\vec{k}, \vec{r}\right) = b_{i\alpha}^{v}\left(\vec{k}, \vec{r}\right) + \sum_{j, \gamma} C_{j\gamma}^{i\alpha} b_{j\gamma}^{c}\left(\vec{k}, \vec{r}\right) \qquad (3.45)$$

Imposing the orthogonality condition:

$$\left\langle b_{j\beta}^{c}\left(\vec{k}, \vec{r}\right) \middle| b_{i\alpha}^{v'}\left(\vec{k}, \vec{r}\right) \right\rangle = \left\langle b_{i\alpha}^{v'}\left(\vec{k}, \vec{r}\right) \middle| b_{j\beta}^{c}\left(\vec{k}, \vec{r}\right) \right\rangle = 0,$$

the expansion coefficients in Eq. (3.45) are given by:

$$C_{j\gamma}^{i\alpha} = -\left\langle b_{j\gamma}^c\left(\vec{k},\vec{r}\right)\Big| b_{i\alpha}^v\left(\vec{k},\vec{r}\right)\right\rangle$$

$$C_{j\gamma}^{i\alpha*} = -\left\langle b_{i\alpha}^v\left(\vec{k},\vec{r}\right)\Big| b_{j\beta}^c\left(\vec{k},\vec{r}\right)\right\rangle \tag{3.46}$$

After orthogonalization, the matrix elements in the lower-right quadrant in Fig. 3.6 are given by:

$$\left\langle b_{i\alpha}^{v'}\left(\vec{k},\vec{r}\right)\Big| b_{j\beta}^{v'}\left(\vec{k},\vec{r}\right)\right\rangle = \left\langle b_{i\alpha}^v\left(\vec{k},\vec{r}\right)\Big| b_{j\beta}^v\left(\vec{k},\vec{r}\right)\right\rangle$$

$$- \sum_{l,\gamma}\left\langle b_{i\alpha}^v\left(\vec{k},\vec{r}\right)\Big| b_{l\gamma}^c\left(\vec{k},\vec{r}\right)\right\rangle\left\langle b_{l\gamma}^c\left(\vec{k},\vec{r}\right)\Big| b_{j\beta}^v\left(\vec{k},\vec{r}\right)\right\rangle$$

$$- \sum_{l,\gamma}\left\langle b_{l\gamma}^c\left(\vec{k},\vec{r}\right)\Big| b_{j\beta}^v\left(\vec{k},\vec{r}\right)\right\rangle\left\langle b_{j\beta}^v\left(\vec{k},\vec{r}\right)\Big| b_{l\gamma}^c\left(\vec{k},\vec{r}\right)\right\rangle$$

$$- \sum_{l,\gamma}\sum_{m,\delta}\left\langle b_{i\alpha}^v\left(\vec{k},\vec{r}\right)\Big| b_{l\gamma}^c\left(\vec{k},\vec{r}\right)\right\rangle\left\langle b_{m\delta}^c\left(\vec{k},\vec{r}\right)\Big| b_{j\beta}^v\left(\vec{k},\vec{r}\right)\right\rangle\delta_{lm}\delta_{\gamma\delta} \tag{3.47}$$

$$\left\langle b_{i\alpha}^{v'}\left(\vec{k},\vec{r}\right)\Big| H\Big| b_{j\beta}^{v'}\left(\vec{k},\vec{r}\right)\right\rangle = \left\langle b_{i\alpha}^v\left(\vec{k},\vec{r}\right)\Big| H\Big| b_{j\beta}^v\left(\vec{k},\vec{r}\right)\right\rangle$$

$$- \sum_{l,\gamma}\left\langle b_{i\alpha}^v\left(\vec{k},\vec{r}\right)\Big| b_{l,\gamma}^c\left(\vec{k},\vec{r}\right)\right\rangle\left\langle b_{l\gamma}^c\left(\vec{k},\vec{r}\right)\Big| H\Big| b_{j\beta}^v\left(\vec{k},\vec{r}\right)\right\rangle$$

$$- \sum_{l,\gamma}\left\langle b_{l\gamma}^c\left(\vec{k},\vec{r}\right)\Big| b_{j\beta}^v\left(\vec{k},\vec{r}\right)\right\rangle\left\langle b_{j\beta}^v\left(\vec{k},\vec{r}\right)\Big| H\Big| b_{l\gamma}^c\left(\vec{k},\vec{r}\right)\right\rangle$$

$$- \sum_{l,\gamma}\sum_{m,\delta}\left\langle b_{i\alpha}^v\left(\vec{k},\vec{r}\right)\Big| b_{l,\gamma}^c\left(\vec{k},\vec{r}\right)\right\rangle\left\langle b_{m\delta}^c\left(\vec{k},\vec{r}\right)\Big| b_{j\beta}^v\left(\vec{k},\vec{r}\right)\right\rangle$$

$$\times \left\langle b_{l\gamma}^c\left(\vec{k},\vec{r}\right)\Big| H\Big| b_{m\delta}^c\left(\vec{k},\vec{r}\right)\right\rangle \tag{3.48}$$

The new secular equation in the orthogonalized space is:

$$\left|\left\langle b_{i\alpha}^{v'}\left(\vec{k},\vec{r}\right)\Big| H\Big| b_{j\beta}^{v'}\left(\vec{k},\vec{r}\right)\right\rangle - \left\langle b_{i\alpha}^{v'}\left(\vec{k},\vec{r}\right)\Big| b_{j\beta}^{v'}\left(\vec{k},\vec{r}\right)\right\rangle E\left(\vec{k}\right)\right| = 0 \tag{3.49}$$

which has its dimension reduced. The main numerical error present in Eq. (3.49) is the assumption of no core–core overlap between the Bloch sums. Test calculations on Si in which the $1s$, $2s$, and $2p$ orbitals are considered to be the core orbitals indicate that the eigenvalues for states near the gap obtained from Eq. (3.49) differ from those obtained from Eq. (3.14) by less than 0.0001 eV (Ching and Lin, 1975b). This is of negligible difference for most electronic structure studies. One way to improve the accuracy is to treat some high lying core states as valence states in the orthogonalization scheme. It is also possible to retain the core states of one or a few atoms in a simulation cell in order to study the electronic structures of the core states specific to those atoms (see Chapter 11).

3.6 Brillouin zone integration

For a regular crystal, the Kohn–Sham equation, Eq. (3.4) or (3.14), has to be solved at many k-points inside the irreducible portion of the BZ. This is usually done by using either a tetrahedral or special k-point method. The tetrahedral method is based on dividing the irreducible wedge of the BZ into space filling tetrahedrons and linearly interpolating the energy eigenvalues within the tetrahedral units (Lehmann and Taut, 1972, Rath and Freeman, 1975). For identical numbers of k-points the tetrahedron method has much higher spectral feature resolution for k-dependent functionals of the crystal that are integrated over the volume of the BZ (e.g. the density of states (DOS)). In the special k-point sampling method (Chadi and Cohen, 1973, Monkhorst and Pack, 1976), special k-points within the BZ with different weighting factors are used which also provide an effective means of integration over the BZ. The number of k-points required for accurate BZ integration depends on the symmetry of the crystal and is inversely proportional to the volume of the unit cell. It also depends on the materials type whose electronic structure is being studied.

For an insulating crystal with a well-defined band gap, the use of a large number of k-points may not be necessary. However, for metallic systems with a Fermi level, a much larger number of k-points are usually required since the precise determination of the Fermi energy is important for fast convergence of the self-consistent potential and charge density. Since the computational cost generally scales with the number of k-points used and the number of atoms in the unit cell, it is important to determine the appropriate number of k-points needed through sensible tests. Most applications of the OLCAO methods are for large complex systems (see later chapters for many types of applications) where large supercells are generally employed, a single k-point at the zone center ($k = 0$) then suffices. In the rare case where more than one k-point is desirable, it is better to use additional k-points at the corner of the reduced BZ where the matrix equations are again real due to symmetry. This is a significant advantage because complex matrix equations can take more CPU time by a factor of four to eight depending on the operation.

In general, the matrix elements in Eq. (3.49) are not sparse because interactions in the present applications of the OLCAO method are sufficiently far-ranging. In dealing with complex systems, the dimension of Eq. (3.49) may still be formidably large, especially if an extended basis is used, so that other time-saving numerical techniques may need to be applied. Recent advances in computer technology have rendered this problem less urgent. Depending on machine design, it is now possible to perform full diagonalization of matrix of the size 30,000 × 30,000 so systems with 5000 or larger atoms can be dealt with if a sp^3 minimal basis is used. This has significantly expanded the types of problems that can be solved to include large biological molecules or embedded nano-particles.

3.7　Special advantages in the OLCAO method

The basic theory and techniques of the OLCAO method have been presented above, but it is easy to lose sight of what it is about the OLCAO approach that gives it special advantages when dealing with the full range of simple to complex systems. Some advantages are derived from the use of atomic orbitals as the basis for expanding the solid state wave function while others come from the extensive use of Gaussian functions. In this section we will comment briefly on many of the different contributing factors that lend advantage to the OLCAO method.

The solid state wave function is expanded from a finite set of atomic orbital functions. This approach makes direct interpretation of the solid state wave function in terms of traditional chemistry concepts very easy and it can be used to great effect when sorting through the data obtained from large and complicated systems. Even for an extended basis set the dimension of the matrix to be diagonalized when solving the secular equation is relatively small and hence it is possible to use direct diagonalization methods to obtain the full spectrum of energy eigenvalues and eigenvectors at a given k point. With such an extended basis set, the energy spectrum can reach relatively high into the unoccupied states, which is important if it is necessary to compute optical transitions at a high photon frequency. The choice of the number of orbitals to include in the basis set is also quite flexible so that one can select the best basis for a given type of calculation considering the time, disk space, and accuracy that is required. For example, an extended basis for optical properties calculations and a minimal basis for Mulliken effective charge and bond order calculations. Further, because certain orbitals can be designated as "core" orbitals and thus orthogonalized out of the secular equation it is possible to isolate specific core orbitals for studying core level transitions or core level photoemission. The basis functions are also transferable to a very large extent so that, generally speaking, they can be pulled from a data base for use in any type of crystal without the need to prepare special functions for different types of studies.

The basis functions themselves are expanded in terms of Gaussian functions. This approach then allows for efficient analytic evaluation of multi-center integrals through the Gaussian transformation technique with no approximations. A further gain in efficiency comes from the use of the same Gaussians for all orbitals of a given element's basis functions where the only modification is the leading coefficient and the exact number of Gaussians to use for each s, p, d, or f orbital. While this does add to the number of Gaussians, it greatly improves the efficiency because so many of the integrals can be reused and the others can be grouped together and computed at once in the algorithm's loop structure.

Each of the different parts of the potential function of the solid state system is in the functional form of atom-centered Gaussians. The complete potential function in each calculation is iterated to self-consistency and the potential provided in the data base is used only as the initial potential. This part is very different from the pseudopotential methods where the choice of which pseudopotential to use must be predetermined and are not always transferable. In the case of the OLCAO method, the form of the potential function for each element is pre-determined but they remain broadly transferrable because of the aforementioned self-consistent iteration of the coefficients. The charge density uses the exact same set of atom-centered Gaussians as the potential function. This also greatly reduces the total number of integrals that need to be computed. Specifically, the electron–electron interactions can be computed using three-center Gaussian integrals instead of four-center integrals. It is also helpful that the atom-centered potential and charge density functions can be used for easy visual presentation on dense three dimensional grids in real space without separate lengthy calculations.

Beyond the specific approach, the execution of the OLCAO code follows a simple model that can be judiciously separated into different stages to suit different machine configurations and computational platforms. The OLCAO method can be easily coupled with other popular electronic structure codes to excellent synergistic effect. For example, the plane wave pseudopotential method as implemented in the VASP package is very efficient in relaxing the structures of materials, whereas the physical properties of the resulting structure can be efficiently analyzed using the OLCAO method. This strategy will be demonstrated in many examples in later chapters. Because of the high efficiency of the OLCAO method where a large number of calculations is not prohibitive, it can be used to study electronic properties under different physical conditions such as temperature, pressure, strain, or with the presence of defect levels and so on as long as the atomic scale structures can be accurately determined by other methods. In principle, the OLCAO method can be applied to any materials with any elements in the periodic table. In reality, this is still a goal to be reached in the future because the physics involving heavy elements with f electrons, or highly correlated systems cannot be adequately described by density functional theory. The OLCAO method represents an excellent balance between efficiency, accuracy, and ease of interpretation so that the results are reliable and they can be obtained and understood for a wide variety of complex systems.

References

Barth, U. V. & Hedin, L. (1972), *Journal of Physics C: Solid State Physics*, 5, 1629.

Becke, A. D. (1986), *The Journal of Chemical Physics*, 85, 7184–7.

Boys, S. F. (1950), *Proceedings of the Royal Society of London. Series A. Mathematical and Physical Sciences*, 200, 542–54.

Boys, S. F. & Shavitt, I. (1960), *Proceedings of the Royal Society of London. Series A. Mathematical and Physical Sciences*, 254, 487–98.

Browne, J. C. & Poshusta, R. D. (1962), *The Journal of Chemical Physics*, 36, 1933–7.

Ceperley, D. M. & Alder, B. J. (1980), *Physical Review Letters*, 45, 566.

Chadi, D. J. & Cohen, M. L. (1973), *Physical Review B*, 8, 5747.

Ching, W. Y. & Lin, C. C. (1975a), *Phys. Rev. Lett.*, 34, 1223–6.

Ching, W. Y. & Lin, C. C. (1975b), *Physical Review B*, 12, 5536.

Clementi, E. & Roetti, C. (1974), *Atomic Data and Nuclear Data Tables*, 14, 177–478.

Frisch, M. J., Trucks, G. W., Schlegel, H. B., et al. (2009), Gaussian 09. Wallingford, CT: Gaussian, Inc.

Gunnarsson, O. & Lundqvist, B. I. (1976), *Physical Review B*, 13, 4274.

Hedin, L. & Lundqvist, B. I. (1971), *Journal of Physics C: Solid State Physics*, 4, 2064.

Huzinaga, S. (1965), *The Journal of Chemical Physics*, 42, 1293–302.

Huzinaga, S. (1984), *Gaussian Basis Sets for Molecular Calculation* (New York: Elsevier).

Lee, C., Yang, W. & Parr, R. G. (1988), *Physical Review B*, 37, 785.

Lehmann, G. & Taut, M. (1972), *physica status solidi (b)*, 54, 469–77.

Monkhorst, H. J. & Pack, J. D. (1976), *Physical Review B*, 13, 5188.

Mulliken, R. S. (1955), *J. Chem. Phys.*, 23, 1833.

Obara, S. & Saika, A. (1986), *The Journal of Chemical Physics*, 84, 3963–74.

Ouyang, L. & Ching, W. Y. (2001), *J. Am. Ceram. Soc.*, 84, 801–5.

Perdew, J. P. & Wang, Y. (1992), *Physical Review B*, 45, 13244.

Perdew, J. P., Burke, K., & Ernzerhof, M. (1996), *Phys. Rev. Lett.*, 77, 3865.

Rath, J. & Freeman, A. J. (1975), *Physical Review B*, 11, 2109.

Schwarz, K. (1972), *Physical Review B*, 5, 2466.

Shavitt, I. & Karplus, M. (1965), *The Jounal of Chemical Physics*, 43, 398–414.

Slater, J. C. (1951), *Physical Review*, 81, 385.

Vosko, S. H., Wilk, L., & Nusair, M. (1980), *Canadian Journal of Physics*, 58, 1200.

Wigner, E. (1934), *Physical Review*, 46, 1002.

Calculation of Physical Properties Using the OLCAO Method

4

In this chapter, we list many of the common physical properties that can be easily calculated using the OLCAO method discussed in Chapter 3. Simple examples obtained by using the OLCAO method are presented here while modern practical applications of such calculations to more complex systems are described in later chapters. Application to core-level spectroscopy will be separately discussed in Chapter 11.

4.1 Band structure and band gap

The band structure $E_n(\boldsymbol{k})$ is the most fundamental electronic property of a crystalline solid and has played a pivotal role in the early development of condensed matter physics (Callaway, 1964). It bears a similarity to the atomic levels of isolated atoms or the molecular orbitals (MO) of molecules except that the energy levels have a momentum (or $\boldsymbol{k}$) dependence due to the periodicity of the crystalline lattice. Diagrams of the band structure are generally restricted to those $\boldsymbol{k}$ values found along high symmetry directions of the Brillouin zone (BZ). From the band structure of a solid, one can immediately know if the crystal is a metal, a semi-metal, a half-metal, a semiconductor, or an insulator based on the presence (or absence) of a band gap between occupied and unoccupied electron states. By sampling $E_n(\boldsymbol{k})$ throughout the whole of the BZ, the density of states (DOS) G(E) of the crystal, defined as the number of energy states per unit energy per crystal cell, can be calculated according to:

$$G(E) = \frac{\Omega}{(2\pi)^3} \sum_n \int d^3k \, \delta \left(E - E_n \left(\vec{k} \right) \right) \tag{4.1}$$

For an amorphous solid where there is no long range order (or lattice periodicity), $\boldsymbol{k}$ is no longer a meaningful quantum number and the band structure concept loses its meaning. Substantial importance then falls to the DOS because it contains all the necessary information related to the electronic structure of the solid. However, electronic structure methods for crystalline solids can still be applied to amorphous solids because such systems are usually represented by a model with periodic boundary conditions that is sufficiently large to make the effects of periodicity negligible. As long as the model is sufficiently large, the corresponding BZ becomes suitably small so that it is only necessary to solve the secular equation at one $\boldsymbol{k}$-point and thereby drop the $\boldsymbol{k}$ dependence of the energy and wave function. The choice of which k point to use is simple because the so-called Γ k point at $k = (0,0,0)$ has the particularly special advantage of a real solution which is much more efficient to obtain computationally than a general k point with its complex valued solution. The same principle applies to a complicated crystal that has a large unit cell and true periodicity, unlike an amorphous model. Many such examples will be given in the later chapters when the calculated results of material systems of recent interest are discussed.

The most important physical quantity for semiconductors and insulators is the band gap, defined as the separation between the top of the occupied valence bands (VB) and the lowest unoccupied conduction band (CB). It is equivalent to the separation between the highest occupied molecular orbital (HOMO) and the lowest unoccupied molecular orbital (LUMO) in MO orbital theory (no k dependence). The band gap is said to be direct if the top of the VB (TVB) and the bottom of the CB (BCB) are at the same k point (usually Γ) or indirect if they are at different k points. Many of the physical properties of semiconductors depend on the size and nature of the band gap. Electric conduction occurs by thermal excitation (if the band gap is small) because electrons are excited from the TVB to the BCB, thus creating holes in the VB and injecting electrons into the CB. Conduction can be achieved at much lower temperatures by doping semiconductors with donor or acceptor elements to generate charge carriers, either holes in the VB or electrons in the CB. As already alluded to in the introductory chapter, it is well known that the band gaps of semiconductors and insulators are usually underestimated by 30–50% within the local density approximation of density functional theory (DFT). This is because DFT is exact only for the ground state calculation whereas the unoccupied CB contains the excited states. Various theories based on a many-body perturbation approach exist to improve the band gap estimation in DFT (Onida et al., 2002). However, they tend to be computationally intensive, if not prohibitive, and hence are restricted to systems that are often considered to be relatively simple ones.

No band gap exists for metallic systems and instead the energy value that separates the occupied and unoccupied states is called the Fermi level (E_F). Only electrons close to the Fermi level can contribute to electric conduction. The DOS at E_F, or N(E_F), is an important physical quantity for metals because the greater in magnitude N(E_F) is, the more charge carries there will be for electric conduction. High conductivity metals such as copper or silver have more free electrons at E_F mostly from the 4s or 5s orbitals.

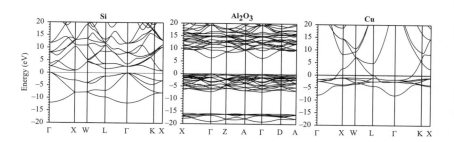

Fig. 4.1.
Band structures of Si, α-Al$_2$O$_3$, and Cu.

Figure 4.1 shows the band structures of Si (a semiconductor), α-Al_2O_3 (an insulator), and Cu (a metal) using the OLCAO method.

4.2 Density of states and its partial components

The total DOS (TDOS) of a crystal can be resolved into partial components, or the partial density of states (PDOS). In the OLCAO method, the partitioning of the TDOS into PDOS of different atomic or orbital components (based on n, l, and m quantum numbers) is very natural because the Bloch functions are expressed in terms of atomic orbitals. The PDOS can be further resolved into spin-dependent components (s quantum number) when spin-polarized calculations are performed (see Section 4.4 below). The PDOS is a very useful quantity that provides a wealth of information because the interactions between different atoms or orbital components are typically manifested by the alignment of peaks in their PDOS spectra. Thus, because the atom-, orbital-, and spin-projected PDOS can be readily obtained it is possible to perform direct interpretation of the nature of electronic bonding and interactions in a solid.

In the OLCAO method, it is convenient to define a fractional charge $\rho_{i\alpha}^m$ for the ith orbital of the αth atom, of the normalized state $\Psi_m(r)$ with energy E_m according to Mulliken's population analysis scheme (Mulliken, 1955):

$$1 = \int |\Psi_m(r)|^2 \, d\vec{r} = \sum_{i,\alpha} \rho_{i,\alpha}^m, \qquad (4.2)$$

$$\rho_{i,\alpha}^m = \sum_{j,\beta} C_{i\alpha}^{m*} C_{j\beta}^m S_{i\alpha,j\beta} \qquad (4.3)$$

Where $C_{j\beta}^m$ is the eigenvector coefficient of the mth state wave function and $S_{i\alpha,j\beta}$ is the overlap matrix Eq. (3.15). The fractional charge is a useful quantity and it is a natural product of methods in which atomic orbitals are used for the basis expansion. As long as the basis functions are reasonably localized, the Mulliken scheme is a simple and effective way to partition an electronic state into orbital species. The decomposition of TDOS into PDOS is done by using $\rho_{i\alpha}^m$ as the projection operator.

In crystalline solids, the presence of van Hove singularities in the DOS is important and special techniques such as the linear analytic tetrahedron method have been widely used for their accurate determination (Lehmann and Taut, 1972, Rath and Freeman, 1975). Such features take on less prominence for amorphous solids represented by a large unit cell or other complex systems with extended defects. For these cases a very small number of k points (or even one) selected with a reasonable sampling method and the application of appropriate broadening will usually be sufficient to obtain a detailed DOS. Figure 4.2 shows the total DOS and PDOS (resolved into atomic and orbital components) of α-Al_2O_3 as an example. Figure 8.3 shows the comparison of the TDOS of crystalline quartz (α-SiO_2) and a large model of amorphous SiO_2 (a-SiO_2) (Huang et al., 1999).

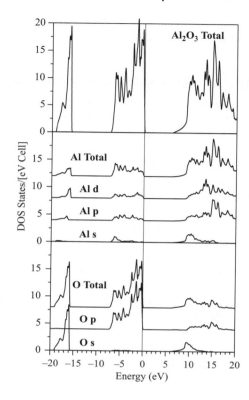

Fig. 4.2.
Total and partial DOS in α-Al_2O_3.

4.3　Effective charges, bond order, and the localization index

From the definition of fractional charge based on the Mulliken scheme in Eq. (4.3), two important quantities of practical use can be easily obtained. They are the effective charge Q_α^* on each atom α and the bond order $\rho_{\alpha\beta}$ (also called the bond overlap population) for each pair of atoms (α, β):

$$Q_\alpha^* = \sum_i \sum_n \sum_{j,\beta} C_{i\alpha}^{m*} C_{j\beta}^m S_{i\alpha,j\beta} \qquad (4.4)$$
$$\text{occ}$$

$$\rho_{\alpha\beta} = \sum_n \sum_{j,\beta} C_{i\alpha}^{*n} C_{j\beta}^n S_{i\alpha,j\beta} \qquad (4.5)$$
$$\text{occ}$$

The charge transfer that a particular atom or group of atoms experiences in a material can be derived from the effective charge by subtracting the computed effective charge from the number of so-called valence electrons that are present in the neutral atom. It should be noted that the calculated effective charges should not be confused with formal charges that are used to describe valence states of a solid. For example, it is traditional to describe the Na and Cl ions in crystalline NaCl as having valence states of $+1$ and -1 denoted by $Na^{+1}Cl^{-1}$. However, the calculated effective charges on crystalline NaCl will never give Na an effective charge of 0 and Cl an effective charge of 8.

The bond order values indicate the strength of the bond between each pair of atoms. Generally speaking, covalent bonds are much stronger than ionic bonds and therefore have larger bond order values. Bond orders tend to scale with bond length but because the quantum mechanical wave function is used to compute $\rho_{\alpha\beta}$ the bond angle also plays a determining role.

In actual calculations of Q_α^* and $\rho_{\alpha\beta}$, a minimal basis set (MB) in the Bloch function expansion is always used because the Mulliken scheme works best with a more localized basis set. It should be noted that any results from the Mulliken scheme are basis-dependent. Direct comparison of Q_α^* and $\rho_{\alpha\beta}$ values for different atoms and pairs of atoms in different crystals are meaningful only if they are calculated using the same method and same basis set. Trends and qualitative features remain comparable with results from other methods though. As an example of this calculation, the Q^* and $\rho_{\alpha\beta}$ values as computed for α-, β-, and γ-Si_3N_4 are tabulated in Table 4.1.

Another quantity that is useful for great physical insight into the electronic states of non-crystalline materials is the localization index (LI). The LI can be defined from the fractional charges of Eq. (4.3) as:

$$L_m = \sum_{i,\alpha} \left[\rho_{i\alpha}^m \right]^2 \qquad (4.6)$$

L_m is a measure of the probability density of the state m at different sites. L_m lies between 1 for a completely localized state, in which the charge is confined to a single orbital, to N^{-1} for a completely delocalized state where N is the total number of electron states in the model. With a sufficiently large number of atoms in the model structure for amorphous solids, Eq. 4.6 enables us to make realistic estimates of the localization of the one-electron state in a disordered

Table 4.1. Comparison of Q^* and $\rho_{\alpha\beta}$ for three phases of Si_3N_4.

	α-Si_3N_4	β-Si_3N4	γ-Si_3N_4
Si1 Q*	2.405	2.419	2.522
Si2 Q*	2.523	—	2.556
N1 Q*	6.160	6.181	6.092
N2 Q*	6.158	6.200	—
N3 Q*	6.172	—	—
N4 Q*	6.127	—	—
ρ_{Si1-N1}	—	0.357, 0.356 × 2	0.369 × 4
ρ_{Si1-N2}	0.336	0.343	—
ρ_{Si1-N3}	0.381, 0.321	—	—
ρ_{Si1-N4}	0.334	—	—
ρ_{Si2-N1}	0.347	—	0.234 × 4, 0.235 × 2
ρ_{Si2-N2}	—	—	—
ρ_{Si2-N3}	0.344	—	—
ρ_{Si2-N4}	0.352, 0.314	—	—

solid across the entire energy spectrum. The reciprocal of the LI is the so-called participation ratio used by others for the same purpose of quantifying the localization of a state in non-crystalline solids. Hence, LI is sometimes also called the inverse participation ratio. The LI of amorphous SiO_2 is shown in Fig. 8.4 as an illustration of the computation.

4.4 Spin-polarized band structures

For magnetic materials, spin-polarized calculations based on the local spin density approximation (LSDA) are necessary. In the LSDA calculation, spin becomes a variable in $V_{exch}(r)$. There are many different forms for $V_{exch}(r)$ including the use of generalized gradient approximations (GGA) as discussed in Chapter 3. The one that was adopted for the OLCAO method is the von Barth–Hedin potential as modified by Moruzzi et al. (Moruzzi et al., 1978) in which the exchange-correlation functional takes the form:

$$\varepsilon_{xc}\left[\rho_\uparrow, \rho_\downarrow\right] = \varepsilon_{xc}^p[r_s] = \left[\varepsilon_{xc}^f(r_s) - \varepsilon_{xc}^p(r_s)\right] f_{xc}(\rho_\uparrow, \rho_\downarrow) \qquad (4.7)$$

The superscripts p and f in Eq. (4.7) stand for paramagnetic and ferromagnetic cases and r_s is defined as $(4/3)\pi r_s^3 = 1/\rho$. ε_{xc}^f and ε_{xc}^p in Eq. (4.7) are ferromagnetic and paramagnetic exchange-correlation energies worked out in a parameterized form by fitting results to the homogeneous electron gas. f_{xc} is given by:

$$f_{xc}(\rho_\uparrow, \rho_\downarrow) = \left[(2\rho_\uparrow/\rho)^{\frac{4}{3}} + (2\rho_\downarrow/\rho)^{\frac{4}{3}} - 2\right]/(2^{4/3} - 2) \qquad (4.8)$$

The spin-up states correspond to the case in which the spins are considered to be aligned along a specific direction (usually taken as the **z** directional component of the crystal lattice) and the spin-down states have spins pointed in the opposite direction. The spin-up and spin-down states are coupled together via the exchange potential, Eq. (4.8). The total charge density, $\rho(\mathbf{r})$ in Eq. (4.8), is the sum of spin-up and spin-down charge densities: $\rho(\vec{r}) = \rho_\uparrow(\vec{r}) + \rho_\downarrow(\vec{r})$. Their difference gives the spin-density function, $\rho_s(\vec{r}) = \rho_\uparrow(\vec{r}) - \rho_\downarrow(\vec{r})$.

The spin-polarized OLCAO method starts by partitioning the total charge $\rho = \sum \rho_A(r - r_A)$ into $\rho_\uparrow$ and $\rho_\downarrow$ components. (For simplicity, we suppress the site index **A** for $\rho_A(r)$ from now on.) To start the self-consistent iterations in the spin-polarized calculation, it is necessary to give the spin-up and spin-down potential representations an initial rigid splitting. Eq. (3.4) is then solved for each spin case and the results are merged to determine the new spin-polarized charge density and potential. In general, convergence toward self-consistency in a spin-polarized calculation is much slower because of the additional spin variable and the need for locating the Fermi level accurately.

The spin-polarized calculation described above is a collinear-spin approximation, because the spin of the magnetic atom is considered to be in a specified direction. In principle, it is possible to have the spin of a magnetic atom oriented in a general direction, or in a non-collinear spin description (Antropov et al., 1995). The non-collinear spin–spin interaction in magnetic materials is very important in many systems. The non-collinear spin calculation has not yet been implemented in the OLCAO method but should be straightforward. The spin-polarized band structure of Fe is illustrated in Fig. 4.3.

4.5 Scalar relativistic corrections and spin-orbit coupling

Electrons in the core states of a heavy element have a sufficiently high velocity that relativistic effects become important and they should be treated by relativistic quantum mechanics. The relativistic correction to the electronic structure of a solid starts with the Dirac equation for the relativistic electron:

$$c\vec{\sigma} \cdot \vec{p}\psi_v + (mc^2 - E - e\phi)\psi_u = 0,$$
$$c\vec{\sigma} \cdot \vec{p}\psi_u - (mc^2 + E + e\phi)\psi_v = 0,$$

(4.9)

In Dirac notation, the wave function has two components denoted as ψ_u (large component) and ψ_v (small component) which each have their own up and down spin components for a total of four components (Bjorken and Drell, 1964). In Eq. (4.9), $\vec{\sigma}$ stands for the Pauli spin matrices, $\vec{p}$ is the momentum operator, ϕ is the scalar potential experienced by the electron, c is the speed of light, and m the mass of electron. In most cases, c is much larger than the average velocity of an electron in a solid, and one can concentrate only on the "large component" ψ_u. The Dirac equation, Eq. (4.9), can be reduced to the following simplified form where the subscript u for the large component has been dropped.

$$(E' + e\phi)\psi = \frac{1}{2}m(\sigma \cdot \vec{P})K(\sigma \cdot \vec{P})\psi$$

(4.10)

The relativistic kinetic energy term K can be approximated as:

$$K = \frac{1}{1 + \left(\frac{E'+e\phi}{2mc^2}\right)} \approx 1 - \frac{E' + e\phi}{2mc^2},$$

(4.11)

And Eq. (4.10) can then be written as

$$(E' + e\phi)\psi = \left[\frac{1}{2m}\vec{p}^2 - \frac{\vec{p}^4}{8m^3c^2} - \frac{e\hbar^2}{8m^2c^2}\vec{\nabla} \cdot \vec{\nabla}\phi - \frac{e\hbar}{4m^2c^2}\vec{\sigma} \cdot \vec{\nabla}\phi \times \vec{p}\right]\psi$$

(4.12)

Eq. 4.12 is similar to the non-relativistic Schrodinger equation with the addition of the last three terms in the Hamiltonian. They are respectively labeled as the mass velocity, Darwin, and spin-orbit coupling terms. Note that the Pauli spin matrices $\vec{\sigma}$ are present in only the spin-orbit coupling term. Therefore, the mass velocity and the Darwin terms have no off-diagonal elements in the two spin components of the wave function. They behave like a scalar and can be separated out to account for the scalar relativistic correction to the non-relativistic potential (Koelling and Harmon, 1977). The separation of the spin dependent and spin independent parts of the relativistic correction is extremely convenient for performing practical calculations because both the

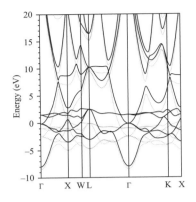

Fig. 4.3.
Spin-polarized band structure of Fe. Solid (dashed) line is for minority (majority) spin.

mass-velocity and the Darwin terms can be treated as corrections to the one-electron potential $V(r)$ in the non-relativistic Schrodinger equation.

The correction for the mass-velocity term can be obtained rather easily by writing

$$\vec{p}^4 = (p_x^2 + p_y^2 + p_z^2) \cdot (p_x^2 + p_y^2 + p_z^2) \tag{4.13}$$

Where p_x, p_y, and p_z are substituted by the corresponding differentiation operators. The correction terms to the Hamiltonian matrix, Eq. (3.16), are evaluated analogously to the momentum operator. This can be evaluated very efficiently by using the generalized Cartesian Gaussian orbitals of the form $x^n y^l z^m \, exp(-\alpha r^2)$ as described in Eq. (3.42) where n, l, m are integers.

To obtain the correction for the Darwin term, it is necessary to obtain the second derivative of the crystal potential which is expressed as a sum of atom-centered atomic potentials $V_A(r)$ cast as a linear combination of GTOs (see Eqs (3.11) to (3.13)).

$$V_A(r) = -\frac{Z}{r} + V_{Coul}(r) + V_{exch}(r) \tag{4.14}$$

The correction for the first term in Eq. (4.14) is simply related to the atomic number of the element since $-\vec{\nabla} \cdot \vec{\nabla}[-Z/r] = 4\pi Z\delta(r)$ while the second term is related to the charge density $\rho(r)$ through Poisson's equation $\nabla^2 V_{Coul}(r) = -4\pi\rho(r)$. The third term, which accounts for less than 10% of $V_{Coul}(r)$, is due to $V_{exch}(r)$ and can be obtained numerically (Zhong et al., 1990). A comparison of the non-relativistic and scalar-relativistically corrected band structures of Ni and Nb, as calculated by the OLCAO method, are shown in Fig. 4.4. The agreement of these calculations with those calculated by using other methods is very good.

The relativistic correction due to the spin-orbit coupling term in Eq. (4.12) is generally an order of magnitude smaller than the scalar relativistic corrections. However, it is still important in magnetic systems and for many other properties. There are different approaches for treating spin-orbit coupling that depend on the system under study and the physical process involved. In semiconductors, one tends to concentrate on states near the band edges at a high symmetry point in the Brillouin zone. Perturbation theory for the degenerate states is usually applied. The more general approach is to first establish a spin-polarized band structure from the spin-resolved calculation and then add a spin-orbit correction to it. The inclusion of a spin-orbit coupling term in the spin-polarized calculation doubles the dimension of the Schrödinger equation because both spins have to be considered simultaneously.

The spin-polarized band structure calculation provides two Hamiltonian matrices, one for the spin-up band (majority spin) and the other for the spin-down band (minority spin), which are diagonalized separately. The spin-orbit interaction couples them together and the matrix size is doubled:

$$\begin{pmatrix} \overleftrightarrow{U} & 0 \\ 0 & \overleftrightarrow{D} \end{pmatrix} \rightarrow \begin{pmatrix} \overleftrightarrow{U}' & \Delta \\ \Delta^+ & \overleftrightarrow{D}' \end{pmatrix} \tag{4.15}$$

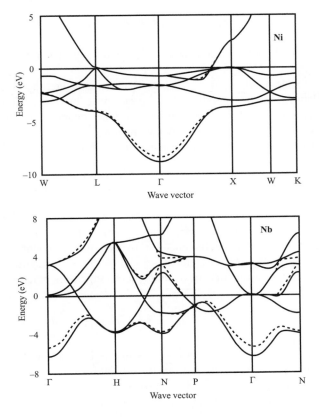

Fig. 4.4.
Scalar relativistic band structure for Ni (upper) and Nb (lower). Dashed lines are for non-relativistic bands.

Source: X.F. Zhong, Y.N. Xu and W.Y. Ching, "Orthogonalized Linear Combination of Atomic Orbitals Method. IV. Inclusion of Relativistic Effects", Phys. Rev. B41, 10545 (1990).

In Eq. (4.9), **U** and **D** are the original Hamiltonian matrices for spin-up and spin-down states. They become **D'** and **D'** after application of the scalar relativistic correction. 0 stands for the null matrix, and Δ is the coupling matrix. The overlap matrices are the same for both spin cases. It is assumed that the scalar relativistic correction has already been included in **U** and **D**. To obtain explicit expressions for Δ, **U'**, and **D'**, we need to evaluate the matrix elements of the spin-orbit coupling term $\frac{e\hbar}{4m^2c^2}\vec{\sigma}\cdot\vec{\nabla}\phi\times\vec{p}$ between spin-polarized Bloch sums. In a truly first-principles approach, explicit expressions can be derived for these matrix elements which involve integrals of the derivatives of the crystal potential and GTOs. A simpler approach is to treat the strength of spin-orbit coupling by a parameter ξ,

$$\xi = \frac{\hbar^2}{2m^2c^2}\int |u_j(\vec{r})|\,\frac{1}{r}\frac{dV_A(r)}{dr}\,d\vec{r} \qquad (4.16)$$

where V_A, $u_j(\mathbf{r})$ are the atomic potential and the atomic orbital state j respectively.

ξ can be obtained either from experiments or accurate atomic calculations. In the parameterized form, the spin-orbit coupling term can be written as:

$$\xi \vec{l} \bullet \vec{s} = \frac{1}{2}\xi(\vec{j}^2 - \vec{l}^2 - \vec{s}^2) \qquad (4.17)$$

where j, l, s are the total, orbital, and spin quantum numbers of the atom and ξ is given by Eq. (4.16).

The matrix elements of Eq. (4.17) between atomic wave functions are quite simple. Using a bracket notation, the atomic wave functions are expressed as direct products $|\alpha, l, m\rangle |s, m_s\rangle$ where $|\alpha, l, m\rangle$ denotes the spatial parts (α designates quantum numbers other than $|l, m_s\rangle$) and $|s, m_s\rangle$ denotes the spin part (with $m_s = \pm 1/2$). The product functions can be transformed into coupling functions:

$$|l, m\rangle |s, m_s\rangle = \sum_{j, m_j} \langle l, s, j, m_j | l, m, s, m_s\rangle |l, s, j, m_j\rangle \qquad (4.18)$$

where $\langle l, s, j, m_j | l, m, s, m_s\rangle$ are obtained from the inverse transformation of Clebsch–Gordon coefficients for two angular momentum coupling (Friedrich, 1998). The matrix elements of Eq. (4.17) become straightforward because the coupling functions are simultaneous eigenfunctions of j, l, s and we have:

$$\xi \langle \vec{l} \bullet \vec{s}\rangle = \frac{1}{2}\xi \left(\vec{j}(\vec{j} + 1) - \vec{l}(\vec{l} + 1) - \frac{3}{4} \right) \qquad (4.19)$$

Test calculations for relativistic corrections in the OLCAO method using the above approach have been carried out for crystalline Ni, Nb, and Ce in the scalar relativistic limit. Full relativistic correction including spin-orbit coupling was tested in the case of ferromagnetic Fe. The results are in good agreement with similar calculations using other approaches (Zhong et al., 1990).

4.6 Magnetic properties

In a spin-polarized calculation, the effective charges of Eq. (4.4) can be further resolved into spin-up and spin-down components.

$$Q_\alpha^* = Q_{\alpha\uparrow}^* + Q_{\alpha\downarrow}^* \qquad (4.20)$$

and their difference gives the spin magnetic moment at the site:

$$\langle M_s\rangle_\alpha = Q_{\alpha\uparrow}^* - Q_{\alpha\downarrow}^* \qquad (4.21)$$

The site-specific effective spin magnetic moment is very important in the study of magnetic properties of materials. The way $\langle M_s\rangle_\alpha$ is evaluated in the OLCAO method does not involve any arbitrarily chosen parameters such as atomic sphere radius and is highly reliable. It is also straightforward to calculate the susceptibility enhancement in magnetic materials by exploiting the electronic states near the Fermi level. (Moruzzi et al., 1978)

With the inclusion of the spin-orbit interaction, it is possible to estimate the orbital moments $\langle M_l\rangle$ for a crystal. In the OLCAO method, $\langle M_l\rangle$ can be expressed as:

$$\langle M_l \rangle = g \sum_{m,k}^{occ} \sum_{i,\alpha} \sum_{j,\beta} C_{i\alpha}^{m*} C_{j\beta}^{m} \left\langle b_{i\alpha}(\vec{k}, \vec{r}) | l_z | b_{j\beta}(\vec{k}, \vec{r}) \right\rangle \quad (4.22)$$

where l_z is the z-component of the angular momentum operator.

The matrix elements in Eq. (4.22) can be expressed as a lattice sum of overlap integrals because l_z operates on only the angular part of the atomic orbitals in the Bloch sum. The difficulty with orbital moment calculations for a crystal lies in the fact that Eq. (4.22) is not invariant under the operation of a wave vector group as in the case of energy eigenstates. Thus the summation over k covers the entire BZ instead of the irreducible portion of it. For amorphous materials or complex materials with a large unit cell, wave-vector k is irrelevant and the orbital moment may actually be easier to evaluate. Recent examples will be discussed further in Chapters 5 and 6.

4.7 Linear optical properties and dielectric functions

The optical properties of solids (insulators or metals) can be calculated using the OLCAO method based on the theory of inter-band optical absorptions within the random phase approximation (RPA) (Ehrenreich and Cohen, 1959). The standard way is to directly evaluate the real part of the frequency-dependent optical conductivity $\sigma_1(\hbar\omega)$:

$$\sigma_1(\hbar\omega) = \frac{2\pi e \hbar^2}{3m^2 \omega \Omega} \sum_{n,\ell} |\langle n|\vec{p}|\ell\rangle|^2 \, f_\ell [1 - f_n] \delta(E_n - E_l - \hbar\omega) \quad (4.23)$$

Here f_ℓ is the Fermi–Dirac function of the occupied band state ℓ and the δ function ensures the conservation of energy in the transition process between the occupied state ℓ and unoccupied state n with energy E_ℓ and E_n respectively; and Ω is the volume of the unit cell.

In crystalline or amorphous metals, the conductivity at low photon frequency is not due to inter-band transitions but rather intra-band transitions which are usually approximated by the Drude formula (Drude, 1952) which is valid for free electron metals:

$$\sigma_D = \frac{Ne^2\tau}{m^* \left(1 + \omega^2\tau^2\right)} \quad (4.24)$$

where τ is the relaxation time and m^* is the effective mass of the conduction electron. In amorphous metals with no translational symmetry, the structure of the glass is usually modeled by a large supercell and the distinction between inter-band and intra-band transitions no longer exists since there is no k dependence for energy. The optical conductivity at low frequency can be calculated exactly using Eq. (4.23) subject only to the size of the model. Similarly, the intra-band transitions in crystalline solids can be obtained by using a large supercell and only one k point at the center of the BZ. The original bands in the regular BZ now fold into many sub-bands (or energy levels) within the smaller BZ. This approach, of course, increases the computational burden when there are a large number of atoms in the supercell. However, with the efficient OLCAO method, this is a very practical way to account for the intra-band transitions in metallic systems.

For insulating systems with a band gap, there are no intra-band transitions and the linear optical properties are described by the complex dielectric function. The imaginary part of the dielectric function $\varepsilon_2(\hbar\omega)$ is related to $\sigma_1(\hbar\omega)$ by

$$\varepsilon_2(\omega) = 4\pi \frac{\sigma_1(\omega)}{\omega}, \tag{4.25}$$

By transforming the summation over the unit cell in Eq. (4.23) to an integral over the BZ in k space, $\varepsilon_2(\hbar\omega)$ takes the following equivalent form:

$$\varepsilon_2(\hbar\omega) = \frac{e^2}{\pi m \omega^2} \int_{BZ} dk^3 \sum_{n,l} \left| \langle \psi_n(\vec{k}, \vec{r}) \right| - i\hbar\vec{\nabla} \left| \psi_l(\vec{k}, \vec{r}) \rangle \right|^2$$

$$f_l(\vec{k})[1 - f_n(\vec{k})]\delta(E_n(\vec{k}) - E_l(\vec{k}) - \hbar\omega). \tag{4.26}$$

The real part of the dielectric function can be extracted from ε_2 through the usual Kramers–Kronig relation (Martin, 1967),

$$\varepsilon_1(\hbar\omega) = 1 + \frac{2}{\pi} P \int_0^\infty \frac{s\varepsilon_2(\hbar\omega)}{s^2 - \omega^2} ds \tag{4.27}$$

The integration limit in Eq. (4.27) is usually replaced by a finite cutoff value, since $\varepsilon_2(\hbar\omega)$ or $\sigma_1(\hbar\omega)$ is calculated only for a finite range of photon energy. This could introduce some uncertainty in the KK-transformed $\varepsilon_1(\hbar\omega)$. As an example, the dielectric properties of α-B_{12} are shown in Fig. 4.5 where the z-axis is aligned with the body diagonal of the rhombohedral cell.

Once the complex dielectric function of an insulator is evaluated, all other related optical constants can be obtained. These include optical absorption power $\alpha(\hbar\omega)$ involving the wave length λ of the photon,

$$\alpha(\hbar\omega) = \frac{\hbar\omega\varepsilon_2(\hbar\omega)}{nc\hbar} \sim \frac{\varepsilon_2(\hbar\omega)}{\lambda} \tag{4.28}$$

the energy loss function $F(\omega)$ and the reflectivity spectrum $R(\omega)$.

$$F(\omega) = IM\{-\frac{1}{\varepsilon(\omega)}\} = \frac{\varepsilon_2(\omega)}{[\varepsilon_1^2(\omega) + \varepsilon_2^2(\omega)]} \tag{4.29}$$

$$R(\omega) = \left| \frac{\sqrt{\varepsilon(\omega)} - 1}{\sqrt{\varepsilon(\omega)} + 1} \right| \tag{4.30}$$

The peak energy in the energy loss function $F(\omega)$ is identified as the plasmon frequency which is the energy of collective excitation of the electrons in a solid and is an experimentally measurable quantity.

In the OLCAO method, the momentum matrix elements in Eqs (4.23) or (4.26) for optical transitions can be easily evaluated as sums of two-center integrals between GTOs as shown in Chapter 3 Section 3.4. The square of the momentum matrix is averaged over the three Cartesian directions for isotropic crystals or amorphous solids. In anisotropic crystals or complex materials with

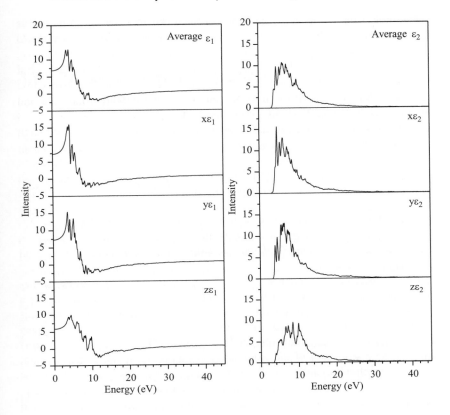

Fig. 4.5.
Real and imaginary dielectric function in α-Boron and its Cartesian components.

directional dependence, the individual components of the momentum matrix elements can be separated into x–x, y–y and z–z components. In certain applications such as in magneto-optical calculations, the off-diagonal matrix elements x–y, y–z, and z–x of the momentum operator may have to be calculated.

4.8 Conductivity function in metals

The optical conductivity function in metallic systems is the key physical quantity to investigate electronic transport properties. For example, metallic glasses show many interesting physical phenomena unique to disordered alloys such as a negative temperature coefficient of resistivity, Mooij correlation for resistivity and thermal power, resistivity saturation, sign reversal of the Hall coefficient, superconductivity, negative magneto-resistance, and so on. Numerical calculation of transport properties in metallic systems at low temperature starts with the evaluation of an energy-dependent conductivity function $\sigma(E = \hbar\omega)$ according to the Kubo–Greenwood formula (Kubo, 1957) as a generalization or extension of inter-band optical conductivity, Eq. (4.23).

$$\sigma(E) = \frac{2\pi\hbar e^2}{3m^2\Omega} \sum_{n,m} |\langle n|\vec{p}|m\rangle|^2 \delta(E_n - E)\delta(E_m - E) \qquad (4.31)$$

The double summation in Eq. (4.31) is over all energy states and the double delta function describes the scattering process of an electron with energy E_n to that of energy E_m. $\langle n|\vec{p}|m\rangle$ is the momentum matrix element described above. For transport property calculations, only states in the vicinity of the Fermi level E_F are important. The momentum matrix elements and the distribution of energy E_m contain all the information about the quantum coherence and multiple scattering associated with the electron transport.

The temperature dependent d.c. conductivity $\sigma(0, T)$ can be obtained from $\sigma(E)$ through

$$\sigma(0, T) = \int \left[\frac{\partial f(E)}{\partial E} \right] \langle \sigma(E) \rangle \, dE, \qquad (4.32)$$

where $f(E)$ is the Fermi distribution function and $\langle \sigma(E) \rangle$ in Eq. (4.32) denotes the configurational average of $\sigma(E)$. In the computational sense, $\langle \sigma(E) \rangle$ is interpreted as the average of calculations over as many independent structural models as possible.

The temperature dependence of $\sigma(0, T)$ in Eq. (4.32) is through the Boltzmann factor in the Fermi distribution function f. At low temperature, f is a step function at E_F and the conductivity of the metal is simply $\sigma(E_F)$ whereas the resistivity ρ (not to be confused with the electron charge density) is given by $1/\sigma(E_F)$. The resistivity obtained according to Eq. (4.32) is purely due to the elastic scattering of the conduction electrons in the solid and does not include the phonon (lattice) scattering. Equation (4.32) is valid only at low temperatures. At higher temperatures, lattice vibrational effects and electron–phonon interactions become important and must be appropriately accounted for.

In the OLCAO calculation of $\sigma(E)$ for metals using a large supercell with a finite number of atoms, the energy spectrum is discrete. The discrete energy spacing implies that the δ function condition in Eq. (4.25) can never be satisfied exactly at all energies. This difficulty can be circumvented by replacing each discrete energy level by a Gaussian of unit area and finite width. The larger the supercell the smaller should be the width of the Gaussian and the more accurate would be the calculation. Hence, the accuracy in determining resistivity ρ or conductivity $\sigma(E_F)$ depends to a large extent on the density of states at E_F which is different for different materials.

The thermopower S(T) can be obtained from $\langle \sigma(E) \rangle$ as:

$$S(T) = -\frac{\pi^2}{3} \frac{k^2 T}{e} \frac{\partial}{\partial E} \log\langle \sigma(E) \rangle \Big|_{E=E_f} \qquad (4.33)$$

Because $S(T)$ involves the derivatives of $\ln\langle \sigma(E) \rangle$, it is more difficult to evaluate accurately.

The transport and optical properties of several metallic glasses have been calculated within this general framework using the OLCAO method and will be discussed more in Chapter 6. (Ching et al., 1990, Zhao et al., 1990) An example of this type of calculation is shown in Fig. 4.6 for amorphous nickel.

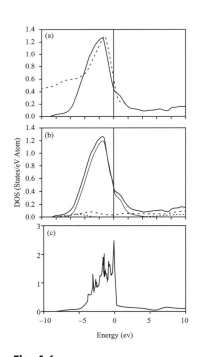

Fig. 4.6.

(a) Total and (b) partial DOS of a-Ni compared with (c) crystalline Ni.

Source: W.Y. Ching, G.L. Zhao and Yi He, "Theory of Metallic Glasses, I: Electronic Structures", Phys. Rev. B42, 10878 (1990).

4.9 Non-linear optical properties of insulators

The linear optical properties of insulators described in Section 4.7 are part of the more general material response to external electromagnetic fields $E(\omega)$. The frequency dependent polarization $P(\omega)$ of the medium is expressed as a power series of the incident field $E(\omega)$:

$$P_i(\omega) = \sum_j \chi_{ij}^{(1)}(\omega)E_j(\omega) + \sum_{jk} \chi_{ijk}^{(2)}(\omega = \omega_1 + \omega_2)E_j(\omega_1)E_k(\omega_2)$$

$$\text{(4.34)}$$

$$+ \sum_{jkl} \chi_{ijkl}^{(3)}(\omega = \omega_1 + \omega_2 + \omega_3)E_j(\omega_1)E_k(\omega_2)E_l(\omega_3) + \cdots,$$

In Eq. (4.34), $\chi^{(n)}(\omega)$ is the nth order frequency dependent complex susceptibility; $\omega_1, \omega_2, \omega_3 \ldots$ are the frequencies of the incident field and ω is the frequency generated by the polarization of the medium. The linear susceptibility $\chi^{(1)}(\omega)$ is related to the linear dielectric function $\varepsilon(\hbar\omega) = \varepsilon_1(\hbar\omega) + i\varepsilon_2(\hbar\omega)$ discussed in Section 4.7 by

$$\chi^{(1)}(\omega) = (1/4\pi)[(\varepsilon(\omega) - 1)] \qquad \text{(4.35)}$$

$\chi^{(2)}(\omega)$ and $\chi^{(3)}(\omega)$ are the second and third order non-linear susceptibilities which are tensors of rank three and four, respectively. When $\omega_1 = \omega_2 = \omega'$ and $\omega = \omega_1 + \omega_2 = 2\omega'$, this results in the simplest second order non-linear optical process called second harmonic generation (SHG). Similarly the simplest third order non-linear process is the third harmonic generation (THG) corresponding to the process when $\omega_1 = \omega_2 = \omega_3 = \omega'$ and $\omega = 3\omega'$. For centrosymmetric crystals with a center of inversion such as Si or Ge, the second order non-linear susceptibility $\chi^{(2)}(\omega)$ vanishes on symmetry grounds and the lowest non-linear optical process is the third order susceptibility $\chi^{(3)}(\omega)$.

The non-linear optical properties of elemental and compound semiconductors and insulators are of great scientific and technological importance with relations to laser technology (Butcher and Cotter, 1990). Non-linear optical techniques have been applied to many diverse disciplines such as atomic, molecular, and solid state physics, materials science, chemical dynamics, surface and interface science, biophysics, and medicine. The OLCAO method has been used to calculate the SHG and THG of an extensive list of semiconductors and insulators. The calculation is similar to but more complicated than the linear optical property calculations. Specific formulae for SHG and THG were derived for the imaginary part of the susceptibility tensor and the real part is obtained through the Kramers–Kronig conversion. The quantity of primary interest is the susceptibility $\chi^{(2)}(0)$ and $\chi^{(3)}(0)$ in the zero frequency limit. The calculation involves not only the momentum matrix elements between the occupied valence band states and the unoccupied conduction band states, but also those between the conduction band states because of the virtual electron process that dominates the excitation. These results will be discussed in more detail in Chapter 5.

4.10 Bulk properties and geometry optimization

Ab initio total energy and force calculations based on density functional theory have been very successful in predicting crystal structures and bulk elastic properties of materials. For crystals with high symmetry, such calculations have become a routine exercise and many highly efficient computational tools are now available, most of them based on plane wave basis expansions. The optimization process entails calculating the forces exerted on each atom under a small displacement which are the gradient of the total energy E_T with respect to the nuclear coordinates of the atom in the crystal. For complex crystals with many internal parameters or for non-crystalline materials with no symmetry at all, it is still a time consuming task to fully optimize the geometry.

Geometry optimization using the OLCAO method, like any other method with a local orbital basis expansion, is not easy. This is because the atomic basis functions in the OLCAO method are nuclear-position-dependent. There is a significant Pulay correction (Pulay, 2002) to the Hellmann–Feynman forces that is difficult to evaluate with local orbital methods. On the other hand, methods based on plane wave expansion can calculate forces very efficiently.

One way to circumvent the above difficulty in directly calculating forces in the OLCAO method is to use an approach based on finite differences of total energy. Let $E_T\,(a, b, c, \alpha, \beta, \gamma, x_1, x_2, x_3 \ldots)$ be the total energy of a general crystal with lattice parameters specified by $a, b, c, \alpha, \beta, \gamma$ and internal parameters by $x_1, x_2, x_3 \ldots$ and so on. The total energy E_T can be calculated very efficiently in the OLCAO method according to:

$$E_T = \sum_{n,k}^{occ} E_n(\vec{k}) + \int \rho(\vec{r}) \left(\varepsilon_{xc} - V_{xc} - \frac{V_{e-e}}{2} \right) d\vec{r} + \frac{1}{2} \sum_{\gamma,\delta} \frac{Z_\gamma Z_\delta}{\vec{R}_\gamma - \vec{R}_\delta},$$

$$(4.36)$$

The energy gradient for each parameter P_i is obtained by the finite difference approach according to:

$$\frac{\partial E_T}{\partial P_i} \approx \frac{E_T(P_i + \Delta P_i) - E_T(P_i - \Delta P_i)}{2 \times \Delta P_i}$$

$$(4.37)$$

A first-order-gradients-only algorithm such as the conjugated gradient or steepest descent is then used to minimize the total energy in the entire parameter space. Such a procedure has been implemented in many popular computational schemes including the General Utility Lattice Program (GULP) (Gale, 1997, Gale and Rohl, 2003), which utilizes crystal symmetry to reduce the number of total computations. The key element in the above simple finite difference in total energy method is that E_T must be calculated accurately and fast. For each energy gradient, a minimum of two additional self-consistent field (SCF) calculations for the ground state are needed to estimate the force for a change in one parameter. Fortunately, the OLCAO method is very efficient and fast, thus making this method practical for simpler crystals. This scheme has been

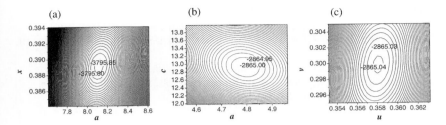

Fig. 4.7.

Energy surface contour as a function of crystal parameters: (a) $MgAl_2O_4$, (b) α-Al_2O_3 and (c) internal parameters in α-Al_2O_3.

Source: L. Ouyang, W.Y. Ching, "Geometry Optimization and Ground State Properties of Complex Ceramic Oxides," J. Amer. Ceram. Soc. 84[4], 801-805 (2001).

Table 4.2. Comparison between calculated and experimental parameters in α-Al_2O_3.

α-Al_2O_3	Calculated	Measured
c (Å)	12.9746	12.9860
a (Å)	4.7901	4.7620
c/a	2.709	2.727
u (Al)	0.3582	0.3520
v (O)	0.3000	0.3060
B (GPa)	247.9	254.4

applied to many crystals. Figure 4.7 and Table 4.2 show the results for two crystals, $MgAl_2O_4$ and α-Al_2O_3. (Ouyang and Ching, 2001).

It should be pointed out that for complex crystals or disordered materials, the finite difference method outlined here is not viable. In systems with no symmetry of any kind, the finite difference method requires the calculation of total energy for the displacement of each atom in three Cartesian directions which renders the calculation too onerous and not worthwhile in spite of the OLCAO method's efficiency. A better strategy is to combine OLCAO with other *ab initio* methods with superior geometry optimization abilities and then to use the OLCAO method for computing physical properties based on the optimized structures to obtain the advantages in interpretation that OLCAO offers. This will be further elaborated on in Chapter 11.

References

Antropov, V. P., Katsnelson, M. I., Van Schilfgaarde, M., & Harmon, B. N. (1995), *Phys. Rev. Lett.*, 75, 729.

Bjorken, J. D. & Drell, S. D. (1964), *Relativistic Quantum Mechanics* (New York: McGraw-Hill).

Butcher, P. N. & Cotter, D. (1990), *The Elements of Nonlinear Optics* (Cambridge: Cambridge University Press).

Callaway, J. (1964), *Energy Band Theory* (New York: Academic Press).

Ching, W. Y., Zhao, G. L., & He, Y. (1990), *Phys. Rev. B*, 42, 10878–86.

Drude, P. K. L. (1952), *Theory of Optics* (New York: Dover).

Ehrenreich, H. & Cohen, M. H. (1959), *Physical Review*, 115, 786.

Friedrich, H. (1998), *Theoretical Atomic Physics* (Berlin: Springer-Verlag).

Gale, J. D. (1997), *Journal of the Chemical Society, Faraday Transactions*, 93, 629–37.

Gale, J. D. & Rohl, A. L. (2003), *Molecular Simulation*, 29, 291–341.

Huang, M.-Z., Ouyang, L., & Ching, W. Y. (1999), *Phys. Rev. B*, 59, 3540–50.

Koelling, D. D. & Harmon, B. N. (1977), *Journal of Physics C: Solid State Physics*, 10, 3107.

Kubo, R. (1957), *Journal of the Physical Society of Japan*, 12, 570.

Lehmann, G. & Taut, M. (1972), *physica status solidi (b)*, 54, 469–77.

Martin, P. C. (1967), *Physical Review*, 161, 143.

Moruzzi, V. I., Janak, J. F., & Williams, A. R. (1978), *Calculated Electronic Properties of Metals* (New York: Pergamon).

Mulliken, R. S. (1955), *J. Chem. Phys.*, 23, 1833.

Onida, G., Reining, L., & Rubio, A. (2002), *Reviews of Modern Physics*, 74, 601.

Ouyang, L. & Ching, W. Y. (2001), *J. Am. Ceram. Soc.*, 84, 801–5.

Pulay, P. (2002), *Molecular Physics*, 100, 57–62.

Rath, J. & Freeman, A. J. (1975), *Physical Review B*, 11, 2109.

Zhao, G. L., He, Y. & Ching, W. Y. (1990), *Phys. Rev. B*, 42, 10887–98.

Zhong, X.-F., Xu, Y.-N., & Ching, W. Y. (1990), *Phys. Rev. B*, 41, 10545–52.

Application to Semiconductors and Insulators

5

The OLCAO method has been applied to a wide variety of materials and with this chapter we will begin to systematically explore them. Because the OLCAO method has been in practice for more than 35 years, some results obtained during the early part of that period will be less accurate and perhaps mundane but they are still important for understanding the present day capabilities of the OLCAO method. Hence, we begin with simple crystals and proceed to more complex systems that have been studied in recent years. In this chapter, the application of the OLCAO method to a very large number of semiconductors and insulators (crystals with band gaps) is summarized. It is difficult to classify these crystals into distinctively different groups and some overlap between them is inevitable regardless of the method of organization. We have adopted the following categories to emphasize both specific capabilities and the range of applicability of the method: elemental and binary semiconductors, binary insulators, oxides, nitrides, carbides, elemental boron and borides, and finally phosphates. Because much has already been said about each of the crystal systems in these groups we will focus our comments on the special points that exemplify the utility and uses of the OLCAO method rather than merely re-presenting the published results.

5.1 Elemental and binary semiconductors

We begin with a discussion of semiconductors formed from the group IV, III–V, and II–VI elements. Silicon, in particular, played an important role in the development of the OLCAO method. In fact, the name, "OLCAO" was coined in 1975 (Ching and Lin, 1975, Ching and Lin, 1977) when the method was first tested on a high pressure phase of silicon, Si-III. As a result, note that from this chapter onward we will often use the word "OLCAO" as a proper noun for the name of the program suite that implements the OLCAO method. The program was subsequently applied to a (111) interface between Si in the diamond and hexagonal wurtzite structures (Huang and Ching, 1983), randomly stacked bi-layers and superlattices of Si (Ching and Huang, 1985, Ching et al., 1984a), and Si_xGe_{1-x} random alloys (Huang and Ching, 1985b). Although these works used a minimal basis and were far less accurate than the present day fully self-consistent *ab initio* calculations, they framed OLCAO as a method that was highly applicable to complex systems and non-crystalline

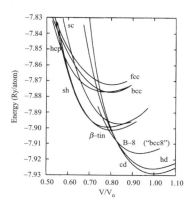

Fig. 5.1.

Calculated total energy per atom of various phases in Si as a function of volume. V_0 is the equilibrium volume of f.c.c., Si.

Source: F. Zandiehnadem and W.Y. Ching, "Total Energy, Lattice Dynamics and Structural Phase Transition of Si by the Orthogonalized Linear Combination of Atomic Orbitals Method", Phys. Rev. B41, 12162-79 (1990).

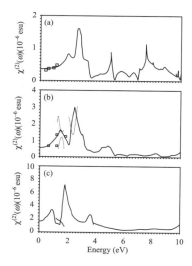

Fig. 5.2.

Calculated second harmonic generation $|\chi^{(2)}(\omega)|$ for (a) GaP, (b) GaAs and (c) GaSb. The symbols are experimental data cited in the referenced paper.

solids. Particularly noteworthy was the calculation of 32 important group IV, III–V and II–VI semiconductors in diamond, zinc blend, wurtzite, and NaCl structures using a minimal basis semi-*ab initio* approach (Huang and Ching, 1985a). This was a simple way to obtain reasonably accurate information on a large number of semiconductor crystals. The band structures, the density of states, and the valence and conduction band effective masses of these 32 crystals were obtained. These results were more accurate than those obtained using tight binding parameters (Harrison, 1970). Such calculations provided better basic materials parameters and were used to estimate band offsets in semiconductor heterojunctions and superlattices (Ruan and Ching, 1986, Ruan and Ching, 1987, Ruan et al., 1988).

When the OLCAO method was made fully self-consistent within the LDA of DFT (Feibelman et al., 1979, Harmon et al., 1982) it was immediately applied to many other crystals including semiconductors. A crucial test for its accuracy was the computation of the phase transitions in Si as a function of pressure (Zandiehnadem and Ching, 1990). This result (Fig. 5.1) shows that proper transitions between 10 Si phases can be obtained using total energy data calculated from the OLCAO method at different volumes for each phase. This indicates that by using the OLCAO method with a localized basis one can obtain reasonably accurate total energies even though direct force calculation is a technical problem that is difficult to overcome.

The calculation of the non-linear optical properties of elemental and compound semiconductors using the OLCAO method was well demonstrated in a series of papers by Huang and Ching (Ching and Huang, 1993, Huang and Ching, 1992, Huang and Ching, 1993b, Huang and Ching, 1993a). There, the band structure and the linear and non-linear optical properties were calculated for 18 cubic semiconductors including group IV semiconductors C, Si, and Ge; III–V compounds AlP, AlAs, AlSb, GaP, GaAs, GaSb, InP, InAs, and InSb; and II–VI semiconductors ZnS, ZnSe, ZnTe, CdS, CdSe, and CdTe. Given the computational resources available at the time this was an extremely rare type of study and its success was attributed to the efficiency of the OLCAO method, especially in regards to the accurate determination of the higher energy conduction band (CB) states involved in the large number of CB–CB optical matrix elements that appear in the expressions for non-linear optical susceptibilities (see Chapter 4). To illustrate, Fig. 5.2 shows the calculated second harmonic generation $|\chi^{(2)}(\omega)|$ for GaP, GaAs, and GaSb showing favorable agreement with experimental data. Fig. 5.3 shows the real and imaginary parts and the absolute value of the dispersion relations for the third harmonic generation $|\chi^{(3)}_{1111}(\omega)|$ and $|\chi^{(3)}_{1212}(\omega)|$ in Si. For centrosymmetric crystals such as Si and Ge, the second harmonic generations vanish and third order non-linearity dominates.

The OLCAO method has been used to calculate the electronic structure, optical, and structural properties of 10 wurtzite compounds (BeO, BN, SiC, AlN, GaN, ZnO, ZnS, CdS, and CdSe) (Xu and Ching, 1993). These are either wide band gap insulators (BeO) or typical semiconductors (CdS and CdSe) that do not belong to the traditional group IV, III–V, and II–VI semiconductors. Comparative studies that include a large number of crystals with the same structure have been a hallmark of the application of the OLCAO method.

It is well known that the band gaps of semiconductors and insulators from density functional calculations tend to be underestimated by between 30 to 50% depending on the crystal, the method used, and the exchange-correlation potential employed. An effective means of gap correction has been at the center of attention of many condensed matter physicists for years. The approaches taken to remedy the band gap problem range from the simplest so-called "scissor operator" (a rigid shift of the conduction bands), to more elaborate many-body corrections such as the GW approach (Hedin and Lundquist, 1969, Hybertsen and Louie, 1985) or the implementation of self-interaction corrections (SIC) (Harrison and et al., 1983, Heaton et al., 1982, Perdew and Zunger, 1981). In general, both GW and SIC approaches are computationally demanding and have been limited to only the most simple crystals until recently. Intermediate between the two approaches are some other computationally efficient and theoretically justifiable schemes. An approximate energy and k-point dependent GW self-energy correction scheme was introduced into the OLCAO method by Gu and Ching based on the Sternes–Inkson model (Sternes and Inkson, 1984). The on-site exchange integrals were evaluated using the LDA Bloch function throughout the BZ. A k-point weighted band gap E_g and a plasmon frequency (ω_p) determined by valence electron density was used to estimate the dielectric constant. Application to diamond, Si, Ge, GaP, GaAs, and ZnS showed that the GW-corrected gap values are generally within 10% of the experimental values (Gu and Ching, 1994).

More recently, there has not been much work using the OLCAO method that is specifically directed towards semiconductors. This is mainly due to the maturity of the field for computing the electronic structures of traditional semiconductors, the existence of many excellent works using other methods, and also the growth direction of OLCAO toward systems with more complex structures that require many more atoms to model. However, with this in mind, there is a good prospect that the OLCAO method will again find useful application in the area of elemental and binary semiconductors when the systems are no longer represented as infinite crystalline solids but are instead modeled as complex polycrystals or nano-particles.

5.2 Binary insulators

Binary insulators are mostly alkali or alkali earth halides and oxides with simple cubic crystal structures, ionic bonding, and large band gaps. They are certainly not the complex structures that this book is devoted to. Still, they constitute an important class of crystals where many of the early methods of band theory were tested and they are related to many complex inorganic crystals. A comprehensive study of these systems using the OLCAO method was published in 1995 (Ching et al., 1995) in which the band structures, and the linear and the non-linear (third harmonic generations) optical properties of 27 alkali halides, alkali-earth fluorides, oxides, and sulfides were reported. They are LiF, LiCl, LiBr, LiI, NaF, NaCl, NaBr, NaI, KF, KCl, KBr, KI, RbF, RbCl, RbBr, RbI, CaF$_2$, SrF$_2$, CdF$_2$, BaF$_2$, MgO, CaO, SrO, BaO, MgS, CaS, and SrS. The work was motivated by a series of papers by Lines (Lines, 1990b,

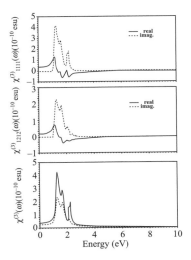

Fig. 5.3.
Calculated third-harmonic generations for Si. Top: $\chi_{1111}^{(2)}(\omega)$; middle: $\chi_{1212}^{(2)}(\omega)$; and bottom: comparison between $|\chi_{1111}^{(2)}(\omega)|$ (solid line) and $|\chi_{1212}^{(2)}(\omega)|$ (dashed line).

Lines, 1990a, Lines, 1991) who used a simple bond orbital theory to study the linear and non-linear optical response in ionic insulators. It was an excellent opportunity to test the accuracy of the first principles OLCAO method for such calculations without the use of empirical parameters. Furthermore, extensive experimental data from (Adair et al., 1989) were available in the form of non-linear refractive indices. These are related to the third harmonic generation (THG) coefficients and can thus be used for comparison with the calculated data. Because these ionic insulators all have large band gaps and the LDA theory underestimates the band gap to a large extent, a "scissor operator" was applied to these calculations to adjust the band gap to the experimentally measured values. The ability to calculate the dispersion relations at finite frequencies for all non-vanishing elements of the non-linear optical susceptibility tensor is a strong point of the OLCAO method as was already demonstrated in the case of elemental and binary semiconductors discussed in the last section. For centrosymmetric binary ionic insulators where the second order non-linearity vanishes, the third order non-linearity is the one to be calculated. The results of these calculations showed that the THG coefficient $\chi^{(3)}(0)$ can vary over two orders of magnitude as the size of the anions change; whereas for a fixed anion, the $\chi^{(3)}(0)$ values have the same order of magnitude and do not scale with the size of the cation. This was computable because the full *ab initio* calculation of the excited CB states provides realistic wave functions that can vary considerably even for isoelectronic crystals with different cations.

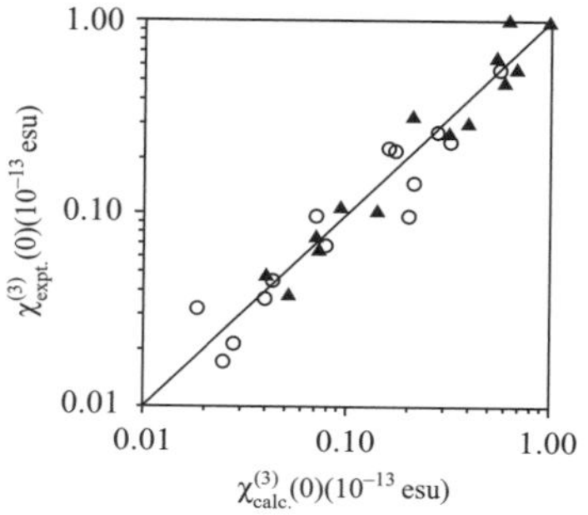

Figure 5.4 shows a comparison of the calculated values of $\chi^{(3)}(0)$ for 13 crystals with experimental data. The agreement is quite impressive and it validates the accuracy of the OLCAO method for non-linear optical calculations. Figure 5.5 shows a comparison of the ratio $|\chi^{(3)}_{1212}(0)/\chi^{(3)}_{1111}(0)|$ for the same 13 crystals. The agreement is less impressive but still considered to be satisfactory because no empirical parameters other than the band gap were involved.

In addition to the above capabilities, a simplified version of the self-interaction correction (SIC) to DFT within the OLCAO method was applied to the calculation of the optical properties of crystalline CaF_2 (Gan et al., 1992). The calculated LDA band gap for CaF_2 is 6.53 eV vs. the measured value of 11.6 eV (Barth et al., 1990). The SIC correction procedure followed the orbital-by-orbital approach proposed by Perdew and Zunger (Perdew and Zunger, 1981) and implemented in the LCAO method by Heaton and Lin (Heaton and Lin, 1982) which involves the construction of the localized Wannier functions. In the simplified version for CaF_2, the time consuming construction of Wannier function was replaced by using the minimal basis atomic orbitals in the OLCAO method, and the orbital SIC potential was replaced by a Mulliken-weighted combination of orbital SIC functions obtained in an iterative fashion. This simplified scheme increased the band gap of CaF_2 from 6.35 eV to 8.20 eV which is sill 29% lower than the experimental gap. The calculated imaginary part of the dielectric function for CaF_2 is shown in Fig. 5.6 and its spectral features show quite good agreement with experiment after aligning the main absorption peak. Although further improvement and refinement of the SIC procedure in the OLCAO method is possible and worth pursuing, it has not been attempted at this point in time.

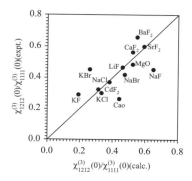

Fig. 5.4.
Calculated vs. measured data of $\chi^{(3)}(0)$ for 13 cubic crystals: LiF, NaF, NaCl, NaBr, KF, KCl, KBr, CaF_2, SrF_2, CdF_2, BaF_2, MgO, and CaO (see cited reference for detail).

Fig. 5.5.
Calculated ratios of $|\chi^{(3)}_{1212}(0)/\chi^{(3)}_{1111}(0)|$ vs. experimental ones for the 13 cubic crystals. (see cited reference for detail).

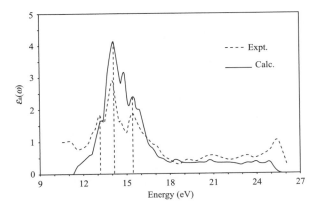

Fig. 5.6.
Imaginary part of the dielectric function of CaF_2 after SIC correction. The solid line is for calculation and the dashed line is experimental data.

5.3 Oxides

The term "oxides" represents a vast variety of solid state materials. They can be semiconductors such as ZnO or large gap insulators such as α-quartz (α-SiO_2), and many of them are metals as will be discussed in later chapters. The OLCAO method has been applied to many oxides ranging from binary cubic crystals such as MgO, ZnO, or BeO (discussed above) to ternary and quaternary crystals. Many of these oxides overlap heavily with other classes of insulators such as the oxynitrides and phosphates, while some of the binary oxides actually have very complex structures such as Y_2O_3 (yttria in the bixbyite structure). Some oxides are bioceramics and they will be discussed in Chapter 7 under the heading of complex crystals.

5.3.1 Binary oxides

Silicon dioxide (SiO_2) is inarguably the most important and abundant oxide on Earth. It has many different polymorphic phases with α-quartz (α-SiO_2) being the most common and thermodynamically stable phase. Early in 1985, Li and Ching used the OLCAO method with a superposition of atomic charge model to calculate the band structures of all the polymorphs of SiO_2. This includes the 4:2 coordinated phases of α-quartz, β-quartz, β-tridynmite, α-crystobalite, β-crystobalite, keatite, coesite, two idealized forms of β-crystobalite, and the 6:3-coordinated high pressure phase stishovite (Li and Ching, 1985). These polymorphs have crystal lattices in hexagonal, tetragonal, cubic, and monoclinic structures. This investigation was a very early example of how the OLCAO method was used for comparative studies of a whole class of crystals that enabled us to correlate the electronic properties with variations in structural parameters such as bond lengths and bond angles. The calculations were later repeated using a much more accurate and fully self-consistent version of the OLCAO code (Ching, 2000, Xu and Ching, 1988). Figure 5.7 shows the band structure of α-quartz compared to that of the high pressure stishovite phase. The study of crystalline SiO_2 is intimately related to the study of other silicate crystals and of amorphous SiO_2 to be discussed in Chapter 8. The non-linear

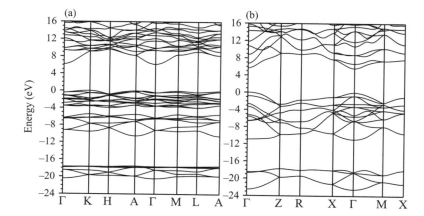

Fig. 5.7.
Comparison of the band structures of (a)
α-SiO₂ and (b) Stishovite.

optical properties of α-SiO$_2$ in the form of the second harmonic generation
have also been studied and show good agreement with experiment on the
calculated value of $|\chi^{(2)}(0)|$ (Huang and Ching, 1994).

Another very important class of binary oxide is alumina (Al$_2$O$_3$) with
α-Al$_2$O$_3$ (corundum) being the most thermodynamically stable phase. The
process of obtaining α-Al$_2$O$_3$ involves many intermediate phases or the so-
called transition alumina (β-, γ-, η-, θ-, κ-, χ-, etc) (Wefer and Misra, 1987).
α-Al$_2$O$_3$ has a rhombohedral (or trigonal) structure with two formula units
per primitive cell. Figure 5.8 shows the crystal structure of α-Al$_2$O$_3$ and the
band structure calculated using the OLCAO method (Ching and Xu, 1994,
Xu and Ching, 1991a). By partitioning the OLCAO charge density in real
space the crystal is shown to be highly ionic with an effective charge formula
of Al$_2^{+2.75}$O$_3^{-1.83}$ which is of course different from the full ionic model
of Al$_2^{+3}$O$_3^{-2}$ routinely used in the interpretation of experimental data. The

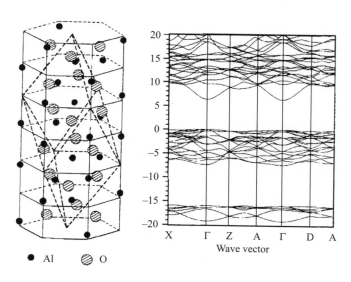

Fig. 5.8.
Hexagonal and rhombohedral unit cell of
α-Al₂O₃ and the calculated band
structure.

calculated direct band gap is 6.31 eV. The experimentally measured gap value in α-Al$_2$O$_3$ is somewhat uncertain due to the presence of an excitonic peak near the absorption edge, but it is estimated to be around 8.1 eV (French, 1990). Calculations using the crystal parameters of α-Al$_2$O$_3$ at both 2000 K and at room temperature showed that the differences in the electronic structure were very small. *Ab initio* calculations of the electronic and optical properties of corundum immediately led to the study of microstructures such as grain boundaries in alumina that will be discussed in Chapter 9.

Among all the transition aluminas, γ-Al$_2$O$_3$ is deemed the most important and controversial because of its important application as a catalyst material or catalytic support. In spite of many experimental and theoretical studies, there have been continuing debates about its structure and properties.

Mo et al. used the OLCAO method to calculate the electronic structure of γ-Al$_2$O$_3$ based on the defective spinel lattice model (Mo et al., 1997). Two 54 atom spinel supercell models were made with different locations for the three Al vacancy sites. The first, denoted by (Al$_5\square_3$)[Al$_{16}$]O$_{32}$, had three Al vacancies selected from the eight available tetrahedral cation sites while the second, denoted by (Al$_8$)[Al$_{13}\square_3$]O$_{32}$, had three Al vacancies selected from the sixteen available octahedral cation sites where $\square$ represents an Al vacancy. Total energy calculations and DOS analysis on these two models favored the model with Al vacancies at the octahedral sites. However, these models are not exactly stoichiometric because the formula for them is Al$_{21.333}\square_{2.667}$O$_{32}$. This defective spinel model was later challenged by other researchers and it was proposed that a low symmetry tetragonal structure (space group I4$_1$/amd) with both octahedrally and tetrahedrally coordinated Al sites was more appropriate (Men et al., 2005). Based on this new model, extensive studies of the physical properties of γ-Al$_2$O$_3$ were carried out including lattice phonons, bulk structural properties, electronic structure and bonding, optical and spectroscopic properties (Ching et al., 2008) using both OLCAO and other DFT based computational programs. The crystal structure of this γ-Al$_2$O$_3$ model and the band structure calculated using the OLCAO method are shown in Fig. 5.9. The calculated LDA band gap for γ-Al$_2$O$_3$ is 4.22 eV compared to that of 6.33 eV for α-Al$_2$O$_3$. The electronic structure of another transition alumina θ-Al$_2$O$_3$ was also studied by the OLCAO method and was shown to have an indirect band gap of 4.64 eV (Mo and Ching, 1998).

Another important binary oxide that is isoelectronic with alumina is yttria (Y$_2$O$_3$). Y$_2$O$_3$ is an important refractory oxide widely used in a variety of applications. It is also well known that a small addition of Y$_2$O$_3$ to α-Al$_2$O$_3$ affects the properties of alumina in the so-called "yttrium effect" which has the beneficial effect of increasing the adhesion of oxide scales in all Al-containing metals. Y$_2$O$_3$ is also used as a laser host material. Unlike alumina, it has only one known phase, the bixbyte phase (space group Ia-3) and its structure is quite complicated. The cubic cell contains 80 atoms with two non-equivalent Y sites (8a and 24d) and one O site (48e). Between α-Al$_2$O$_3$ and Y$_2$O$_3$, there are three intermediate ternary oxides, YAlO$_3$ (YAP), Y$_3$Al$_5$O$_{12}$ (YAG), and Y$_2$Al$_2$O$_9$ (YAM) which will be discussed in the next subsection. The electronic structure and optical properties of Y$_2$O$_3$ were studied using the OLCAO method, first

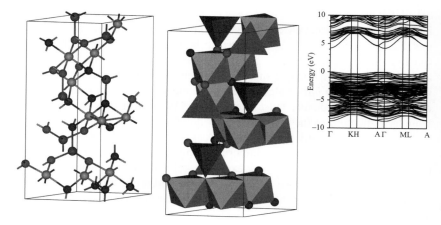

Fig. 5.9.

Crystal structure and the calculated band structure of γ-Al_2O_3.

in 1990 (Ching et al., 1990) and then later in a more detailed fashion with the addition of mechanical properties calculations in 1997. (Xu et al., 1997). From the charge density analysis, it was shown that the partial charge description for Y_2O_3 is $Y_2^{+2.16}O_3^{-1.44}$. In this respect, Y_2O_3 is less ionic than α-Al_2O_3. The calculated band gap for Y_2O_3 is 4.54 eV, less than that of α-Al_2O_3. Such differences were traced to the larger size of the Y ion and, via orbital decomposition of the wave function, to the character of the 4d electrons in Y.

Other binary oxides in which the OLCAO method has been applied include the three phases of ZrO_2 (Zandiehnadem et al., 1988), Cu_2O, CuO, (Ching et al., 1989), V_2O_3, and V_2O_5 (Parker et al., 1990a, Parker et al., 1990b), three phases of TiO_2 (Mo and Ching, 1995), and B_2O_3 (Ouyang et al., 2002). The early calculations on the band structures of cubic, tetragonal, and monoclinic phases of ZrO_2 showed that they have band gaps of 3.84, 4.11, and 4.51 eV respectively. The cubic phase of ZrO_2 is hyperphysical because it has to be stabilized by adding Y_2O_3. The calculated optical conductivities for the three phases of ZrO_2 showed reasonable agreement with the data from vacuum ultraviolet (VUV) spectroscopy (French et al., 1994).

Cuprous oxide (Cu_2O) is a well studied semiconductor crystal in which Cu is in a Cu^{2+} state while cupric oxide (CuO) with Cu in the Cu^{1+} state is a Mott insulator where LDA calculations generally fail. For Cu_2O, the calculated intrinsic band gap is only 0.8 eV whereas the calculated optical gap is on the order of 2.0 to 2.3 eV. This is because the optical transition at the Γ point is dipole forbidden. This work also underscores the importance of distinguishing the intrinsic band gap as obtained from the first-principles band structure calculations and the optical band gap extracted from experiments. The calculated band structure for CuO within the one electron approximation is quite interesting. It shows the presence of a Fermi surface with intrinsic holes at the top of the valence band (VB) and a semiconductor-like gap of 1.60 eV at Γ, similar to that of the $YBa_2Cu_3O_7$ superconductor (to be discussed in the next chapter). The calculated electronic structure and optical properties for VO_2 and V_2O_5 show very good agreement with absorption spectra obtained from ellipsometry measurements (Parker et al., 1990a, Parker et al., 1990b). This is somewhat unexpected given the complexity of the interactions in transition metal oxides.

The OLCAO method was also used to calculate the electronic structure, mechanical, and optical properties of the three phases of TiO_2, rutile, anatase, and brookite (Mo and Ching, 1995). This is another well studied binary oxide with extensive applications in many industries. An important conclusion from that study was the revelation of subtle differences in the optical anisotropy in these three phases. This can be traced to the variations in the inter-atomic bond lengths and bond angles in the three phases which result in slightly different electronic structures. Figure 5.10 shows the calculated total DOS and the O-2s, O-2p, and Ti-3d components of the partial DOS in the three crystals.

In contrast, the vitreous B_2O_3 glass and the crystalline phases of B_2O_3 were less well studied. There are two crystalline phases for B_2O_3; the low pressure phase (B_2O_3-I) whose structure consists of infinitely linked B-O_3 triangular units and the high pressure phase (B_2O_3-II) which contains interconnected B-O_4 tetrahedral units with 6- and 8-membered rings. The OLCAO method was applied to study the electronic structure and the optical properties of these two phases (Li and Ching, 1996a). Both are wide band gap insulators with

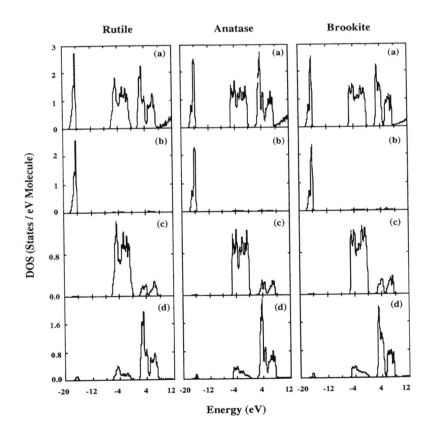

Fig. 5.10.
Calculated total and partial DOS of three phases of TiO_2: rutile, anatase and brookite: (a) total, (b) O_{-2s}, (c) O_{-2p}, and (d) Ti_{-3d}.

calculated gap values of 6.20 eV and 8.85 eV respectively while the calculated static dielectric constants for the two were quite similar and close to 2.30. It was concluded that B_2O_3-I is less ionic than B_2O_3-II and its properties should be very close to that of the vitreous B_2O_3.

5.3.2   Ternary oxides

The OLCAO method was applied to ternary alkali silicates (Na_2SiO_3, Na_2SiO_5, Li_2SiO_3, and Li_2SiO_5) as early as the mid 1980s (Ching et al., 1983, Ching et al., 1984b). This was the first time that it had been applied to crystals with fairly complicated structures. A rather simple potential with a minimal basis set was used in these calculations. The calculated DOS of the VB showed good agreement with experimental XPS spectra. These early calculations motivated us to make more effective use of the OLCAO method and shaped many of its later applications.

Many of the most important ferroelectric crystals are perovskites. In the cubic or tetragonal form, their structures are relatively simple. The self-consistent OLCAO method was applied to three cubic perovskites $SrTiO_3$, $BaTiO_3$, and $KNbO_3$ (Xu et al., 1990, Xu et al., 1994) to investigate their band structure and optical properties and the results showed favorable agreement with experimental optical data. Similar calculations were then extended to $LiNbO_3$ which is another important ferroelectric and non-linear optical crystal. Unlike $SrTiO_3$ or $BaTiO_3$, $LiNbO_3$ has a rhombohedral unit cell similar to alumina. The electronic structure of $LiNbO_3$ was calculated with self-interaction correction (SIC) similar to that applied to CaF_2 discussed above (Ching et al., 1994). The SIC correction increased the direct band gap from the LDA value of 2.62 eV to 3.56 eV in much closer agreement with experiment. Additionally, the calculated optical properties were also significantly improved showing the difference from a simple rigid shift of the conduction band states or the "scissor operator" practice. Additionally the calculated optical bire-fringence data in the form of ordinary (n_o) and extraordinary (n_e) refractive indices showed excellent agreement with the measured values. These results are illustrated in Fig. 5.11.

Another interesting ferroelectric crystal which is distinctively different from the ones discussed above is $NaNO_2$ which has a very simple body-centered orthorhombic structure. $NaNO_2$ is classified as a two-dimensional ferroelectric in which the positive Na ion and the negative NO_2 ion form a dipole with the moment pointing in the y-direction. Polarization reversal is achieved by the rotation of the NO_2 molecule about the axis normal to the mirror plane. The electronic structure and optical absorption spectrum of $NaNO_2$ in the ordered phase were calculated by using the OLCAO method. The occupied energy levels and the first unoccupied band are very flat which is typical of molecular solids (Jiang et al., 1992). The effect of disorder in the paraelectric phase of $NaNO_2$, associated with random orientation of the NO_2 molecule, was later investigated and shown to have a slight reduction in the band gap and a weakening of the dielectric screening by disorder scattering (Zhong et al., 1994).

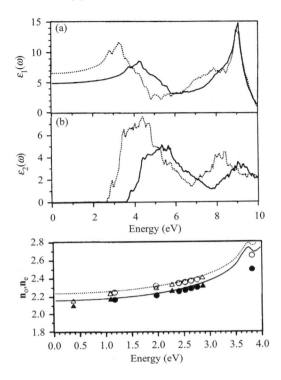

Fig. 5.11.
Top: Calculated $\varepsilon_1(\omega)$ (a) and $\varepsilon_2(\omega)$ (b) for LiNbO$_3$. Solid line, self-energy-corrected result; dashed line, LDA result. Bottom: Calculated n_e (solid line) and n_o (dashed line). Triangles and circles are experimental data (see cited reference for details).

In the late 1970s, Chen proposed an anionic-group theory model to describe and predict non-linear optical properties of the borate crystals (Chen and Liu, 1986). This was one of the few examples in which theory led the discovery of important inorganic crystals with excellent optical properties and materials qualities. Among them, β-BaB$_2$O$_4$ (BBO) and LiB$_3$O$_5$ (LBO) are the most outstanding. The OLCAO method was applied to investigate the electronic structure and optical properties of these two crystals (Xu and Ching, 1990, Xu et al., 1993). Both LBO and BBO have fairly large unit cells and complex crystal structures but somewhat different local bonding characterizations. LBO has an orthorhombic cell with four formula units per cell (space group Pn2$_1$a) in which two of the B atoms are three-fold bonded as in B$_2$O$_3$, and one B site is at the center of a distorted tetrahedron; whereas BBO has a trigonal structure with six formula units per cell (space group R3c) with alternating and closely separated layers of (B$_3$O$_6$)$^{-3}$ and Ba ions. The OLCAO calculation showed that LBO has a much larger LDA band gap (direct gap at $\Gamma = 7.37$ eV) than BBO (direct gap at $\Gamma = 5.61$ eV and indirect gap $= 5.52$ eV). The differences in the electronic structure of BBO and LBO are attributed not only to the different local environments of the ionic units but also to the unique role played by the Ba-5p orbitals in the inter-atomic bonding. The calculated optical properties in these two crystals are in good agreement with the experimental data.

Spinel oxides constitute an important class of ceramic oxides. In particular, magnesium spinel (MgAl$_2$O$_4$) has a combination of many desirable properties suitable for a variety of applications. There are two types of spinels having an f.c.c. structure (space group $Fd\text{-}3m$): normal spinel and inverse spinel. In the normal spinel, all trivalent (Al^{3+}) ions are in an octahedral coordination and all divalent (Mg^{2+}) ions are in a tetrahedral coordination. It can be conveniently designated by the formula $(Mg_8)[Al_{16}]O_{32}$ in the 54-atom cubic cell description. In the inverse spinel, all the tetrahedral sites are occupied by Al^{3+} and the octahedral sites are occupied by both Al^{3+} and Mg^{2+} in equal proportions denoted by $(Al_8)[Mg_8Al_8]O_{32}$. In many oxide systems, voids can be treated as a trivalent ion to form what is called the defect inverse spinel as in $\gamma\text{-}Fe_2O_3$ and $\gamma\text{-}Al_2O_3$. Between the normal and inverse spinels, there are intermediate phases with random cation distributions characterized by a disorder parameter λ ranging from 0 to 0.5. λ is simply the fraction of octahedral sites occupied by Mg^{2+} so that the inverse spinel can be described by the formula $(Mg_{8-16\lambda}\,Al_{16\lambda})[Mg_{16\lambda}\,Al_{16-16\lambda}]O_{32}$ with $\lambda = 1/2$ corresponding to a completely random distribution. The electronic structure and the ground state properties of the normal, inverse, and partially inverse $MgAl_2O_4$ spinels were studied in considerable detail by the OLCAO method (Mo and Ching, 1996, Xu and Ching, 1991a). For the normal spinel, charge analysis showed its ionic description to be $Mg^{+1.79}Al_2^{+2.63}O_4^{-1.76}$ compared to $Mg^{+1.83}O^{-1.83}$ and $Al_2^{+2.63}O_3^{-1.75}$ for MgO and $\alpha\text{-}Al_2O_3$ crystals. Thus, $MgAl_2O_4$ is less ionic than MgO but similar to $\alpha\text{-}Al_2O_3$ and it has a band gap of $5.80\,\text{eV}$ which is smaller than $\alpha\text{-}Al_2O_3$. Optical properties calculations of these three crystals showed good agreement with measured VUV data except near the absorption edge where the experimental data showed strong excitonic peaks in all three crystals. An interesting and intriguing point was that the calculated optical spectra were in good agreement with the measured optical absorption peaks in the range above $10\,\text{eV}$ without needing to apply corrections to the band gap values to account for the LDA gap underestimation.

The OLCAO calculation of the inverse and partially inverse spinels showed a reduced band gap E_g due to disorder with the E_g minimum at $5.80\,\text{eV}$ when $\lambda = 4/16$. The variation in the total DOS with λ is illustrated in Fig. 5.12. Detailed inspection of the atom and orbital decomposed PDOS of Al and Mg at the octahedral and tetrahedral sites revealed strong effects of disorder related to local bonding (Mo and Ching, 1996). The OLCAO method has also been used to study optical spectra changes caused by isolated O vacancy defect structures in $MgAl_2O_4$ as well as Fe substitutional impurities at the Mg and Al sites. This will be discussed in detail in Chapter 9.

The other major ternary oxides are yttrium aluminates, yttrium silicates, and aluminum silicates. These crystals all have large unit cells, exhibit different types of bonding, contain different compositions where the number of bonds per atom varies, and have variations in charge transfer. Most importantly, these are all materials with important applications. The OLCAO method is ideally suited for studying the electronic, optical, and bonding properties of these crystals.

Between the two important and stable ceramic crystals, alumina and yttria, there are three congruently melted compounds with different Al to Y ratios: $Y_3Al_5O_{12}$ (YAG), $YAlO_3$ (YAP), and $Y_4Al_2O_9$ (YAM). Together

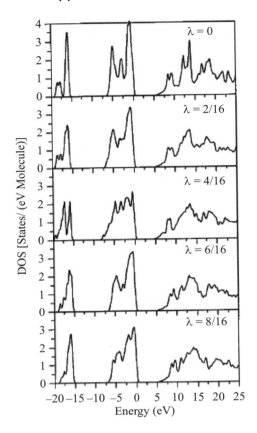

Fig. 5.12.
Calculated total DOS of the normal spinel ($\lambda = 0$) and partially inverse spinel ($\lambda \neq 0$).

they constitute the Y-Al-O system with distinct and well defined local atomic coordinations. YAG or yttrium alumina garnet is the most important solid state laser host material and it is also known for its creep resistance and application in high temperature ceramic composites. It has a complicated bcc structure of space group Ia3d with 160 (80) atoms in the cubic (primitive) cell. The Y ions are dodecahedrally bonded to eight O ions. There are two different sites for Al, the octahedrally bonded Al_{oct} at the 16(a) site and the tetrahedrally bonded Al_{tet} at the 24(d) site. YAP is also an important laser host crystal with a $GdFeO_3$ type structure (space group *Pbnm*) which is slightly distorted from the perfect perovskite. Both Y and Al are octahedrally coordinated in YAP with two non-equivalent O sites. YAM has a very complicated monoclinic structure (space group $P2_1/c$). All four Y sites, two Al sites, and nine O sites are non-equivalent resulting in a plethora of local bonding environments. The electronic structures, bonding, and bulk structural properties in YAG, YAP, and YAM were calculated using the OLCAO method and are discussed in conjunction with the end members α-Al_2O_3 and Y_2O_3 in the Y-Al-O series (Ching and Xu, 1999, Xu and Ching, 1999).

Similar to the Y-Al-O series, there are two well-established yttrium silicates between SiO_2 and Y_2O_3. They are Y-orthosilicate Y_2SiO_5 and Y-pyrosilicate

$Y_2Si_2O_7$, both have a monoclinic structure of space group C2/c and $P2_1/c$ respectively. Y_2SiO_5 is another well known laser host crystal whereas $Y_2Si_2O_7$ is much less studied and is mostly recognized as a precipitated phase in the ceramic interlayer between Si_3N_4 and Si_2N_2O. The electronic structure, the bulk structure properties, and the optical absorptions in these two crystals were studied in great detail by the OLCAO method (Ching et al., 2003a). An important conclusion was that neither can be represented as the weighted sum of α-SiO_2 and Y_2O_3. This is because the Si-O bond in the two ternary compounds tends to be weaker than that in SiO_2 while the Y-O bond appears to be stronger than that in Y_2O_3 crystal, thus underscoring the importance of *ab initio* calculation of individual crystals rather than taking the average of the properties of simpler crystals that compose them.

Another group of common ternary oxides is the aluminosilicates Al_2SiO_5. More recently, the electronic structure and spectroscopic properties of three aluminosilicate polymorphs (sillimanite, andalusite, and kyanite) were studied using the OLCAO method (Aryal et al., 2008). Sillimanite and andalusite have orthorhombic crystal structures with space group *Pnnm* and *Pbnm* respectively while kyanite has a very low symmetry triclinic structure of space group *P*-1. The calculated band gaps for the three crystals are 5.01 eV, 5.05 eV, and 5.80 eV respectively. As with the yttrium silicates, the electronic structure of the aluminosilicates cannot be approximated as the sum of Al_2O_3 and SiO_2. Also calculated were the XANES spectra of these three crystals which will be discussed in Chapter 11.

A more general class of the ternary aluminosilicates with a long history is the Mullite system (Shepherd and Rankin, 1909). It has a chemical formula of $Al_{4+4x}Si_{2-2x}O_{10-x}$ (x = 0 to 1.0) and is typically considered to be a mixture of $nAl_2O_3 \bullet mSiO_2$, where the most common mullite phases are 3/2 (x = 0.25, n = 3, m = 2), 2/1 (x = 0.40, n = 2, m = 1), 4/1 (x = 0.667, n = 4, m = 1), and 9/1 (x = 0.826, x = 9, m = 1). Sillimanite is a precursor of mullite and it represents the x = 0 end member of the mullite series. The other end member with x = 1 is the elusive ι-Al_2O_3 phase of alumina (Foster, 1959). Mullite is an excellent ceramic and refractory material because it has a low thermal expansion, low thermal conductivity, and it remains strong at high temperatures. Furthermore, it has excellent creep resistance and an outstanding stability under harsh chemical environments. These properties make it a ceramic par excellence that is ideal for structural and functional materials applications (Schneider and Komarneni, 2005).

There exists essentially no electronic structure information on the different mullite phases due to uncertainty in the precise atomic scale structure. The experimental data on the structure of mullite phases are always presented in the context of a perfect crystal with partially occupied sites and so large scale modeling is required to produce an accurate representation. In this sense, mullite phases can be considered to be defective aluminosilicates obtained by creating an O vacancy along with the replacement of two tetrahedrally coordinated Si by two Al atoms which can be either tetrahedrally or octahedrally bonded to O such that the charge neutrality condition is maintained. The OLCAO method is well suited for such systems and some work has been done in this direction by first making supercell models for mullite and then using the OLCAO method to study the electronic structure and its variation with x (Aryal, 2007). They are all large band gap insulators similar to the crystalline aluminosilicates but with the presence of defect-like states at the band edges. This knowledge is important for understanding how the physical properties of mullite changes with changing alumina content.

5.3.3   Laser host crystals

Many of the important laser crystals have the garnet structure such as the YAG crystal already discussed above. The garnet structure has a cubic cell with 8 formula units of the general form $A_3B'_2B''_3O_{12}$ where the A sites (24(c)) are for rare earths, the B' and B'' sites (16(a) and 24(d)) are for transition metal or metalloid ions, and where the O ions occupy the 96(h) site. The garnet structure can be viewed as interconnected dodecahedrons (at the A site), octahedrons (at the B' site), and tetrahedrons (at the B'' site) with shared O at the corners of the polyhedra. The three main classes of synthetic garnets are the yttrium aluminum garnets $Y_3Al_5O_{12}$ (YAG), iron garnets (to be discussed in the next chapter), and gallium garnets with $Gd_3Sc_2Ga_3O_{12}$ (GSGG), $Gd_3Sc_2Al_3O_{12}$ (GSAG), and $Gd_3Ga_5O_{12}$ (GGG) being the most common. Laser operation depends on the doping of the trivalent rare earth or transition metal ions with elements such as Nd^{3+} or Cr^{3+} at the B' site or Cr^{4+} at the B'' site so that the crystal can produce the desired localized levels within the gap for emission or absorption at specific frequencies. Resistance to radiation is also an important issue for laser host materials. Most theoretical studies on laser crystals concentrate on spectroscopy of the energy levels of the doped ions in the local environment of the host crystal using the framework of ligand field theory. In these studies, the atomic multiplet levels are calculated using crystal field parameters fixed by experimental data while the physical properties of the host materials are usually ignored. As a result, the electronic structure and bonding in many of the laser crystals has not been studied at all.

The OLCAO method was applied to the GSGG, GSAG, and GGG crystals in addition to YAG to obtain information on the electronic structure, inter-atomic bonding, and mechanical strength in these crystals (Xu et al., 2000). All four crystals have similar band structures and inter-atomic bonding except for GSGG and GSAG where the Sr atom at the octahedral site showed a greater covalent character and stronger bonds compared to Ga or Al in GSGG and YAG. This could have implications on the stability of the Cr^{3+} ion in GSGG and its resistance to radiation. The calculated effective charges on each ion and the bond order values for these four crystals are listed in Table 5.1 and 5.2.

The other two important laser crystals that do not have the garnet structure that have also been studied by the OLCAO method are $BeAl_2O_4$ and $LiYF_4$ (LYF) (Ching et al., 2001c). $BeAl_2O_4$ is the mineral chrysoberyl and when doped with Cr bears the name of alexandrite. $BeAl_2O_4$ has an orthorhombic cell (space group *Pnma*) with two non-equivalent octahedrally coordinated Al sites and three non-equivalent O sites. The Be atom is tetrahedrally coordinated with relatively short Be-O bond lengths. LYF has a tetragonal cell (space group

Table 5.1. Calculated effective charge Q_α^* in the four garnet crystals.

	GSGG	GSAG	GGG	YAG
A site				
Y (exc. 4p)				2.033
Gd (exc. 4f)	1.895	1.903	1.424	
B′ site				
Sc	2.291	2.276		
Al				1.939
Ga (exc. 3d)			1.599	
B″ site				
Al		1.893		1.839
Ga (exc. 3d)	1.999		1.965	
O site				
O	6.645	6.672	6.748	6.708

Table 5.2. Calculated bond order $\rho_{\alpha,\beta}$ in the four garnet crystals. Interatomic distances in parenthesis (Å)

	GSGG	GSAG	GGG	YAG
A site				
Y-O				0.075(2.432)
				0.081(2.303)
Gd-O	0.066(2.477)	0.066(2.479)	0.069(2.473)	
	0.062(2.392)	0.064(2.371)	0.069(2.358)	
B′ site				
Sc-O	0.112(2.088)	0.113(2.083)		
Al-O				0.096(1.937)
Ga-O			0.086(2.006)	
B″ site				
Al-O		0.132(1.775)		0.133(1.761)
Ga-O	0.121(1.854)		0.121(1.848)	
O site				
O-O	0.011(2.815)	0.014(2.705)	0.011(2.808)	0.014(2.658)
	0.005(2.808)	0.006(2.815)	0.013(2.705)	0.014(2.696)
	0.006(2.808)	0.004(2.852)	0.006(2.848)	0.007(2.837)

$14_1/a$) with four formula units per cell. The Li ions are four-fold bonded whereas the Y ion is eight-fold bonded to F with fairly large Y-F bonds. Results from the OLCAO calculation showed that $BeAl_2O_4$ ($LiYF_4$) has a band gap of 6.45 eV (7.54 eV) and a bulk modulus of 217.2 GPa (90.0 GPa). Thus, LYF is much softer than $BeAl_2O_4$. This can be easily explained by the higher ionic bonding character in LYF than in $BeAl_2O_4$ which has a significant amount of covalent mixing. This is consistent with the calculated effective charges and bond order values in these two crystals.

5.3.4 Quaternary oxides and other complex oxides

The OLCAO method has also been applied to several complex crystals and quaternary oxides. One example is the calculation of the structure and properties of the so-called EST-10 microporous titanosilicate (Ching et al., 1996). Microporous titanosilicates are a class of synthetic inorganic framework materials with potential applications as molecular sieves. The structure of EST-10 was determined by an ingenious combination of several experimental techniques and molecular modeling (Anderson et al., 1994). The monoclinic unit cell (space group C2/c) contains 16 Si_5TiO_{13} units for a total of 304 atoms. The porous structure is composed of corner sharing SiO_4 tetrahedra and TiO_6 octahedra linked by bridging oxygen (see Fig. 5.13). The framework itself is electron deficient and can be stabilized by adding alkali ions such as Na. The OLCAO calculation showed that by intercalating with 32 Na ions, the system becomes a semiconductor with a direct band gap of 2.33 eV. Total energy calculations indicate that the Na ions are located inside the 7-member ring pore adjacent to the one-dimensional Ti-O-Ti-O chain. This calculation was another early example of the OLCAO method being used to study materials with complex structures.

Another example of the application of OLCAO to complex structures is the study of the electronic structure and bonding in crystalline $(Na_{1/2}Bi_{1/2})TiO_3$ (NBT) and its ordered solid solution with $BaTiO_3$ (Xu and Ching, 2000). It was shown that large single crystals of (Na, Bi)TiO_3 can be grown after doping with $BaTiO_3$, which results in a solid solution with a large piezoelectric strain of up to 0.85% (Chiang et al., 1998, Godlewski et al., 1998). Such a promising

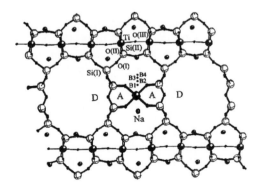

Fig. 5.13.
Projection of crystal structure of ETS-10 along the [110] direction. For clarity, the smallest spheres are the O atoms and the large shaded (open) spheres are the Ti (Si) atoms (see cited reference for detail).

Pb-free A-site relaxor material with multiple ferroelectric phases and superior piezoelectric actuation is extremely attractive for large scale relaxor materials applications. OLCAO was applied to NBT and its solid solution with 6% of $BaTiO_3$ using a supercell of 320 atoms in an ordered superstructural arrangement (Xu and Ching, 2000). The results showed that the Na ions are not fully ionic and the Bi ions have a high degree of covalent character due to the paired 6s electrons in Bi. This may also account for the increased bulk modulus over that of the single crystal perovskites $PbTiO_3$ and $PbZrO_3$ which are B-site relaxors with extensive applications.

ZrW_2O_8 is an interesting ternary oxide with a negative thermal expansion due to the rotational flexibility of the ZrO_6 octahedra and the WO_4 tetrahedra in the interconnected framework structure. There are two known crystalline forms of ZrW_2O_8. The low pressure phase α-ZrW_2O_8 transforms into γ-$ZrWO_4$ at a relatively low pressure of 0.21 GPa. The crystal structures of both phases are quite complicated. α-ZrW_2O_8 has a cubic cell (space group $P2_13$) with 44 atoms and γ-ZrW_2O_8 has a low symmetry orthorhombic cell (space group $P2_12_12_1$) with 132 atoms. The OLCAO calculation of the band structures for both crystals showed that α-ZrW_2O_8 has an indirect gap of 2.84 eV whereas γ-ZrW_2O_8 has a direct gap of 2.17 eV. The calculated bulk modulus of 63 GPa for α-ZrW_2O_8 is in good agreement with experiment. It was predicted that γ-ZrW_2O_8 should be a harder material on the basis of a larger calculated total crystal bond order value due to the formation of additional O–W bonds at higher pressure and reduced volume.

5.4 Nitrides

5.4.1 Binary nitrides

The earliest application of the OLCAO method to nitrides was to the quasi-one-dimensional crystal polysulfur nitride (S_2N_2) using a simple superposition of atomic potentials model (Ching et al., 1977). The calculated DOS was in good agreement with x-ray photoemission data and a band gap of 1.52 eV was obtained. The same method was then used to calculate the band structure of β-Si_3N_4 and α-Si_3N_4 in 1981 with the inclusion of Si-3d orbitals in the basis expansion (Ren and Ching, 1981). The calculations were extended to the oxynitride crystals Si_2N_2O and Ge_2N_2O (Ching and Ren, 1981). α- and β-silicon nitrides have a hexagonal cell while the Si_2N_2O cell is orthorhombic. These early calculations were considered to be the first time that the band structures of these "complex" structural ceramics were studied. A decade later, the electronic structure and optical properties of these crystals were recalculated using the fully self-consistent OLCAO method and a larger basis set and the results were compared with those of α-SiO_2 (Xu and Ching, 1995). Effective charge calculations for these crystals suggested ionic formulae of α-$(Si^{+2.52})_3(N^{-1.89})_4$, β-$(Si^{+2.50})_3(N^{-1.87})_4$, $(Si^{+2.54})_3(N^{-1.90})_4(O^{-1.25})$, and α-$(Si^{+2.60})(O^{-1.30})_2$ which show a significant covalent bonding character for the nitrides. These numbers are consistent with the calculated LDA band gaps of 4.63 eV, 4.96 eV, 5.59 eV, and 5.20 eV for these same crystals. The band

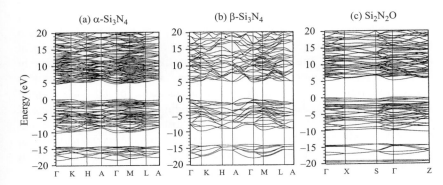

Fig. 5.14.
Band structures of (a) α-Si_3N_4, (b) β-Si_3N_4, and (c) Si_2N_2O.

structures of α-Si_3N_4, β-Si_3N_4, and Si_2N_2O are shown in Fig. 5.14. Optical properties calculations show good agreement with experimental data. All four crystals have intrinsic absorption tails because their wave functions near the conduction band minimum and valence band maximum are such that direct optical transitions are symmetry-forbidden. This emphasizes the important but often overlooked point of comparing the calculated band gap (the intrinsic gap) and the experimentally measured gap (usually extrapolated from optical absorption data above the threshold).

The ground state total energies for α-Si_3N_4 and β-Si_3N_4 from the OLCAO calculations were used as input data to obtain parameters for the pair-wise potentials for lattice dynamic and simplified phonon calculations in these two crystals (Chiang et al., 1998). These detailed crystalline silicon nitride calculations paved the way for later *ab initio* studies of microstructures in polycrystalline ceramics (see Chapter 8).

AlN is another important binary nitride with many applications. At the time when the fully self-consistent OLCAO method was implemented, it was first applied to AlN in the wurtzite structure (Ching and Harmon, 1986). The wurtzite structure has a hexagonal lattice characterized by two basic parameters, the c/a ratio and an internal parameter μ, which can deviate from the ideal values of $c/a = 1.63333$ and $\mu = 0.375$. Many important semiconductors and insulators have their ground state in the wurtzite structure. Although the calculated LDA band gap of 4.4 eV for AlN is a 30% underestimation of the experimental band gap of 6.3 eV, the calculated lattice constant, internal parameters, bulk modulus, bond strength, and pressure dependent frequency of the A1 mode of the transverse optical (TO) phonons were in good agreement with experiment. This gave a vote of confidence to the subsequent vast application of the OLCAO method to many other crystals. The calculated optical properties of AlN in the form of the complex dielectric function was in good agreement with data obtained from a VUV study up to the transition energy of 40 eV (Loughin et al., 1993). Similar calculations were also carried out for other binary nitrides in the wurtzite structure (GaN and InN) together with other stable wurtzite crystals (BeO, SiC, ZnO, ZnS, CdS, and CdSe) (Xu and Ching, 1993). Also presented in these calculations were the bulk modulus obtained from total energy data fitted to the Murnaghan equation of state (Murnaghan, 1944) with results in good agreement with experimental data.

In 1989, Liu and Cohen (Liu and Cohen, 1989) predicted that β-C_3N_4 (which is isostructural with β-Si_3N_4) may be a superhard material with a bulk modulus exceeding that of diamond based on a simple semi-empirical rule (Cohen, 1985). This exciting and provocative suggestion immediately attracted a lot of attention and prompted many attempts to synthesize this elusive material that are continuing even to this day. The OLCAO method was used to study the electronic structure, bonding, and optical properties of the hypothetical β-C_3N_4 crystal based on the structure suggested by Liu and Cohen (Yao and Ching, 1994). The calculated results on the crystal parameters, band gap, and bulk modulus were similar to those obtained by Liu and Cohen (Liu and Cohen, 1990). The optical absorption spectrum showed two major peaks at 7.9 and 13.9 eV which can be viewed as a prediction because no such measurement has since been conducted on validated samples of β-C_3N_4.

Boron nitride (BN) is an important wide gap insulator with many fascinating properties such as extreme hardness, light weight, low dielectric constant, and so on that make it highly competitive in many industrial applications. It is the second hardest material known after diamond. BN exists in three crystalline forms: hexagonal (h-BN), cubic zinc blend (c-BN), and wurtzite (w-BN). The normal phase of h-BN is a layered structure that is isostructural with graphite. At high temperature and pressure, h-BN may transform into w-BN. The c-BN has also been synthesized in laboratory experiments under pressure. The electronic structure, ground state total energy, and linear optical properties of these three phases of BN were studied by using the OLCAO method (Xu and Ching, 1991b). They all have large band gaps with calculated indirect band gaps of 4.07 eV, 5.18 eV, and 5.81 eV, and direct minimal band gaps of 4.2 eV (at H), 8.7 eV (at Γ), and 8.0 eV (at Γ) respectively for h-BN, c-BN, and w-BN. The effective masses for the VB and the CB were also obtained as well as the Mulliken effective charges which showed an increased B to N charge transfer in going from h-BN to c-BN to w-BN, consistent with increased ionicity in the series. The calculated equilibrium lattice constants, the bulk modulus and their derivatives, and the cohesive energies were in good agreement with experiments. From the total energy vs. volume data for the three crystals, the possible phase transitions from h-BN to c-BN and then to w-BN were investigated and the transition pressures estimated. The calculated optical properties for h-BN showed excellent agreement with the inelastic electron energy scattering measurements which gave credence to the results for the other two phases where reliable optical data were lacking.

The binary compound ZrN is known to be a metallic superconductor; however, Zr_3N_4 is an insulator which exists in the N-rich limit of ZrN_x films. Photoemission and optical data on ZrN_x films showed dramatic changes from x = 1 to x = 1.333 (Sanz et al., 1998), pointing to the existence of an insulating Zr_3N_4 compound. While the exact structure of Zr_3N_4 is unknown it is believed to be a vacancy-stabilized cubic structure (denoted as $Zr_3 \square N_4$). Lerch et al. (Lerch et al., 1996) reported the synthesis of a crystalline Zr_3N_4 with an orthorhombic structure (space group *Pnam*) o-Zr_3N_4. It was also speculated that Zr_3N_4 may exist in the spinel structure similar to that of cubic spinel nitrides γ-Zr_3N_4 (see next subsection). The OLCAO method was used to identify the likely structure of Zr_3N_4 by investigating the properties of $Zr_3 \square N_4$, γ-Zr_3N_4, and o-Zr_3N_4. The calculations confirmed that the ordered vacancy stabilized $Zr_3 \square N_4$ model is the most likely structure. All three crystals are insulators with small indirect band gaps. The total energy per formula unit follows the order $Zr_3 \square N_4 < o$-$Zr_3N_4 < \gamma$-Zr_3N_4, and $Zr_3 \square N_4$ has the largest bulk modulus at 299 GPA. Comparison between the experimental XPS data and the calculated DOS of the upper VB clearly showed that the best matching is with $Zr_3 \square N_4$ as illustrated in Fig. 5.15. It was also pointed out that previous calculations by others on models of $Zr_3 \square N_4$ showed it to be metallic due to the failure to properly relax the structure.

5.4.2 Spinel nitrides

The 1999 discovery of the cubic spinel phase of Si_3N_4 (denoted as c-Si_3N_4 or γ-Si_3N_4), in addition to the other two well known silicon nitride forms (α-Si_3N_4 and β-Si_3N_4), under a high pressure of 15 GPa and at a temperature exceeding 2000 K had a major impact on the search for new nitride compounds (Zerr et al., 1999). This crystal was believed to be the first time that Si was found in an octahedral coordination with N. γ-Si_3N_4 was expected to have many physical properties that were very different from the α- and β- phases of Si_3N_4. The OLCAO method was immediately applied to investigate these properties (Mo, 1999). It was found that γ-S_3N_4 has a direct band gap of 3.45 eV similar to GaN, an electron effective mass of 0.51 m_e, and a static dielectric constant of 4.70. The calculated high bulk modulus of 280 GPa confirmed it to be a superhard material. All this led to the high expectations of many special applications based on its unique properties. Unfortunately synthesis of a large quantity of pure γ-Si_3N_4 remains elusive and in the absence of large single crystals, many of its predicted properties remain unverified. Figure 5.16 shows the calculated band structure for γ-Si_3N_4 and the hypothetical γ-C_3N_4.

The experimental and theoretical investigations on γ-Si_3N_4 were immediately followed by the search for other single and double spinel nitrides. It was

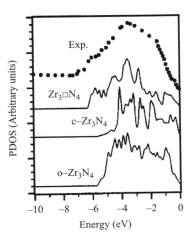

Fig. 5.15.

Comparison of the upper valence band DOS of (a) $Zr_3 \square N_4$, (b) c-Zr_3N_4 and (c) o-Zr_3N_4 with the experimental XPS data (dotted line).

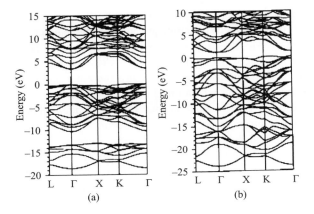

Fig. 5.16.

Calculate band structure of: (a) c-Si_3N_4; (b) c-C_3N_4.

soon confirmed that γ-Ge$_3$N$_4$ and γ-(Si, Ge)$_3$N$_4$ (Leinenweber et al., 1999, Serghiou et al., 1999) and later γ-Sn$_3$N$_4$ were also existent (Shemkunas et al., 2002). Furthermore, the OLCAO calculations showed that the electronic structure of γ-Sn$_3$N$_4$ was quite different from that of γ-Si$_3$N$_4$ and γ-Ge$_3$N$_4$ with a smaller direct band gap of 1.40 eV and an attractive small electron effective mass of 0.17 m_e (Ching and Rulis, 2006). The theoretical calculations were extended to 20 other single and double cubic spinel nitrides with cation elements from the groups IVA (Ti, Zr, Hf) and IVB (C, Si, Ge, Sn). Later, this was further extended to cover 39 such crystals in various combinations (Ching et al., 2001a, Ching et al., 2000, Ching et al., 2001b). We denoted them as c-AB$_2$N$_4$ or c-BA$_2$N$_4$ in reference to the cubic spinel phase but this has the same meaning as the γ phase. Results on the crystal structure parameters, band gaps, effective charges, bond order values, stability of double nitrides relative to single nitrides, bulk modulus, and so on were reported (Ching et al., 2002a). Of the 32 double nitrides, only 9 were predicted to be energetically favorable. Among the potentially stable double nitrides, the most interesting ones were c-CSi$_2$N$_4$ with exceptionally strong covalent bonds and large bulk modulus, c-SiGe$_2$N$_4$ with a favorable direct band gap of 1.85 eV and c-SiTi$_2$N$_4$ which is metallic. One of the more contentious issues of debate was the precise location for the cations in the double nitrides. For example, in c-[Si, Ge]$_3$N$_4$, should Ge occupy the tetrahedral sites or the octahedral sites of the spinel lattice or are they randomly distributed over both sites (Dong et al., 2003, Soignard et al., 2004)? Such issues can only be resolved by more accurate calculations and with experimental input.

An interesting point in the study of single and double spinel nitrides is that c-Ti$_3$N$_4$ is a narrow gap semiconductor and it may be a likely structure for the defect model of TiN$_x$ films. Because c-Si$_3$N$_4$ is a wide gap semiconductor and c-SiTi$_2$N$_4$ is metallic, by doping Ti at the octahedral sites of c-Si$_3$N$_4$, the direct band gap can be adjusted and a metal-insulator transition in c-Si[Si$_{1-x}$Ti$_x$]$_2$N$_4$ should occur. This was indeed the case. An insulator to metal transition was predicted to occur at x = 0.44 when a sufficient amount of Si was replaced by Ti while forming a solid solution (Ching et al., 2000). It was found that as x increases, the size of the band gap tends to shrink. This is due to the growth of the unoccupied "impurity bands," which have significant Ti3d$_{x^2-y^2}$ and 3d$_{3z^2-r^2}$ (e$_g$) components. With a reduced and adjustable band gap, an increase in the static dielectric constant, a closer lattice match with TiN, and a stable structure with strong covalent bonds, c-Si[Si$_{1-x}$Ti$_x$]$_2$N$_4$ could potentially find a variety of applications both as a structural ceramic and as a specialized material for semiconductor technology.

There were many other extensions and innovative ideas related to the synthesis, structure determination, property characterization, and practical applications of crystalline γ-Si$_3$N$_4$ as well as other closely related hard materials. Experimental determination of the band gap of γ-Si$_3$N$_4$ was attempted and values were obtained that were close to theoretical prediction (Leitch et al., 2004). These were discussed in a comprehensive review article (Zerr et al., 2006).

5.4.3   Ternary and quaternary nitrides and oxynitrides

The OLCAO method has been applied to several ternary and quaternary nitrides. This includes the metallic nitrides that will be separately discussed in the next chapter. The compounds we discuss in this subsection are mostly oxynitrides and can also be considered as insulating oxides. Silicon oxynitride Si_2N_2O has already been discussed in Section 5.4.1 in the comparative study of α-Si_3N_4, β-Si_3N_4, Si_2N_2O, and α-SiO_2 (Xu and Ching, 1995). An expected conclusion was that the electronic structure of Si_2N_2O could not simply be taken as a weighted average of the electronic structures of Si_3N_4 and α-SiO_2 (Xu and Ching, 1995).

An important class of quaternary structural ceramics is the Si-Al-O-N solid solution system sometimes simply denoted as SiAlON. There are two forms, α-SiAlON and β-SiAlON, obtained by simultaneous substitution of the (Si,N) pair by the (Al,O) pair in α-Si_3N_4 and β-Si_3N_4 respectively. The charge-neutral crystal is generally represented by the formula α-, β-$Si_{6-z}Al_zO_zN_{8-z}$ for $z = 1,2,3,4$. Transformation between α-SiAlON and β-SiAlON has been a subject of continuing interest. Similar pair substitution in Si_2N_2O results in what is called orthorhombic SiAlON, or the O'-SiAlON. SiAlON materials are easier to densify and more ductile than Si_3N_4 at high temperatures due to the replacement of the stronger Si-N covalent bonds by the more ionic Al-O bonds. The OLCAO method was used to study the electronic structure and bonding in β-$Si_{6-z}Al_zO_zN_{8-z}$ for $z = 1,2,3,4$ using a 14-atom hexagonal unit cell (Xu et al., 1997). Also calculated was the bulk modulus as a function of z by fitting the total energy data near the equilibrium volume to the Murnaghan equation of state. The bulk modulus and the overall bond strength decreases only slightly with an increase in z. Charge density plots show only a slight decrease in the covalent character of the solid solution as z increases. A rather surprising finding was that the substitution of the (Si,N) pair by an (Al,O) pair actually strengthened the remaining bonds because of the effective redistribution of the charges. This was seen as a potential reason for the superior mechanical properties of the SiAlON system. The calculation of the SiAlON system was later extended to the spinel lattice or the c-SiAlON with $z = 1$, or with just one pair substitution (Ouyang and Ching, 2002). It was found that Al prefers the octahedral site of the spinel lattice. The lowest energy configuration among several solid solution models has a direct band gap of 2.29 eV and it retains the same strong covalent bonding as in c-Si_3N_4.

Another class of important quaternary oxynitrides that have been studied in considerable detail by using the OLCAO method are from the Y-Si-O-N system (Ching, 2004, Ching et al., 2004, Ching and Rulis, 2008, Ching et al., 2003b, Ouyang et al., 2004, Xu et al., 2005). These quaternary crystals have rather complicated crystal structures. $Y_2Si_3N_4O_3$ (M-mullite), $Y_4Si_2O_7N_2$ (N-YAM), $YSiO_2N$ (wallastonite), $Y_{10}(SiO_4)_6N_2$ (N-apatite), and a more recently synthesized crystal, $Y_3Si_5N_9O_3$, with a high N content together with the binary and ternary compounds that have been described in previous sections constitute the most well known phases of the SiO_2-Y_2O_3-Si_3N_4-YN phase diagram (see Fig. 5.17). These quaternary crystals have a plethora of

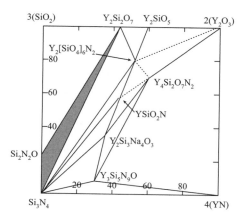

Fig. 5.17.
The SiO_2-Y_2O_3-Si_3N_4-YN phase equilibrium diagram.

local cation-anion bonding configurations that is due to O/N disorder. The electronic structures of these complex crystals have never been studied before. A comprehensive study of the electronic structure and bonding in these crystals in the form of atom-resolved partial DOS, Mulliken effective charges and bond order values together with spectroscopic investigations on the complex frequency-dependent dielectric functions and core level spectroscopy was most revealing (Ching, 2004). It was shown that Y-O and Y-N bonding were not negligible in spite of their fairly long bond lengths. These studies contributed to the understanding of the formation and properties of intergranular glassy films (IGFs) that exist in polycrystalline ceramics such as Si_3N_4 and helped motivated the direct modeling of the IGF structures that will be discussed later in Chapter 8.

5.4.4 Other complex nitrides

Recently, two new three-dimensional quaternary lithium alkaline earth nitridosilicate crystals $Li_2CaSi_2N_4$ and $Li_2SrSi_2N_4$ were synthesized at 900° C (Zeuner et al., 2010). The crystal parameters were fully determined by x-ray diffraction. Both have a complex cubic structure (space group Pa-3 and Z = 12) with 108 atoms in the unit cell. The electronic structure, inter-atomic bonding, optical properties, and core level spectroscopic XANES/ELNES in the form of Li-K, Ca-K, Ca-$L_{2,3}$, Sr-K, Sr-$L_{2,3}$, Si-K, Si-$L_{2,3}$, and N-K for all crystallographically non-equivalent sites were calculated using the OLCAO method. Both crystals are large gap semiconductors with a direct band gap of 3.48 eV and slightly different refractive indices of 2.05 and 2.12 respectively. The XANES/ELNES spectra for the two non-equivalent Ca (Sr) sites and the two N sites showed noticeable differences, reflecting the dependence on local environment. The calculated Li-K edge was in good agreement with the measure EELS spectrum. Figure 5.18 shows the calculated Li-K edge in $Li_2CaSi_2N_4$ which agrees with the experimental curve very well. All the above results were obtained in less than two calendar days using a local computer of

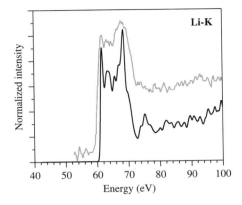

Fig. 5.18.
Comparison of the calculated (lower curve) and experimental (upper curve) Li-K edge in $Li_2CaSi_2N_4$.

very modest power. This demonstrates the efficiency of the OLCAO method for studying the fundamental properties of complex crystals.

5.5 Carbides

Silicon carbide, together with silicon nitride, is one of the two most important prototypes of non-oxide ceramics. In this subsection, we discuss carbides as a class of insulators by itself. SiC in the wurtzite structure has been studied with other wurtzite crystals and that of β-C_3N_4 and c-C_3N_4 were discussed under nitrides. Application to grain boundaries in SiC (Rulis et al., 2004) will be separately discussed later in Chapter 9. Here we focus on the various polymorphs of SiC and other more recent complex Al-Si-C compounds.

5.5.1 SiC

Whereas Si_3N_4 has only three known forms of crystalline structures, α-, β-, and γ-Si_3N_4, SiC exists in many polymorphic structures where the bonding between Si and C is always tetrahedral. The simplest one is cubic SiC in the zinc blend structure or β-SiC; the other common polymorphs have hexagonal structures and are conveniently labeled as 2H-SiC (the wurtzite phase), 4H-SiC, and 6H-SiC which are collectively known as α-SiC. The more rare and complex ones are those with a rhombohedral structure denoted as 15R-SiC, 21R-SiC, and so on. The main difference between these polymorphs is the stacking sequence of Si-C bi-layers along the c-axis of the crystal. This is illustrated Fig. 5.19 for the six polymorphs of SiC.

The OLCAO method was applied to study the electronic structure, bonding, and spectroscopic properties of the polymorphs of SiC (Ching et al., 2006). The different stacking sequences in SiC polymorphs give rise to crystallographically non-equivalent sites with almost identical nearest and next nearest neighbors but different intermediate range ordering. Mulliken effective charges and bond order calculations revealed that there are minute but recognizable differences in these polymorphs with an average charge transfer from Si to C in the range of 0.85 to 0.89 electrons. The band structures of β-SiC, 2H-SiC,

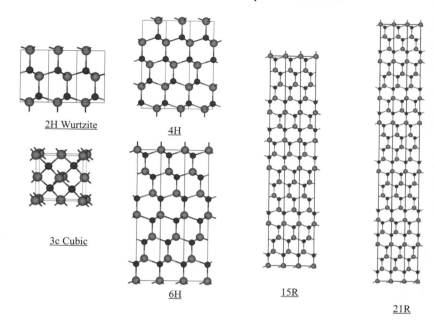

Fig. 5.19.
Sketches of the crystal structures of the
six polymorphs of SiC.

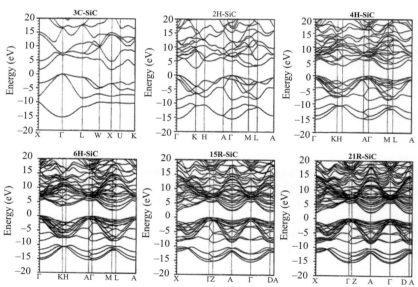

Fig. 5.20.
Band structures of the six polymorphs of
SiC.

4H-SiC, 6H-SiC, 15R-SiC, and 21R-SiC are shown in Fig. 5.20. Four of these
polymorphs (β-, 2H-, 4H-, 6H-) have indirect (direct) band gaps of 1.60, 2.47,
2.43, and 2.20 eV (2.39, 3.33, 3.27, and 3.02 eV) whereas 15R-SiC and 21R-
SiC have direct bands gap of 3.57 eV and 3.30 eV. The calculated optical
properties of the non-cubic phases of SiC show considerable anisotropy for
components in the c-axis and in the plane perpendicular to the c-axis.

5.5.2 Other carbides

The electronic structure and the spectroscopic properties of two ternary aluminum silicon carbides Al_4SiC_4 and $Al_4Si_2C_5$ were also studied by using the OLCAO method (Hussain et al., 2008). They have a hexagonal lattice with different layered structures. The unit cell of Al_4SiC_4 (space group $P6_3mc$) contains 18 atoms with nine non-equivalent sites (4 Al, 1 Si, and 4 C), and $Al_4Si_2C_5$ (space group $R3m$) contains 33 atoms with eleven non-equivalent sites (4 Al, 2 Si, and 5 C). Unlike the so-called MAX phase compounds (M = Transition metal, A = Al or Si, X = C or N) which are metallic layered compounds (to be discussed in Chapter 6), Al_4SiC_4 and $Al_4Si_2C_5$ are semiconductors with small band gaps of 1.05 and 1.02 eV respectively. Because of the difference in the layered structure of these two crystals, their calculated electron effective masses in the plane perpendicular to the c-axis ($m_\perp$) are similar (0.62 m_e for Al_4SiC_4 and 0.619 m_e for $Al_4Si_2C_5$) but are drastically different in the direction parallel to the c-axis ($m_\parallel$) (0.55 m_e for Al_4SiC_4 and 3.06 m_e for $Al_4Si_2C_5$). The hole effective masses in both crystals are very large due to the flat bands at the top of the valence band. The layered structure leads to a large optical anisotropy in the calculated optical properties. Al_4SiC_4 and $Al_4Si_2C_5$ crystals are highly covalent crystals. Mulliken effective charge calculations showed that in Al_4SiC_4, Al and Si lose on average 0.90 and 0.84 electrons to C which gains 1.11 electrons. In $Al_4Si_2C_5$, Al and Si lose on average 0.93 and 0.85 electrons to C which gains 1.08 electrons.

The Al-Si-C system is interesting because of its superior wear resistance properties, low weight, high strength, and high thermoconductivity which could lead to many applications in the modern electronics and power industry. These properties are intimately related to their fundamental electronic structures. It is conceivable that other layered carbides may be discovered that will have different physical properties for specific applications.

5.6 Boron and boron compounds

5.6.1 Elemental boron

Boron is one of the most fascinating low atomic number (Z = 5) elements in the periodic table. In elemental form, it has many allotropes, each with the distinctive feature of having a B_{12} icosahedron with triangular three-center bonds as the fundamental building unit. While the structures of some are well understood, others remain controversial. The relatively higher strength of the inter-icosahedral bonds compared to the intra-icosahedral bonds identifies these materials as inverted molecular solids (Emin, 1987). This helps give rise to spectacular variations in structure and crystal complexity. The phase diagram of elemental B is shown in Fig. 5.21.

The most thermodynamically stable phase of elemental boron is α-B_{12} with a single B_{12} icosahedron in the rhombohedral unit cell. The icosahedron is connected with other icosahedra of the neighboring cells via a number of different inter-icosahedral bonds. To understand the bonding we identify six

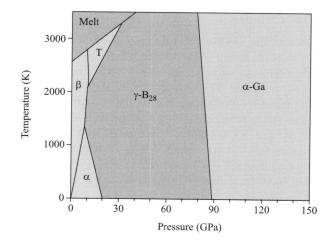

Fig. 5.21.
Phases diagram of elemental B.

of the twelve B atoms as polar sites (B_{polar}) and the remaining six as equatorial sites (B_{equat}). The B_{polar} form two-center two-electron (2c-2e) bonds with the B_{polar} of neighboring icosahedra while the B_{equat} participate in planar three-center two-electron (3c-2e) bonds with the B_{equat} of neighboring icosahedra. β-boron is another low pressure form of elemental boron with a rhombohedral lattice whose crystal structure is not completely known due to the presence of partially occupied sites (Slack et al., 1988). In this respect, β-boron can be considered a disordered form of boron. The tetragonal boron phase (t-Boron) is a high temperature phase. Early work suggested it to have 50 atoms in the tetragonal cell with four B_{12} icosaheda and 2 isolated four-fold bonded B atoms (Hoard et al., 1958). However, other forms of t-Boron with even more complex structures have also been suggested (Vlasse et al., 1979). A new phase of elemental boron, γ-B_{28} formed at higher pressure was discovered only recently and its structure was resolved by using a combination of state-of-the-art experimental and computational techniques (Oganov et al., 2009). It has a di-boron pair in addition to the B_{12} icosahedron in the tetragonal unit cell. An α-Ga-type phase of boron is predicted to exist at extremely high pressures, but has only been studied by theoretical calculations (Ma et al., 2004).

The electronic structure, bonding, charge distribution, and optical properties of α-B_{12} and t-B_{50} were investigated by using the OLCAO method in the 1990s (Li et al., 1992). α-B_{12} is a semiconductor with a gap of 1.70 eV while t-B_{50} is a metal due to electron deficiency at the top of the VB which is separated from the CB above with a sizable gap. The calculations on α-B_{12} were repeated more recently giving a value of 2.61 eV for the indirect band gap, which is larger than the earlier calculations. This is attributed to the more accurate potential model used. The γ-B_{28} phase was also recently studied together with the XANES/ELNES spectral calculations of the elemental B phases (Rulis et al., 2009). An indirect band gap of 2.1 eV for γ-B_{28} was obtained.

The calculation of β-Boron is far more challenging because of the complexity of the crystal structure for this phase. Many different models have been suggested containing 105, 106, 107, or 320 atoms per unit cell. One model for β-boron was suggested in 2007 by van Setten et al. (Van Setten et al., 2007) based on crystal information obtained by Slack et al. in 1988 (Slack et al., 1988). This model was later structurally optimized via the VASP code and studied by the OLCAO method. The model (named β-B_{106}) has a rhombohedral cell with 106 B atoms arranged in a very complicated structure consisting of B_{12} icosahedra on the cell vertices and edges, complex clusters of face sharing icosahedra, and a few non-icosahedral B atoms present in interstitial locations between icosahedra. The electronic structure and the XANES/ELNES spectra of this β-B_{106} model were calculated using the OLCAO method (Wang, 2010). It is a semiconductor with an indirect (direct) band gap of 1.75 (1.83) eV. There is a defect-like acceptor level about 0.21 eV above the VB edge which originates from the defect-like structure. However, these results remain preliminary at the time of this writing.

5.6.2 B_4C

The most important and well studied boron rich compound is boron carbide (B_4C). Boron carbide has been used in a variety of applications including, abrasive tools, neutron absorbers and shielding materials, body armor, and thermoelectric devices. Laboratory samples of boron carbide are not always stoichiometric and the C content can be anywhere between 8 and 20% with the composition tending to be closer to B_5C. They may even be amorphous or contain significant quantities of H. Thus, the term boron carbide actually represents a family of boron rich compounds with highly complex structures and compositions that could influence their specific applications and make definite characterization particularly difficult. The study of boron carbide should therefore start with the ideal stoichiometric B_4C phase in a rhombohedral cell (space group R-$3m$). It is now well accepted that the structure of stoichiometric B_4C consists of 12 atom icosahedra centered on the unit cell vertices like α-B_{12} but which are then interconnected with a three atom chain along the rhombohedral body diagonal (see Fig. 5.22). However, the exact locations of the three C within this structure are somewhat controversial because it is experimentally difficult to distinguish B from C due to their similar scattering cross-sections. Detailed experimental and theoretical investigations have concluded that the preferred structure is one with a C atom at a polar site of the B_{12} icosahedron along with a C-B-C chain (denoted as B_{11}C-CBC). There are also claims that a three atom-C-C-C chain (B_{12}-CCC) could also be a viable structure for B_4C. There were also other studies on possible structures of boron carbides such as $B_{13}C_2$. The $B_{13}C_2$ crystal has an intrinsic hole at the top of the VB because it is missing one electron and is therefore unable to satisfy the ideal bonding configuration (Li and Ching, 1995). Thus, $B_{13}C_2$ is expected to be a p-type semiconductor.

The OLCAO method was applied to study the electronic structure and optical properties of B_4C and other B_{12}-based crystals in the early 1990s (Li and Ching, 1995). It was confirmed that the proper structure for B_4C is B_{11}C(CBC). The charge density distribution and the nature of inter-atomic bonding were

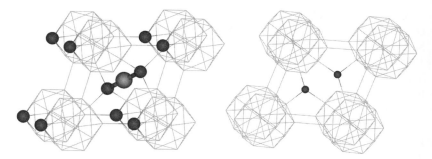

Fig. 5.22.

Crystal structures of B_4C and B_6O.

discussed. More recently, the investigation of B_4C was repeated and extended to include lattice dynamics, phonon calculations, and mechanical properties, but these results are as yet unpublished. The OLCAO method was also used to calculate the XANES/ELNES spectra with results in excellent agreement with experiment (to be discussed further in Chapter 11). The electronic structure and other bonding information obtained were more accurate than those obtained in the earlier study because of the availability of more computational power to apply more stringent calculation criteria.

One of the many applications of B_4C is as a strong lightweight protection material for military equipment and personnel. However, it is known that B_4C undergoes amorphization under uniaxial compression and that it loses its strength when the stress exceeds the Hugoniot elastic limit (HEL) of about 22GPa. The microscopic mechanism of B_4C amorphization under high velocity impact is not fully understood and this topic is a subject of current study.

5.6.3   Other boron compounds

Other than α-B_4C, there are many other B-rich compounds with the same rhombohedral structure such as $B_{12}P_2$, $B_{12}As_2$, $B_{12}O_2$, models of Si-containing B_{12} crystals, etc. The electronic structures and optical properties of these crystals have been calculated using the OLCAO method (Ching and Li, 1998, Li and Ching, 1995, Li and Ching, 1996b, Li et al., 1992). They are all semiconductors with indirect band gaps. Of all these B-rich crystals, the most interesting one is boron suboxide B_{12}-O_2 or B_6O. Boron suboxide has received great attention in recent years because of a combination of several favorable properties. It is a semiconducting optical material but is also one of the hardest substances known. The stoichiometric $B_{12}O_2$ has a rhombohedral unit cell with one B_{12} icosahedral unit and two O ions located in a line along the body diagonal similar to the CBC chain of B_4C (see Fig. 5.22). The B_{polar} has five nearest neighbor (NN) intra-icosahedral bonds and one inter-icosahedral bond to a B_{polar} of a neighboring icosahedron as in α-B_{12}. The B_{equat} also has 5 intra-icosahedral bonds and another short bond (1.493 Å) to one of the center line O atoms. The two O atoms are separated by 3.01 Å along the rhombohedral body diagonal and they are not considered to be bonded. The physical properties of boron suboxide including elastic, mechanical, vibrational, thermodynamic, electronic, optical, and spectroscopic properties have been obtained and are presently in preparation for future publication by S. Aryal et al. The calculated electronic structure and optical absorption spectra are similar to the earlier calculation (Li and Ching, 1996b) but with better accuracy. The calculated indirect band gap is 2.94 eV and the direct band gap of 5.44 eV is at the Γ site. The top of the VB is at the L point and the bottom of the CB is at H, both have reasonable effective masses. More importantly, the calculated XANES/ELNES spectra of the B-K edge is in good agreement with experiment and while the agreement for the O-K edge is less impressive, the mismatch is attributed to the O-deficient samples in the measurement. This will be further discussed in Chapter 11.

5.6.4   Other forms of complex boron compounds

There are many other complex forms of B and B compounds such as alkali-doped B or rare earth doped boron compounds (Douglas and Ho, 2006) including YB_{66} (Perkins et al., 1996) and many yet to be discovered or confirmed forms of elemental B. Of particular recent interest is a variety of B-clusters in the form of cages, ribbons, core shell structures, nano-particles, and so on. These are very interesting structures and their existence underscores the complex bonding that is possible between B atoms. Their properties can certainly be explored by using the OLCAO method.

5.7   Phosphates

Phosphates are oxides. We treat them as a separate class of insulators because of the rather unique role of the $(PO_4)^{-3}$ anion in these crystals. Also, a large portion of bioceramic materials, such as hydroxyapatite or tri-calcium phosphate, are complex phosphates that will be separately discussed in Chapter 7.

5.7.1   Simple phosphates: $AlPO_4$

$AlPO_4$ is a subclass of network oxides with a structure in close analogy to α-quartz. There are three known crystalline phases of $AlPO_4$, α-$AlPO_4$ or berlinite is a well studied crystal with a trigonal cell (space group $P3_221$) and a structure similar to α-SiO_2 in which both Al and P are tetrahedrally coordinated. At a pressure of around 13 GPa, α-$AlPO_4$ transforms into a new orthorhombic phase (o-$AlPO_4$) with a space group of $Cmcm$ in which the Al are octahedrally coordinated. More recently, it was reported that at a still higher pressure of 97.5 GPa, o-$AlPO_4$ transforms into a monoclinic phase (m-$AlPO_4$) of space group p2/m in which both Al and P are octahedrally coordinated (Pellicer-Porres et al., 2007). The fact that these three $AlPO_4$ crystals have the same chemical formula and the same percentage of different atomic species but distinctively different local atomic coordinations offers a very unique opportunity to systematically investigate their structure/property relationship. The OLCAO method was applied to calculate the electronic structure, bonding and XANES/ELNES spectra of the Al-K, Al-L, P-K, P-L, and O-K edges of all crystallographically non-equivalent sites in these three crystals. The calculated band gaps are 5.71, 6.32, and 4.01 eV for α-$AlPO_4$, o-$AlPO_4$, and m-$AlPO_4$ respectively. Mulliken effective charge calculation showed an increasingly covalent character and the crystal bond order values increased in going from α- to o- to m-$AlPO_4$ as the density increased and the inter-atomic bond length decreased. The most important conclusion in this study was that the prevailing notion of using finger printing for XANES/ELNES spectral interpretation is not always valid or is sometimes overly simplified. Among these three crystals, elements with the same number of nearest neighbors can have very different XANES/ELNES spectra. This point will be further discussed in Chapter 11.

5.7.2   Complex phosphates: KTP

KTP or $KTiOPO_4$ belongs to a family of inorganic crystals with a formula unit of $MTiOXO_4$ where M can be K, Rb, or Tl, and X can be either P or As. KTP was found to be a non-linear optical material with large second harmonic generation in Nd-doped lasers (Bierlein et al., 1987). Together with its strong optical damage threshold and other promising materials properties, KTP emerged as a leading material for electro-optical applications similar to the LiB_3O_5 or BaB_2O_4 crystals discussed earlier in subsection 5.3.2. KTP is a fairly complex crystal with an orthorhombic cell (space group $Pna2_1$) characterized by localized TiO_6 and PO_4 units. The electronic structure and the linear optical properties of KTP were investigated by the OLCAO method in 1991 (Ching and Xu, 1991). This early success was instrumental in the application of the OLCAO method to many other complex crystals and structures in later years. A direct band gap of 4.9 eV at Γ was obtained. It was found that the presence of short Ti-O bonds in this crystal has a profound effect on its electronic structure. The calculated optical absorption near the threshold showed considerable anisotropy which may have some implications on its non-linear optical properties. Unfortunately, the calculation was not extended to its non-linear optical properties as were done for some of the binary semiconductors and insulators.

5.7.3   Lithium iron phosphate: $LiFePO_4$

$LiFePO_4$ has become a very important phosphate crystal because of its application in Li ion batteries which could have an important role in addressing present day energy issues (Papike and Cameron, 1976). $LiFePO_4$ belongs to a general class of polyanion compounds in which the structural units are strong covalently bonded polyanions $(XO_4)^{-n}$ (X = P, S, As, Mo, or W) and transition metal octahedrons MO_6 (M = Mn, Fe, Co, Ni etc.) (Padhi et al., 1997). $LiFePO_4$ has an orthorhombic cell of olivine structure (space group $Pnma$) with 28 atoms in it. The structure can be viewed as interconnected LiO_6 and FeO_6 octahedra and PO_4 tetrahedra with shared edges (see Fig. 5.23). A remarkable feature in this crystal is the very short O-O distances in the PO_4 unit.

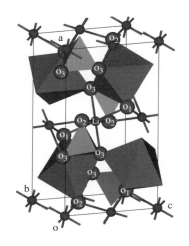

Fig. 5.23.
Crystal structure of $LiFePO_4$.

The electronic structure of $LiFePO_4$ was studied using the spin polarized version of the OLCAO method (Xu et al., 2004b). In principle, with transition metals such as Fe, intra-atomic correlation effects need to be considered but this was ignored at the time. Interestingly, the LSDA calculation showed $LiFePO_4$ to be a half-metal, having a large electron effective mass and a much smaller and highly anisotropic hole effective mass suggesting that the hole-doped composition should have a larger electric conductivity due to increased carrier mobility. Although direct connection to experimentally measured conductivity data was difficult, some insights related to the fundamental electronic structure were quite useful (Hunt et al., 2005, Xu et al., 2004b). The calculated DOS of $LiFePO_4$ and $FePO_4$ crystals appear to be consistent with the data obtained from the resonant inelastic x-ray scattering experiments even though the electron correlation effect was not taken into account (Hunt et al., 2006). Similar spin-polarized calculations were extended to Li_3PO_3, $LiMnPO_4$, $LiCoPO_4$, and $LiNiPO_4$ polyanion compounds (Xu et al., 2004a). $LiPO_3$ has no transition metal ions in it and is an insulator with a large band gap of 5.75 eV. For other crystals with transition metal elements, only $LiFePO_4$ and $LiCoPO_4$ are 100% half metals. $FePO_4$ and $LiMnPO_4$ are insulators with band gaps of 0.27 eV and 1.91 eV respectively between the occupied majority spin band and the unoccupied minority spin band. In the case of $LiNiPO_4$, the calculated Fermi level is in a small gap separating the t_{2g} and e_g minority bands. All of these are very interesting results yet the understanding remains that possible electron correlation effects may change some of these observations considerably.

References

Adair, R., Chase, L. L. & Payne, S. A. (1989), *Physical Review B*, 39, 3337.

Anderson, M. W., Terasaki, O., Ohsuna, T., Philippou, A., Mackay, S. P., Ferreira, A., Rocha, J. & Lidin, S. (1994), *Nature*, 367, 347–51.

Aryal, S., Rulis, P., & Ching, W. Y. (2012), *Journal of the American Ceramic Society*, (In press).

Aryal, S., Rulis, P., & Ching, W. Y. (2008), *Am. Mineral.*, 93, 114–23.

Barth, J., Johnson, R. L., Cardona, M., Fuchs, D., & Bradshaw, A. M. (1990), *Physical Review B*, 41, 3291.

Bierlein, J. D., Ferretti, A., Brixner, L. H., & Hsu, W. Y. (1987), *Applied Physics Letters*, 50, 1216–18.

Chen, C. & Liu, G. (1986), *Annual Review of Materials Science*, 16, 203–43.

Chiang, Y.-M., Farrey, G. W., & Soukhojak, A. N. (1998), *Applied Physics Letters*, 73, 3683–85.

Ching, W. Y. & Lin, C. C. (1975), *Physical Review B*, 12, 5536.

Ching, W. Y., Harrison, J. G., & Lin, C. C. (1977), *Phys. Rev. B*, 15, 5975.

Ching, W. Y. & Lin, C. C. (1977), *Phys. Rev. B*, 16, 2989.

Ching, W. Y. & Ren, S. Y. (1981), *Phys. Rev. B*, 24, 5788–95.

Ching, W. Y., Murray, R. A., Lam, D. J., & Veal, B. W. (1983), *Phys. Rev. B Condens. Matter*, 28, 4724–35.

Ching, W. Y., Huang, M. Z., & Huber, D. L. (1984a), *Phys. Rev. B*, 29, 2337–40.

Ching, W. Y., Song, L. W., & Jaswal, S. S. (1984b), *J. Non-Cryst. Solids*, 61–62, 1207–12.

Ching, W. Y. & Huang, M. (1985), *Superlattices Microstruct.*, 1, 141–5.

Ching, W. Y. & Harmon, B. N. (1986), *Phys. Rev. B*, 34, 5305–8.

Ching, W. Y., Xu, Y. & Wong, K. W. (1989), *Phys. Rev. B*, 40, 7684–95.

Ching, W. Y. & Xu, Y. N. (1990), *Phys. Rev. Lett.*, 65, 895–98.

Ching, W. Y. & Xu, Y. N. (1991), *Phys. Rev. B*, 44, 5332–35.

Ching, W. Y. & Huang, M. Z. (1993), *Phys. Rev. B*, 47, 9479–91.

Ching, W. Y., Gu, Z.-Q. & Xu, Y.-N. (1994), *Physical Review B*, 50, 1992.

Ching, W. Y. & Xu, Y. N. (1994), *J. Am. Ceram. Soc.*, 77, 404–11.

Ching, W. Y., Gan, F. & Huang, M.-Z. (1995), *Phys. Rev. B*, 52, 1596–611.

Ching, W. Y., Xu, Y.-N. & Gu, Z.-Q. (1996), *Phys. Rev. B*, 54, R15585–9.

Ching, W. Y. & Li, D. (1998), *Phys. Rev. B*, 57, 3737–40.

Ching, W. Y., Xu, Y.-N., Gale, J. D. & Ruhle, M. (1998), *J. Am. Ceram. Soc.*, 81, 3189–96.

Ching, W. Y. & Xu, Y.-N. (1999), *Phys. Rev. B*, 59, 12815–21.

Ching, W. Y. (2000), First principles calculation of the electronic structures of crystalline and amorphous forms of Sio_2. *In:* Rod, D., Durand, J.-P. & Dooryhee, E. (eds.) *Structure and Imperfections Amorphous Crystalline Sio_2* Chichester: John Wiley & Sons, Ltd., Chichester).

Ching, W. Y., Mo, S.-D., Ouyang, L., Tanaka, I., & Yoshiya, M. (2000), *Phys. Rev. B*, 61, 10609–14.

Ching, W. Y., Mo, S.-D., & Ouyang, L. (2001a), *Phys. Rev. B*, 63, 245110/1–245110/7.

Ching, W. Y., Mo, S.-D., Tanaka, I., & Yoshiya, M. (2001b), *Phys. Rev. B*, 63, 064102/1–064102/4.

Ching, W. Y., Xu, Y.-N., & Brickeen, B. K. (2001c), *Phys. Rev. B*, 63, 115101/1–115101/7.

Ching, W. Y., Mo, S.-D., Ouyang, L., Rulis, P., Tanaka, I., & Yoshiya, M. (2002a), *J. Am. Ceram. Soc.*, 85, 75–80.

Ching, W. Y., Xu, Y.-N., & Ouyang, L. (2002b), *Phys. Rev. B*, 66, 235106/1–235106/10.

Ching, W. Y., Ouyang, L., & Xu, Y.-N. (2003a), *Phys. Rev. B*, 67, 245108/1–245108/8.

Ching, W. Y., Xu, Y.-N., & Ouyang, L. (2003b), *J. Am. Ceram. Soc.*, 86, 1424–26.

Ching, W. Y. (2004), *Journal of the American Ceramic Society*, 87, 1996–2013.

Ching, W. Y., Ouyang, L., Yao, H., & Xu, Y. N. (2004), *Phys. Rev. B*, 70, 085105/1–085105/14.

Ching, W. Y. & Rulis, P. (2006), *Phys. Rev. B*, 73, 045202/1–045202/9.

Ching, W. Y., Xu, Y.-N., Rulis, P., & Ouyang, L. (2006), *Mater. Sci. Eng., A*, A422, 147–56.

Ching, W. Y., Ouyang, L., Rulis, P., & Yao, H. (2008), *Phys. Rev. B*, 78, 014106/1–014106/13.

Ching, W. Y. & Rulis, P. (2008), *Phys. Rev. B*, 77, 035125/1–035125/17.

Cohen, M. L. (1985), *Physical Review B*, 32, 7988.

Dong, J., Deslippe, J., Sankey, O. F., Soignard, E., & Mcmillan, P. F. (2003), *Physical Review B*, 67, 094104.

Douglas, B. E. & Ho, S.-M. (2006), *Structure and Chemistry of Crystalline Solids* (New York: Springer).

Emin, D. (1987), *Physics Today*, 40, 55–62.

Feibelman, P. J., Appelbaum, J. A., & Hamann, D. R. (1979), *Physical Review B*, 20, 1433.

Foster, J. P. A. (1959), *Journal of The Electrochemical Society*, 106, 971–75.

French, R. H. (1990), *Journal of the American Ceramic Society*, 73, 477–89.

French, R. H., Glass, S. J., Ohuchi, F. S., Xu, Y. N., & Ching, W. Y. (1994), *Phys. Rev. B Condens. Matter*, 49, 5133–42.

Gan, F., Xu, Y. N., Huang, M. Z., & Ching, W. Y. (1992), *Phys. Rev. B*, 45, 8248–55.

Godlewski, M., Goldys, E. M., Phillips, M. R., Langer, R., & Barski, A. (1998), *Applied Physics Letters*, 73, 3686–88.

Gu, Z.-Q. & Ching, W. Y. (1994), *Phys. Rev. B*, 49, 10958–67.

Harmon, B. N., Weber, W., & Hamann, D. R. (1982), *Physical Review B*, 25, 1109.

Harrison, J. G. et al. (1983), *Journal of Physics B: Atomic and Molecular Physics*, 16, 2079.

Harrison, W. A. (1970), *Solid State Theory* (New York: McGraw Hill).

Heaton, R. A., Harrison, J. G., & Lin, C. C. (1982), *Solid State Communications*, 41, 827–9.

Heaton, R. A. & Lin, C. C. (1982), *Physical Review B*, 25, 3538.

Hedin, L. & Lundquist, S. (1969), *Solid State Physics* (New York: Academis).

Hoard, J. L., Hughes, R. E., & Sands, D. E. (1958), *Journal of the American Chemical Society*, 80, 4507–15.

Huang, M. Z. & Ching, W. Y. (1983), *Solid State Commun.*, 47, 89–92.

Huang, M. Z. & Ching, W. Y. (1985a), *J. Phys. Chem. Solids*, 46, 977–95.

Huang, M. Z. & Ching, W. Y. (1985b), *Superlattices Microstruct.*, 1, 137–39.

Huang, M. Z. & Ching, W. Y. (1992), *Phys. Rev. B*, 45, 8738–41.

Huang, M. Z. & Ching, W. Y. (1993a), *Phys. Rev. B*, 47, 9464–78.

Huang, M. Z. & Ching, W. Y. (1993b), *Phys. Rev. B*, 47, 9449–63.

Huang, M. Z. & Ching, W. Y. (1994), *Ferroelectrics*, 156, 105.

Hunt, A., Moewes, A., Ching, W. Y., & Chiang, Y. M. (2005), *J. Phys. Chem. Solids*, 66, 2290–4.

Hunt, A., Ching, W. Y., Chiang, Y. M., & Moewes, A. (2006), *Phys. Rev. B*, 73, 205120/1–205120/10.

Hussain, A., Aryal, S., Rulis, P., Choudhry, M. A., & Ching, W. Y. (2008), *Phys. Rev. B*, 78, 195102/1–195102/9.

Hybertsen, M. S. & Louie, S. G. (1985), *Phys. Rev. Lett.*, 55, 1418.

Jiang, H., Xu, Y. N., & Ching, W. Y. (1992), *Ferroelectrics*, 136, 137–46.

Leinenweber, K., O'keeffe, M., Somayazulu, M., Hubert, H., Mcmillan, P. F., & Wolf, G. H. (1999), *Chemistry—A European Journal*, 5, 3076–78.

Leitch, S., Moewes, A., Ouyang, L., Ching, W. Y., & Sekine, T. (2004), *J. Phys: Condens. Matter*, 16, 6469–76.

Lerch, M., Füglein, E., & Wrba, J. (1996), *Anorg. Allg. Chem.*, 622, 367–72.

Li, D., Xu, Y. N., & Ching, W. Y. (1992), *Phys. Rev. B*, 45, 5895–905.

Li, D. & Ching, W. Y. (1995), *Phys. Rev. B*, 52, 17073–83.

Li, D. & Ching, W. Y. (1996a), *Phys. Rev. B*, 54, 13616–22.

Li, D. & Ching, W. Y. (1996b), *Phys. Rev. B*, 54, 1451–4.

Li, Y. P. & Ching, W. Y. (1985), *Phys. Rev. B*, 31, 2172–9.

Lines, M. E. (1990a), *Physical Review B*, 41, 3383.

Lines, M. E. (1990b), *Physical Review B*, 41, 3372.

Lines, M. E. (1991), *Physical Review B*, 43, 11978.

Liu, A. Y. & Cohen, M. L. (1989), *Science*, 245, 841–2.

Liu, A. Y. & Cohen, M. L. (1990), *Physical Review B*, 41, 10727.

Loughin, S., French, R. H., Ching, W. Y., Xu, Y. N. & Slack, G. A. (1993), *Appl. Phys. Lett.*, 63, 1182–4.

Ma, Y., Tse, J. S., Klug, D. D. & Ahuja, R. (2004), *Physical Review B*, 70, 214107.

Menéndez-Proupin E. & Gutiérrez G. (2005), *Physical Review B*, 72, 035116.

Mo, S.-D. & Ching, W. Y. (1995), *Phys. Rev. B* 51, 13023–32.

Mo, S.-D. & Ching, W. Y. (1996), *Phys. Rev. B*, 54, 16555–61.

Mo, S.-D., Xu, Y.-N., & Ching, W. Y. (1997), *J. Am. Ceram. Soc.*, 80, 1193–97.

Mo, S.-D. & Ching, W. Y. (1998), *Phys. Rev. B*, 57, 15219–28.

Mo, S.-D., Ouyang, L., Ching, W. Y., Tanaka, I., Koyama, Y., & Riedel, R. (1999), *Phys. Rev. Lett.*, 83, 5046–49.

Murnaghan, F. D. (1944), *Proceedings of the National Academy of Sciences*, 30, 244–47.

Oganov, A. R., Chen, J., Gatti, C., et al. (2009), *Nature*, 457, 863–67.

Ouyang, L. & Ching, W. Y. (2002), *Appl. Phys. Lett.*, 81, 229–31.

Ouyang, L., Xu, Y. N., & Ching, W. Y. (2002), *Phys. Rev. B*, 65, 113110/1–113110/4.

Ouyang, L., Yao, H., Richey, S., Xu, Y. N., & Ching, W. Y. (2004), *Phys. Rev. B*, 69, 094112/1–094112/6.

Padhi, A. K., Nanjundaswamy, K. S., & Goodenough, J. B. (1997), *Journal of The Electrochemical Society*, 144, 1188–94.

Papike, J. J. & Cameron, M. (1976), *Rev. Geophys.*, 14, 37–80.

Parker, J. C., Geiser, U. W., Lam, D. J., Xu, Y., & Ching, W. Y. (1990a), *J. Am. Ceram. Soc.*, 73, 3206–8.

Parker, J. C., Lam, D. J., Xu, Y. N., & Ching, W. Y. (1990b), *Phys. Rev. B*, 42, 5289–93.

Pellicer-Porres, J., Saitta, A. M., Polian, A., Itie, J. P., & Hanfland, M. (2007), *Nat Mater*, 6, 698–702.

Perdew, J. P. & Zunger, A. (1981), *Physical Review B*, 23, 5048.

Perkins, C. L., Trenary, M., & Tanaka, T. (1996), *Phys. Rev. Lett.*, 77, 4772.

Ren, S.-Y. & Ching, W. Y. (1981), *Phys. Rev. B Condens. Matter*, 23, 5454–63.

Ruan, Y. C. & Ching, W. Y. (1986), *J. Appl. Phys.*, 60, 4035–38.

Ruan, Y. C. & Ching, W. Y. (1987), *J. Appl. Phys.*, 62, 2885–97.

Ruan, Y. C., Wu, N., Jiang, X., & Ching, W. Y. (1988), *J. Appl. Phys.*, 64, 1271–73.

Rulis, P., Ching, W. Y., & Kohyama, M. (2004), *Acta Mater.*, 52, 3009–18.

Rulis, P., Wang, L., & Ching, W. Y. (2009), *physica status solidi (RRL)—Rapid Research Letters*, 3, 133–35.

Sanz, J. M., Soriano, L., Prieto, P., Tyuliev, G., Morant, C., & Elizalde, E. (1998), *Thin Solid Films*, 332, 209–14.

Schneider, H. & Komarneni, S. (2005), *Mullite* (Weinheim: Wiley-VCH).

Serghiou, G., Miehe, G., Tschauner, O., Zerr, A. & Boehler, R. (1999), *The Journal of Chemical Physics*, 111, 4659–62.

Shemkunas, M. P., Wolf, G. H., Leinenweber, K., & Petuskey, W. T. (2002), *Journal of the American Ceramic Society*, 85, 101–4.

Shepherd, E. S. & Rankin, G. S. (1909), *Am J Sci*, s4–28, 293–333.

Slack, G. A., Hejna, C. I., Garbauskas, M. F., & Kasper, J. S. (1988), *Journal of Solid State Chemistry*, 76, 52–63.

Soignard, E., Mcmillan, P. F., & Leinenweber, K. (2004), *Chemistry of Materials*, 16, 5344–49.

Sterne, P. A. & Inkson, J. C. (1984), *Journal of Physics C: Solid State Physics*, 17, 1497.

Van Setten, M. J., Uijttewaal, M. A., De Wijs, G. A., & De Groot, R. A. (2007), *Journal of the American Chemical Society*, 129, 2458–65.

Vlasse, M., Naslain, R., Kasper, J. S., & Ploog, K. (1979), *Journal of Solid State Chemistry*, 28, 289–301.

Wang, L. Y. (2010), *Electronic Structure of Elemental Boron*. MS Thesis, University of Missouri—Kansas City.

Wefer, K. & Misra, C. (1987), Oxides and Hydroxides of Aluminium. ALCOA technical paper No. 19 (revised), Alcoa Laboratory, Pittsburgh, USA.

Xu, Y. N. & Ching, W. Y. (1988), *Physica B*, 150, 32–6.

Xu, Y. N. & Ching, W. Y. (1990), *Phys. Rev. B*, 41, 5471–4.

Xu, Y. N., Ching, W. Y. & French, R. H. (1990), *Ferroelectrics*, 111, 23–32.

Xu, Y. N. & Ching, W. Y. (1991a), *Phys. Rev. B*, 43, 4461.

Xu, Y. N. & Ching, W. Y. (1991b), *Phys. Rev. B*, 44, 7787–98.

Xu, Y. N. & Ching, W. Y. (1993), *Phys. Rev. B*, 48, 4335–51.

Xu, Y. N., Ching, W. Y., & French, R. H. (1993), *Phys. Rev. B* 48, 17695–702.

Xu, Y. N., Jiang, H., Zhong, X.-F., & Ching, W. Y. (1994), *Ferroelectrics*, 153, 787–92.

Xu, Y. N. & Ching, W. Y. (1995), *Phys. Rev. B*, 51, 17379–89.

Xu, Y. N., Gu, Z.-Q., & Ching, W. Y. (1997), *Phys. Rev. B*, 56, 14993–5000.

Xu, Y. N. & Ching, W. Y. (1999), *Phys. Rev. B*, 59, 10530–35.

Xu, Y. N. & Ching, W. Y. (2000), *Philos. Mag. B*, 80, 1141–51.

Xu, Y. N., Ching, W. Y., & Brickeen, B. K. (2000), *Phys. Rev. B*, 61, 1817–24.

Xu, Y. N., Ching, W. Y., & Chiang, Y.-M. (2004a), *J. Appl. Phys.*, 95, 6583–85.

Xu, Y. N., Chung, S.-Y., Bloking, J. T., Chiang, Y.-M., & Ching, W. Y. (2004b), *Electrochem. Solid-State Lett.*, 7, A131–4.

Xu, Y. N., Rulis, P., & Ching, W. Y. (2005), *Phys. Rev. B*, 72, 113101/1–113101/4.

Yao, H. & Ching, W. Y. (1994), *Phys. Rev. B*, 50, 11231–34.

Zandiehnadem, F., Murray, R. A., & Ching, W. Y. (1988), *Physica B*, 150, 19–24.

Zandiehnadem, F. & Ching, W. Y. (1990), *Phys. Rev. B*, 41, 12162–79.

Zerr, A., Miehe, G., Serghiou, G., et al. (1999), *Nature*, 400, 340–42.

Zerr, A., Riedel, R., Sekine, T., Lowther, J. E., Ching, W. Y., & Tanaka, I. (2006), *Advanced Materials*, 18, 2933–48.

Zeuner, M., Pagano, S., Hug, S., Pust, P., Schmiechen, S., Scheu, C., & Schnick, W. (2010), *European Journal of Inorganic Chemistry*, 2010, 4945–51.

Zhong, X.-F., Jiang, H., & Ching, W. Y. (1994), *Ferroelectrics*, 153, 799–804.

6 Application to Crystalline Metals and Alloys

Compared to the collection of semiconducting and insulating materials that the OLCAO method has been applied to, as described in Chapter 5, the range of its application to metallic systems is somewhat smaller. However, a separate chapter for the discussion of metals is warranted because there are many complex metallic materials and one of the main goals of the OLCAO method is to be applicable to complex materials in general. In fact, the OLCAO method was the first method applied to a very complex metallic permanent magnet crystal, $Nd_2Fe_{14}B$, in the mid-1980s even before the OLCAO method was fully mature. In this chapter, we discuss the application of OLCAO to metals and alloys where the magnetic property is the main topic of interest. Naturally, most of the compounds to be discussed involve Fe because it is one of the most important transition metal elements with a pervasive presence. In the organization of the chapter, some overlap between different material classifications is unavoidable and so some arbitrary choices must be made. For example yttrium iron garnet (YIG), which is a complex oxide of iron, is considered in this chapter, while $LiFePO_4$ was discussed in Chapter 5 under the heading of phosphates and TiN and ZrN were discussed in Chapter 5 under the heading of binary nitrides. On the other hand, the high T_c superconductors, which are metallic oxides, are covered in this chapter even though the OLCAO calculations on these crystals were not spin-polarized. The early contribution of the OLCAO method to the study of oxide superconductors was a timely attempt to understand the electronic structure and its implications for the novel form of superconductivity just discovered. Another early application of the OLCAO method was to metallic glasses which will be discussed separately in Chapter 8 under the heading of non-crystalline solids. The most prominent groups for which classification is relatively easy are the elemental metals and simple alloys, hard permanent magnets, high temperature superconducting metals, and complex metals and alloys.

6.1 Elemental metals and alloys

6.1.1 Elemental metals

In the early stages of the development of the LCAO method, which later led to the OLCAO method, it was applied to simple alkali metals such as Li,

Na, and K (Ching and Callaway, 1973, Ching and Callaway, 1974, Ching and Lin, 1975, Lafon and Lin, 1966). These calculations used a linear combination of single Gaussian orbitals and demonstrated high accuracy not only for the band structures, but also for the wave functions used in optical absorption and Compton profile calculations. Similar LCAO calculations were also used to study transition metals in the spin-polarized mode (Rath et al., 1973, Singh et al., 1975, Wang and Callaway, 1974) to explore their magnetic properties. Although these early works on magnetic metals were not directly made by the OLCAO method as discussed in this book, they were made by closely related methods and they substantially influenced the later development of OLCAO. The current version of the spin-polarized OLCAO method uses the collinear approximation in which the z and -z directions are used to identify the majority and minority spin directions.

An important step in the development of OLCAO with particular application to metallic systems was the inclusion of f electron orbital interactions (Li et al., 1985). This extended the applicability of the method to a large number of crystals with rare earth elements that contain many core states. This extension involved the laborious derivation of multi-center integral formulas for the GTOs as discussed in Chapter 3. Test calculations on γ-Ce (Li et al., 1985) showed good agreement with results obtained using the APW method (Pickett et al., 1981). The Ce atom has a ground state configuration of $[Xe]6s^2 4f^2 5d^0 6p^0$ so the basis set used in this calculation for the valence and unoccupied states included 6s, 7s, 6p, 5d, and 4f orbitals. After orthogonalization to the core, the dimension of the secular equation to be solved was only 17 (2 for the 6s and 7s, 3 for 6p, 5 for 5d, and 7 for 4f), which demonstrates the very economic basis expansion used in the OLCAO method.

The spin-polarized OLCAO method has been used in the calculation of many elemental metals including Fe, Co, Ni, Y, and Nd at various levels of accuracy in conjunction with the studies of different compounds. They will not be separately discussed.

6.1.2   Fe borides

The electronic structure and magnetic properties of three inter-metallic iron compounds with relatively simple crystal structures (FeB, Fe_2B, and Fe_3B) were studied using the self-consistent spin-polarized OLCAO method. FeB has an orthorhombic cell (space group *Pnma*), Fe_2B has a body centered tetragonal cell (space group I4/mcm), and Fe_3B is orthorhombic (space group *Pbnm*). The comparative study of these three compounds and b.c.c. Fe was quite revealing of trends that followed the increasing Fe content. These three crystals have different inter-atomic Fe-Fe, Fe-B, and B-B separations. The calculated spin-projected DOS of these four ferromagnetic crystals are shown in Fig. 6.1. The DOS value at the Fermi level E_F is larger for the minority spin than it is for the majority spin ($N(E_F)_\downarrow > N(E_F)_\uparrow$) except for b.c.c. Fe. The total DOS at E_F for FeB and Fe_2B are comparable (7.54 and 7.46 states/[eV cell]) but they are only about half the value obtained for Fe_3B (13.3 states/eV-cell), and are much larger than the value obtained in b.c.c. Fe (1.22 states/eV-cell). The calculated

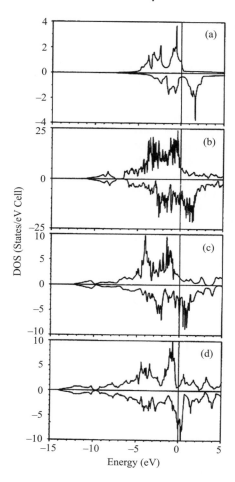

Fig. 6.1.
Spin-projected DOS: (a) fcc Fe; (b)
Fe_3B; (c) Fe_2B; (d) FeB. Above (below)
0 for the majority (minority) spin bands.

spin magnetic moments per Fe in FeB, Fe_2B, and Fe_3B are 1.26 μ_B, 1.95 μ_B, and 1.96 μ_B, which are all less than the moment in b.c.c. Fe (2.15 μ_B). These values are in good agreement with experimentally determined values. These calculations enabled us to compare the trends in the average Fe moment with that of amorphous $Fe_{1-x}B_x$ films. They too are in good agreement. The data for the amorphous films is only slightly lower than the values from crystalline alloys of different B content.

6.1.3   Fe nitrides

Iron nitride has a very complex phase diagram (Jack, 1948). Most of the phases are iron rich compounds and are ferromagnetically ordered. The OLCAO method in the local spin density approximation was applied to the α''-$Fe_{16}N_2$ crystal (Huang and Ching, 1995). α''-$Fe_{16}N_2$ has a body centered tetragonal cell (space group *I4/mmm*) with three different Fe sites, Fe-I (4e), Fe-II (8h), and Fe-III (4d) with different Fe-Fe and Fe-N inter-atomic separations

(Jack, 1951). N sits at the center of an octahedron formed by four Fe-II in the x-y plane and two Fe-I above and below it. This crystal attracted considerable attention in the early 1990s because of reports of it having a giant magnetic moment, but this was never confirmed. Many theoretical calculations were made using different methods including the OLCAO method; all concluded that there was no giant magnetic moment in this crystal. The OLCAO calculation showed that the site decomposed Fe moments were in agreement with more recent experiments and that the N atom had a negative polarization that was obtained by forming partially covalent bonds with the neighboring Fe atoms.

Motivated by the existence of spinel nitrides such as c-Si_3N_4 (discussed in Chapter 5) where Si can be octahedrally coordinated with N, an exploratory study was conducted to see if c-Fe_3N_4 in the spinel lattice could exist and what its electronic and magnetic properties would be (Ching et al., 2002, Xu et al., 2002). The results turned out to be quite interesting. Computations indicate that c-Fe_3N_4 is a ferromagnetic metal and is at least metastable with a magnetic moment of 3.26 μ_B per formula unit, and a large bulk modulus of 304 GPa that was attributed to strong covalent bonds between Fe and N. The magnetic moment per Fe was 2.28 μ_B at the octahedral site but only 0.42 μ_B at the tetrahedral site. Table 6.1 summarizes the calculated results for c-Fe_3N_4 and compares them with the binary FeN in the zinc blend structure. The two crystals have very different properties. FeN is anti-ferromagnetic with a relatively low bulk modulus. This calculation indicated that the influence of N coordination with Fe can drastically change the electronic structure and magnetic properties of Fe nitrides. So far, the existence of c-Fe_3N_4 in the spinel structure has not been confirmed experimentally.

Table 6.1. Comparison of the structure and properties of FeN and c-Fe_3N_4. The numbers in parentheses indicate the number of bonds present.

Crystal	FeN	c-Fe_3N_4
Space group	F-43m (No. 216)	Fd-3m (No. 227)
Lattice constant (Å)	a = 4.307	a = 7.896, u = 0.379
Fe-N bond (Å)	1.865 (4)	FeI: 1.774 (4) FeII: 1.937 (6)
Fe-Fe bond (Å)	3.046	FeI-FeII: 3.273 (4) FeII-FeI: 3.273 (2) FeII-FeII: 2.791 (3)
Total moment (μ_B/F.u.)	0	3.26
Fe moment(μ_B/Fe)	+ 0.028 (FeI), −0.028 (FeII)	0.42 (FeI), 2.28 (FeII)
N moment (μ_B/N)	0	−0.43
Bulk modulus (GPa)	144.2	303.7
B'	4.04	4.28
$PN(E_F)$ (states/eV atom)	0.65	0.88

A similar spin-polarized calculation was also applied to Li_3FeN_2 in connection with studies of the $LiFePO_4$ crystal discussed in Chapter 5. Li_3FeN_2 was considered to be capable of enhancing the reversible capacity in rechargeable Li batteries. The OLCAO calculation in the local spin density approximation (Ching et al., 2003b) showed that stoichiometric Li_3FeN_2 is nearly a half-metal with a high degree of spin-polarization (close to 67%). The calculated Fe magnetic moment of $1.5\,\mu_B$ is slightly lower than the reported experimental value of $1.7\,\mu_B$. Both Li and N were found to be slightly polarized with moments on Li $(-0.27\,\mu_B)$ aligned in the direction opposite to the Fe moment. Charge and spin density maps indicated strong bonding not only between Fe and N in the tetrahedral unit but also between Li and Fe which was not expected. It was speculated that other half-metals may exist in the large family of ternary Fe nitrides with slightly different crystal fields and Fe–Fe interactions.

6.1.4   Yttrium iron garnet

Yttrium iron garnet ($Y_3Fe_5O_{12}$) or YIG is an important ferrimagnetic oxide having a complex garnet structure. It has many important applications in modern technological areas such as microwave communication, non-volatile memory devices, magneto-optics, and so on. Unlike YAG and other laser host materials discussed in Chapter 5, which are large gap insulators, YIG has a much smaller gap and more complex electronic structure associated with the presence of the intra-atomic correlations of the 3d electrons in Fe. The Fe ions in crystalline YIG form two magnetic sublattices, one at the octahedral 16(a) site and the other at the tetrahedral 24(d) site, which are ferrimagnetically ordered with unequal spin-magnetic moments that are oppositely directed. In spite of its prominent role in a variety of applications and interesting fundamental physics, the electronic structure of YIG was not well studied theoretically for many years.

The electronic structure and ferrimagnetic properties of YIG were studied using the spin-polarized OLCAO method in the local spin density approximation (LSDA) (Xu et al., 2000). The Fe 3d bands fall within a large gap similar to that in the YAG crystal. The calculated Fe moments were $-0.62\,\mu_B$ at the octahedral site and $+1.56\,\mu_B$ at the tetrahedral site, confirming the ferrimagnetic ordering in YIG crystal. The source of the ferrimagnetic ordering can be traced to different crystal field effects on the two Fe sublattices that result in different distributions of the minority and the majority spin bands from the Fe e_g and t_{2g} levels.

While the above LSDA calculation on YIG provided reasonable results to explain the ferrimagnetic ordering, it showed YIG to be a metal with the Fermi level E_F resting within the Fe 3d band in contrast to the experimental observation that YIG is an insulator with a photon energy of around 3 eV for optical absorption. The reason for this discrepancy is, of course, the LSDA's neglect of intra-atomic correlations among the 3d electrons. To account for this effect, an ad hoc LSDA $+$ U approach was used (Ching et al., 2001) following the procedure of (Anisimov et al., 1991), and the YIG calculation was repeated.

Table 6.2. Results of magnetic moments (in μ_B) in $Y_3Fe_5O_{12}$ (YIG).

	LSDA	LSDA + U
Fe (16a)	−0.616	−3.266
Fe (24d)	+1.560	+3.939
Y (24c)	+0.016	−0.316
O (96h)	+0.063	+0.106
Total moment/unit cell	17.01	5.07
Fe moment/F.U.	3.45	5.28
Experimental value	4.25–5.0	7.9

In this approach, the orbital dependent on-site potential was added to the usual LSDA potential for Fe 3d electrons in the Hamiltonian.

$$V_{im\sigma} = V^{LDA} + U \sum_{m'} (n_{im'\sigma} - n^0) + (U - J) \sum_{m' \neq m} (n_{im'\sigma} - n^0) \quad (6.1)$$

In Eq. (6.1), the Coulomb repulsion parameter U (Hubbard U) and the exchange parameter J were treated as adjustable parameters and the values of $U = 3.5$ eV and $J = 0.8$ eV were used; m and m′ label the Fe 3d orbitals, i and σ specify the atom and the spin, $(n_{im'\sigma} - n^0)$ is the deviation of the electron number in the m'^{th} orbital of the ith atom from the average value of n^0 (equal to 6/10 in Fe). The revised calculation showed a significant improvement over the LSDA calculation as shown in Table 6.2. Not only was a band gap of 2.88 eV obtained, but the calculated spin-magnetic moments on the Fe sites were in better agreement with the experiment too. Based on the electronic structure using the LSDA + U approach, the optical absorption spectrum in YIG was also calculated, with results in reasonably good agreement with the experiment. This calculation underscores the importance of treating intra-atomic correlations in transition metal oxides with 3d electrons. This effect will be even more important in the highly correlated heavy element systems with localized 4f electrons.

6.2 Permanent hard magnets

Magnetic materials play an important role in many sectors of modern industry especially those related to energy and power distribution. In the early to mid-1980s, there was exciting news of the discovery of a new ternary compound with composition $R_2Fe_{14}B$ where R is a rare earth element (usually neodymium) which had exceptional magnetic properties such as high coercivity and a large energy product. The theoretical limit of this material's energy product (107 MGOe) was far larger than traditional permanent magnets based on $SmCo_5$ and Sm_2Co_{17}. Additionally, Fe was also much cheaper than Co. Over the next quarter century, $Nd_2Fe_{14}B$ has had a huge worldwide economic impact, especially in the automobile industry and for medical systems using nuclear magnetic resonance. Study of the structures and electronic properties of $Nd_2Fe_{14}B$ and related compounds is important for the fundamental understanding of their outstanding properties. In this subsection, we describe the early application of the OLCAO method to permanent magnets as a part of the method development and its demonstrated potential for application to complex materials of technological interest.

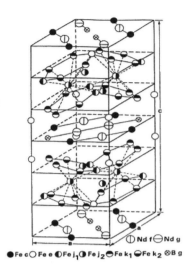

Fig. 6.2.
Crystal structure of $Nd_2Fe_{14}B$.

6.2.1 Application to $R_2Fe_{14}B$ crystals

After the discovery of the $Nd_2Fe_{14}B$ permanent magnet, its crystal structure was determined by neutron scattering (Herbst et al., 1984) (see Fig. 6.2). It has a tetragonal cell with 68 atoms (space group $P4_2/mnm$) with two Nd sites labeled as f and g, six Fe sites (labeled as c, e, j_1, j_2, k_1, and k_2) and one B site which occupies a distinctive position at the center of a trigonal prism. The crystal can be viewed as having layers of Nd, B, Fe(c) atoms at the $z = 0$ and $z = 0.5$ planes with the Fe atoms at the other sites forming packed hexagonal nets in between. The crystal structures for the other $R_2Fe_{14}B$ were found to be similar. The complexity of the crystal structure prevented the detailed calculation of the electronic structure of $R_2Fe_{14}B$ crystals after its discovery.

The OLCAO method was applied to $Y_2Fe_{14}B$ and $Nd_2Fe_{14}B$ (Ching and Gu, 1987, Gu and Ching, 1986, Gu and Ching, 1987a) soon after its structure was determined. $Y_2Fe_{14}B$ was attempted first because Y has no f electrons, which facilitated the calculation. These calculations were done in spin-polarized form using a minimal basis set and a superposition of atomic charge model with carefully constructed atomic potentials obtained from separate calculations on elemental metals. However, the self-consistent scheme was not fully developed at that time and a simplified orbital charge self-consistent scheme based on the calculated Mulliken charges on each atom was adopted (Ching and Huang, 1986). The calculation was then applied to $Nd_2Fe_{14}B$ with the Nd 4f states included in the basis (Ching and Gu, 1987). It is well known that the 4f electrons in rare earth metals are highly localized and an itinerant treatment of the highly correlated f electrons is not feasible. To circumvent this difficulty, the Nd ion was assumed to retain the Nd^{3+} configuration with the full occupation of the three 4f electrons. The Fermi level was then determined by subtracting the 4f part of the DOS from the total DOS (or the frozen f electron approach) and then integrating to the total number of valence electrons in the crystal without the 4f electrons. A similar approach was used to calculate the electronic structures and the magnetic properties of $Y_2Co_{14}B$, $Nd_2Co_{14}B$, Co-substituted $Y_2Fe_{14}B$ (Gu and Ching, 1987a, Gu and Ching, 1987b), and $Gd_2Fe_{14}B$ (Ching and Gu, 1988). Results on the Mulliken effective charges, site-decomposed spin-magnetic moments, spin-resolved partial DOS, and density of states at the Fermi level $N(E_F)$ for both spins were obtained and compared. Charge density maps and spin-density maps on the basal plane and the [110] plane were presented to ascertain the inter-atomic bonding and spin interactions. Although these early calculations were based on many assumptions and were less accurate by today's standards, several important conclusions were obtained. The calculated total and site decomposed magnetic moments were in reasonable agreement with the data from neutron

Below Fig 6.2 legend: ● Fe c ○ Fe e ◑ Fe j_1 ◐ Fe j_2 ◓ Fe k_1 ◒ Fe k_2 ⊗ B g ; ◐ Nd f ⊖ Nd g

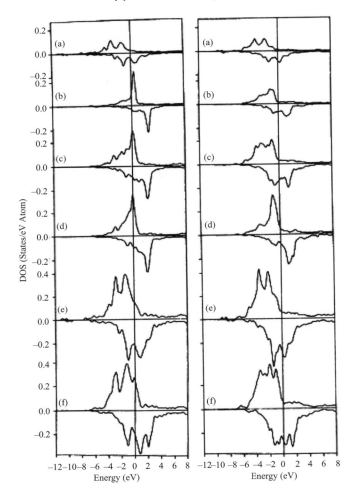

Fig. 6.3.
PDOS at different Fe sites in $Y_2Fe_{14}B$ (left-hand panel) and $Nd_2Fe_{14}B$ (right-hand panel): (a) 4 e, (b) c, (c) j_1, (d) 8 j_2, (e) 16 k_1, and (f) 16 k_2.

scattering. The relative magnitudes of the moments on each site are different for different crystals and were in less good agreement with experiments. The charge density map showed the formation of multi-center bonds on the B atom in the trigonal prism which may contribute to the stability of the crystal structure. The spin-density map showed an interesting network-like structure parallel to the c-axis. There were considerable differences and also similarities between the site-decomposed and spin-resolved PDOS among these crystals, reflecting the variations in local atomic structures. Figure 6.3 shows the comparison of the PDOS at the six different Fe sites in $Y_2Fe_{14}B$ and $Nd_2Fe_{14}B$.

6.2.2 Further applications to $Nd_2Fe_{14}B$

The calculation of the electronic structure and magnetic properties of the $R_2Fe_{14}B$ crystals described above formed the foundation for further studies on the properties specifically related to their application as permanent magnets.

The first such study included the development of a new approach for calculating the crystal field parameters (CFPs) based on the band structure results. CFPs are important for explaining many experimental results related to local electrons in solids. The crystal electric field at a given site is expressed as:

$$V(r, \theta, \phi) = B_{20}O_{20} + B_{22}O_{22} + B_{40}O_{40} + \cdots\cdots + B_{66}O_{66} \qquad (6.2)$$

Where $B_{20}, B_{22}, B_{40}\ldots\ldots B_{66}$ are CFPs and O_{nm} are the corresponding Stevens operators related to the environmental charge and the atomic coordinates (Hutchings, 1964). Traditionally, calculation of the CFPs in solids is based on the so called "point charge model" (PCM) which is somewhat primitive. The CFP results obtained by using PCM can differ from measured values by orders of magnitude and may even have the wrong sign. Some remedial procedures, so-called "screening factors," were then introduced to account for the covalent bonding effect but this ad hoc approach amounted to no better than an arbitrary adjustment. Usually, the CFPs were obtained by fitting to selective experimental data for the best agreement but no fundamental understanding at the microscopic level could be obtained. A new approach using the electronic structure results was introduced to calculate the CFPs at the Nd sites in $Nd_2Fe_{14}B$ hard magnets (Zhong and Ching, 1988, Zhong and Ching, 1989b). In this method, the real space charge distributions from the Bloch functions of the band structure calculation were used to numerically evaluate various contributing parts to the CFP expression. Although the second order parameters, B_{20} and B_{22}, are the most important, CFPs of up to the sixth order could be obtained. The calculated values of -3.77K and -2.85 K for Nd (f) and Nd (g) respectively for B_{20} were in good agreement with experimental values deduced from magnetization and Mössbauer measurements. The same CFPs obtained by using the point charge model were larger by a factor of more than 10. The fourth order and sixth order crystal field parameters were also calculated but they were much smaller than the second order parameters.

One of the most important parameters for hard magnets is the anisotropic energy at the rare earth site. The anisotropy energy in magnetization is due to the combined effect of the crystal field and the exchange field with a basic magnetic Hamiltonian given by $H_m = H_{CF} + H_{ex}$. In calculating the anisotropic energy for $Nd_2Fe_{14}B$, the H_{CF} part of the Hamiltonian was obtained using the CFPs at the Nd site as described above. The H_{ex} part is much more complicated. The Stoner approximation (Stoner, 1938) was used to obtain the Stoner parameter from the spin-polarized DOS (Fig. 6.3) in the rigid band approximation. Together with the calculated local magnetic moments at the Nd(f) and Nd(g) sites, the strength of the molecular field H_m at the two sites was estimated to be 216.9 K and 214.8 K respectively. The magnetic Hamiltonian was diagonalized within the ground state multiplet $4I_{9/2}$ of the Nd^{3+} ion which has 10 multiplet wave functions. The matrix elements of Stevens operators in H_{CF} are easily calculated on this basis since they are expressed in terms of the total angular momentum operator J. The stabilization energy in the $Nd_2Fe_{14}B$ crystal was calculated as the difference between the ground state energies of the 10 x 10 Hamiltonian with the H_{ex} field being parallel or perpendicular to the tetragonal axis of the crystal (Zhong and Ching, 1989a). The stabilization energy $E_A(0)$ at zero temperature is related to the anisotropic coefficients K_1 and K_2 by $E_A(0) = K_1 \sin^2 \theta + K_2 \sin^4 \theta$ where θ is the angle between the magnetization direction and the c-axis. The calculated values for K_1 at the Nd(f) and Nd(g) sites were $18.1 \times 10^7 \text{erg/cm}^3$ and $16.4 \times 10^7 \text{erg/cm}^3$ respectively which were in good agreement with experiments. It was found that H_{ex} is more important than H_{CF} by a factor of approximately 5.8. Thus, the exchange field is much stronger than the crystal field for determining the anisotropy in $Nd_2Fe_{14}B$ which is consistent with the fact that the spin-density map in $Nd_2Fe_{14}B$ shows a network-like structure parallel to the c-axis. Although the calculation of the anisotropy described here is valid only at zero temperature mainly because of the use of the Stoner model, the work was a good example of connecting the fundamental electronic structure with measurable properties in $Nd_2Fe_{14}B$.

In the previous section on the calculation of the magnetic properties of $Nd_2Fe_{14}B$, it was mentioned that although the calculated total spin-magnetic moments were in reasonable agreement with experiment, the site-decomposed Fe moments were not. One possible reason was the neglect of the contribution from orbital moments at each Fe site. The calculation was later extended to include the local orbital moments in $Nd_2Fe_{14}B$ (Zhong and Ching, 1990) by adding the spin-orbit interaction in the Hamiltonian. The magnetic anisotropy in $Nd_2Fe_{14}B$ cannot be understood solely by the local spin moments which are not related to any spatial direction in the absence of an external magnetic field. It is through the spin-orbit interaction and the directionally dependent orbital angular moment that the local spin-moment will have an effect on the magnetic anisotropy. For simplicity, the atomic values for the spin-orbit coupling constants ξ for the p- and d-orbitals of Nd and Fe were used. The steps for such a calculation were described briefly in the method section 4.5 in Chapter 4. The orbital moment was calculated as the expectation value of the z-component of ℓ, the angular momentum operator on the ion between the Bloch functions. The Bloch functions from the diagonalization of the complex matrix equation have to be solved at the star of k points that span the entire Brillouin zone, not just the irreducible portion of the Brillouin zone. This is because the matrix element is no longer invariant under the operations of the wave vector group as in the case of energy eigenvalues. By using more k points in this calculation, which also improved the accuracy of the spin angular moments reported earlier, the total moments at each of the Fe sites were calculated (see Table 6.3). The total orbital moments (as a sum of the p- and d-components) are much smaller than the spin-magnetic moments but they were not negligible and were found to have considerable fluctuations between different sites with the same symmetry. The calculated total moments (as a sum of spin and orbital components) at each Fe site were now in better agreement with experiments. In particular, the moment at the j2 site was greatly improved and it was correctly predicted to have the largest moment among all the Fe sites in $Nd_2Fe_{14}B$.

Another application to $Nd_2Fe_{14}B$ using the results of the fundamental electronic structure is the calculation of the spin-reorientation temperature T_{sr} in the framework of the single ion anisotropy theory using the magnetic Hamiltonian $H_m = H_{CF} + H_{ex}$ introduced earlier. The temperature dependence of

Table 6.3. Total local moments (spin and orbital) in units of μ_B.

Site	e	c	j1	j2	k1	k2
Spin	1.99	2.97	2.48	3.40	2.15	2.28
Spin	2.30	3.07	2.38	3.28	2.15	2.32
Orb.	0.04	0.04	0.01	0.07	0.03	0.02
Total	2.34	3.11	2.39	3.35	2.18	2.34
Exp.	2.10	2.75	2.30	2.85	2.60	2.60
Exp.	1.12	2.23	2.71	3.51	2.41	2.41

H_{ex} is approximated by replacing the molecular field by the Brillouin function (Callaway, 1991) and H_{CF} is considered to be temperature independent at low temperatures. The magnetic Hamiltonian was diagonalized within the ground state of the multiplets of the Nd ion. The eigenstates of the magnetic Hamiltonian were used to construct the partition function for the free energy $F(T, \theta, \phi)$ where (θ, ϕ) specify the direction of the magnetic moment at the Nd site. The angle θ is between the magnetic moment and the c-axis and ϕ is the angle in the basal plane. Using the crystal field parameters calculated for $Nd_2Fe_{14}B$, the search for the lowest free energy showed that the magnetic anisotropy direction at the Nd(f) site was canted away from the c-axis by an angle of about 12° at a T_{sr} of about 140 K. This result was in qualitative agreement with experimental observation (Zhong et al., 1991). The main experimental observation was that the preferential direction of magnetization is a cone with an angle of 25° to 35° against the c-axis while below the spin-reorientation temperature of about 150 K. The physics behind this is the crucial interplay between the CF and the molecular field of the Nd ion as the temperature varies. The first order magnetic process and magnetic phase transitions in $Nd_2Fe_{14}B$ based on the free energy surface and the global properties thus calculated are further discussed in reference (Zhong and Ching, 1993).

6.2.3 Application to Re_2Fe_{17} and related phases

The three most important quantities for high performance permanent magnets with superior applications are the saturation moment, the coercivity, which is related to the anisotropic field in the crystal, and the Curie temperature T_c. $Nd_2Fe_{14}B$ satisfies the first two criteria but its Curie temperature is relatively low. This is the main reason for Co substitution in $Nd_2Fe_{14}B$ in order to increase its Curie temperature. In this respect, the binary rare earth iron compound R_2Fe_{17} (the 2–17 phase) received renewed interest because its T_c can be significantly increased through lattice expansion by interstitial doping with N or C, or by partially replacing Fe with Al, Ga, or Si. This is the so-called magneto-volume effect. The OLCAO method was used to study the electronic structure and magnetic moments in doped crystals of $Nd_2Fe_{17}N$ (Gu et al., 1992, Gu et al., 1993), $Y_2Fe_{17}N_3$, $Y_2Fe_{17}C_3$ (Ching et al., 1994), and the substituted crystals of $Nd_2Fe_{17-x}M_x$, (M = Al, Si, Ga) (Huang and Ching, 1994, Huang and Ching, 1996, Huang et al., 1997). The conclusion from these calculations was that an increase in volume, which increases the Fe–Fe separations, was not a major factor in the enhancement of the Fe moment that would lead to the increase in T_c. For example, based on the supercell calculation of $Nd_2Fe_{17-x}Si_x$ for different amounts of Si substitution (Huang et al., 1997), it was found that the average Fe moment saturates at x = 3. This correlates well with the experimental dependence of T_c on x, but the crystal volume in the Si substituted samples actually decreases in contradiction to the concept of the magneto-volume effect. Obviously, a more refined theory based on the fundamental electronic structure of the crystal is needed.

One of the theories for calculating the Curie temperature was advanced by Mohn and Wohlfarth (Mohn and Wohlfarth, 1987), and it was quite successful for ferromagnetic metals such as Fe, Co, and Ni. T_c is solved from the quadratic equation involving the Stoner–Curie temperature T_c^s and a characteristic temperature T_{sf} based on a spin-fluctuation model.

$$\frac{T_c^2}{T_c^s} + \frac{T_c}{T_{sc}} = 1 \qquad (6.3)$$

Using the values of exchange splitting, the density of states at the Fermi level for both spins, and the equilibrium moment of the crystal as calculated by the OLCAO method for eight crystal systems where experimental data on T_c were available, T_c values were calculated according to Eq. 6.3 and compared with the experimental values (Ching and Huang, 1996). The results are shown in Fig. 6.4. The calculated T_c using the Mohn and Wohlfarth model provides only a qualitative account for T_c enhancement and shows that the magneto-volume effect is an inadequate explanation. Clearly, accurate determination of T_c from first principles for these complicated inter-metallic compounds is a difficult task and a more elaborate theory is necessary.

The OLCAO method has also been applied to other Nd-Fe compounds with much more complicated crystal structures such as Nd_3Fe_{29}, $Nd_3Fe_{28}Ti$ (Ching et al., 1997, Ching and Xu, 2000), and Nd_5Fe_{17} (Gu et al., 2000). These compounds have very low symmetry and many different Fe sites. The

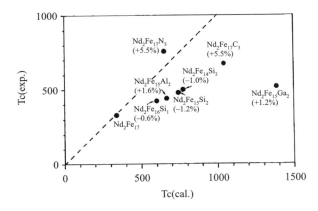

Fig. 6.4.

Calculated and measured T_C for some Nd_2Fe_{17} based compounds. There are uncertainties in the experimental T_C values. (See original reference for details.)

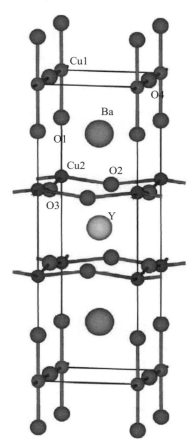

Fig. 6.5.
Crystal structure for $YBa_2Cu_3O_7$.

hexagonal cell of Nd_5Fe_{17} (space group $P6_3/mcm$) contains a total of 264 atoms with 14 different Fe sites and 7 different Nd sites. The idea was to use the calculated moments at each Fe site to determine if there is an obvious statistical correlation of the Fe moments with the number of the Fe–Fe bonds or the average Fe–Fe bond lengths. Although the total moments from these calculations were in reasonable agreement with the measured data, there were no obvious correlations with the Fe–Fe distances or with the Fe–Fe nearest neighbor numbers.

6.3   High T_c superconductors

6.3.1   YBCO superconductor

The discovery in the mid-1980s of the high temperature superconducting oxide, $La_{2-x}Ba_xCuO_4$, with a reported superconducting temperature above 30 K (Bednorz and Müller, 1986) had a revolutionary impact on condensed matter physics and materials science. Within a year, the report of the discovery of a superconductor in the $YBa_2Cu_3O_{7-\delta}$ (YBCO or Y123) compound with a T_c of 92 K (Wu et al., 1987), above the boiling point of nitrogen, raised the real hope that a room temperature superconductor with its immeasurable impact on technology and society was around the corner. The YBCO compound has a perovskite structure with layers of CuO_2 and CuO_4 planes separated by Ba and Y between the planes (see Fig. 6.5). The stoichiometric $YBa_2Cu_3O_7$ has an orthorhombic cell with 13 atoms (space group P_{mmm}) (Beno et al., 1987). There are two Cu sites, Cu(1) and Cu(2), four O sites O(1), O(2), O(3), and O(4), and one site each for Ba and Y. Cu(2), O(2), and O(3) form a buckled Cu–O plane whereas Cu(1) and O(1) form a Cu-O chain. Cu(1) and Cu(2) are connected with O(4) in the z-direction.

Because the electronic structure of a crystal is the basis for all of its properties, the calculation of the band structure of YBCO was immediately attempted using different methods including the OLCAO method (spin-non-polarized) (Ching et al., 1987). The electronic structure is also useful for the formulation of several models for high T_c superconductivity mechanisms. Figure 6.6 shows the calculated band structure and DOS of the YBCO crystal. The most interesting part is the presence of exactly four intrinsic holes on the top of the valence band separated from the higher conduction band with a semiconductor-like band gap of 1.54 eV at the S point of the Brillouin zone. However, the conduction band minimum is at the Γ point (indirect gap = 1.06 eV). The Fermi level E_F resides at the steep edge of the DOS with $N(E_F) = 5$ states/(eV cell). From the band structure, the electron effective mass at S and Γ were computed to be 0.49 m and 1.17 m and the hole effective mass at S was −0.57 m. In addition, a static dielectric constant of ε_0 of 12.9 was estimated. The calculated optical conductivity showed a high anisotropy in the planer and the z-directions due to the two-dimensional characteristic of the band structure (Zhao et al., 1987). The calculated DOS for YBCO was resolved into different site-specific orbital-projected components (Ching et al., 1991). Such calculations provided a quantitative description of the

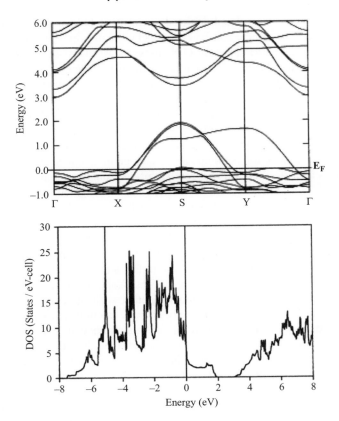

Fig. 6.6.
Calculated band structure (upper) and DOS of $YBa_2Cu_3O_7$.

Cu_{3d}-O_{2p} bonding information between different atoms in the YBCO crystal. Of particular importance are the states of the Cu-3d hole (those of the intrinsic hole-states below the semiconductor-like gap) and those near the Fermi level. They come predominately from the planer orbital $3d_{(x^2-y^2)}$ of Cu(2) and the vertical orbital $3d_{(3z^2-r^2)}$ of Cu(1). The planer orbitals hybridize with the p_x and p_y orbitals of O(2) and O(3), and the vertical orbital interacts strongly with the p_z of O(4). The apical O, or O(4), contributes significantly to the DOS at E_F, supporting the contention that it plays an important role in the superconducting mechanism. The quantitative information obtained from the standard band theory using density functional theory was used in the interpretation of many different types of experimental observations and in the formulation of different types of theories for superconductivity in the high T_c oxides (Kresin et al., 1993, Pickett, 1989).

The spin-non-polarized OLCAO calculation on the YBCO crystals was soon extended to other YBCO based crystals in order to further explore their properties in relation to various experimental observations. These include the fluorine-substituted YBCO (Xu et al., 1988), and the Ga- and Zn- substituted YBCO (Xu et al., 1990). In the F-substituted calculations, substitution of O by one, two, or three F ions at various O sites was performed. The extra electron in F raised the Fermi level and depleted the intrinsic-hole portion of the upper valence band. The conclusion was that O(1) was the most likely site to be replaced by F, and O(4) in the Cu-O chain was least likely. For the Ga and Zn substitution, either the chain site Cu (1) or the plane site Cu(2) were replaced by Ga or Zn. This work was motivated by the experimental finding that Ga-substituted and Zn-substituted YBCO samples have dramatically different superconducting properties. The OLCAO calculation showed that substitution of Cu(1) by Ga has the least effect on the electronic structure. In the other cases, extra peaks appeared in the semiconductor-like band gap or near the bottom of the conduction band due to the change in the electronic interactions as a result of substitution. More recently, the electronic structure of the YBCO crystal with O vacancies at various O sites was investigated using a supercell approach in relation to the calculation of O-K edge XANES spectra (Ching et al. unpublished).

The electronic structure results on $YBa_2Cu_3O_{7-\delta}$ shown above were instrumental in the formulation of the excitonic enhancement mechanism (EEM) to explain the superconducting transition in this and other related compounds (Ching et al., 1987, Wong and Ching, 1989a, Wong and Ching, 1989b). The EEM was based on a two-band model together with the concept of the off-diagonal long-range order of positively charged excitonic quasi-particles (or exciton-like charged clouds). The result was a simultaneous excitonic condensation via phonon-induced coupling. Although the EEM theory was able to produce quantitative fits to normal phase properties such as the Hall Effect, resistivity, thermoelectric power, and several other experimental observations (Wong and Ching, 2004), the theory has not received much recognition so far, but this is also true of many other competing theories about the origin of high T_c superconductivity. It is interesting to note that many of the applications that require high T_c superconductors are already on the market now using YBCO and other similar superconductors.

6.3.2   Other oxide superconductors

In addition to the YBCO superconductor, several other layered oxide superconductors with higher T_c were discovered such as Bi-Ca-Sr-Cu-O, Tl-Ca-Ba-Cu-O, and Hg-Ca-Ba-Cu-O (Maeda et al., 1988, Schilling et al., 1993, Sheng and Hermann, 1988). This shows that the presence of rare earth atoms is not the essential requirement for high T_c superconductivity. Most of these compounds have much more complex structures with stacked layers and higher superconducting transition temperatures compared to YBCO. The T_c for $Bi_2CaSr_2Cu_2O_8$ (Bi2122), $Tl_2CaBa_2Cu_2O_8$ (Tl 2122), and $Tl_2Ca_2Ba_2Cu_3O_{10}$ (Tl 2223) were reported to be 85 K, 110 K, and 125 K respectively and there is evidence that T_c increases with the increased stacking of layers in the crystal unit cell, and in some cases with an increase in hydrostatic pressure. The reported superconducting transition temperature in $HgBa_2CaCu_2O_{6+x}$ with two CuO_2 layers was as high as 133 K.

The OLCAO method has also been applied to calculate the band structure, DOS, and optical conductivities of Bi2021, Bi2122, Bi2223, Tl2021, Tl2122, and Tl2223 crystals (Ching et al., 1989b, Ching et al., 1989a, Zhao and Ching, 1989) and V-substituted Bi2122 (Fung et al., 1990). The results showed that the features of the intrinsic holes on the top of the valence band and the semiconductor-like gap above it remain valid though the specific parameters in the calculated electronic structures change. In particular, the band structures near the CB minima above the valence band with intrinsic holes were very different from those of YBCO. This shows that certain common features are shared by different superconducting oxides. The band structures of Tl2122 and Tl2223 crystals are shown in Fig. 6.7 as illustrations.

Although the electronic structure of the YBCO crystal and other high T_c superconducting oxides calculated by the OLCAO method were used to explain various experimentally observed phenomena, one particular experiment deserves special attention. This is positron annihilation spectroscopy (PAS). PAS is an effective probe for detecting changes in local charge distribution above and below T_c. In a series of experiments on many types of high T_c superconductors (Bharathi et al., 1990a, Bharathi et al., 1990b, Sundar et al., 1990, Sundar et al., 1991), it was found that T_c depends on the positron life time (or equivalently the positron annihilation rate λ) which can be vastly different depending on which region it is in the crystal that the position and electron wave functions have a significant overlap. The site-

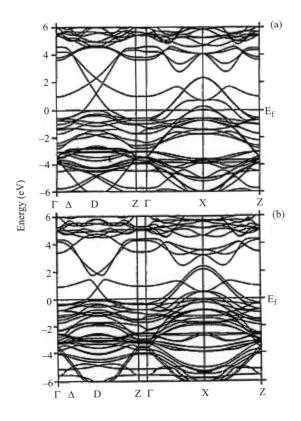

Fig. 6.7.
Band structure of (a) $Tl_2CaBa_2Cu_2O_8$ and (b) $Tl_2Ca_2Ba_2Cu_3O_{10}$.

decomposed total charge density from the OLCAO calculations was used to evaluate the electron-position correlation potential using the scheme proposed by Arponen and Pajanne (Arponen and Pajanne, 1979). This data is necessary for the solution of the positron wave functions and its overlap with the electron wave functions for the calculation of the positron annihilation rate λ. Comparison of the calculated positron density distribution in various planes of the crystal and the experimental values of the annihilation rate provided much useful information on the charge transfer model of the superconductivity mechanism.

6.3.3   Non-oxide superconductors

The OLCAO method has also been applied to the study of the electronic structures of other superconductors. Here we present some results for a class of conventional BCS superconductors (Bardeen et al., 1957b, Bardeen et al., 1957a) in the transition metal ternary compounds. Other classes of superconductors, such as the organic superconductors and alkali-doped C_{60} crystals, will be discussed in the next chapter under application to complex crystals.

The ternary transition metal superconductors with a chemical formula of TT' X (T, T' are 3d or 4d transition metals and X represents Si, Ge, or P) are interesting compounds for exploring the relationship between the crystal structure, electronic structure, and superconducting properties. There are two structural types for TT'X compounds with the same atomic compositions; the orthorhombic (o-) anti-PbCl type (space group *Pnma*) and the hexagonal (h-) Fe_2P-type (space group *P6–m2*). Past studies indicated that TT' X compounds with the hexagonal structure have a higher $T_c (> 10K)$ than those with the orthorhombic structure ($T_c < 5K$). It was therefore astonishing that Shirotani et al. (Shirotani et al., 2000) reported a T_c of 15.5 K for o-MoRuP, higher than h-ZrRuP ($T_c = 13K$). It was quite surprising that there were no serious calculations of the electronic structures of these compounds. The OLCAO method was used to calculate the electronic structure of the four compounds: o-MoRuP, h-MoRuP, o-ZrRuP, and h-ZrRuP. The crystal structure of o-MoRuP had been accurately determined and refined by the x-ray Rietveld technique (Wong-Ng et al., 2003). However, the existence of h-MoRuP was not confirmed and its structure was theoretically modeled by the total energy minimization scheme (Ouyang and Ching, 2001) described in Chapter 3. The calculated band structures for these four crystals are shown in Fig. 6.8 and the calculated results on electronic structure and bulk mechanical properties are listed and compared in Table 6.4.

Based on the trends in the calculated DOS value at the Fermi level $N(E_F)$, it was predicted that the T_c for the hypothetical h-MoRhP crystal could be as high as 21.1 K. This simple prediction is based on the McMillian formula (McMillian, 1968) that T_c is related to the electron-phonon coupling constant which is proportional to $N(E_F)$. This example illustrates the usefulness of accurate electronic structure calculations for predicting physical properties of unknown phases based on a systematic approach.

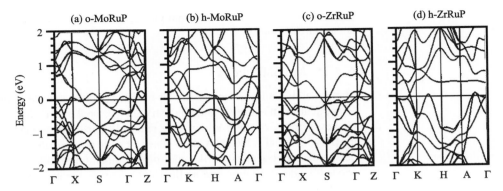

Fig. 6.8.
Calculated band structures of: (a) o-MoRuP, (b) h-MoRuP, (c) o-ZrRuP, and (d) h-ZrRuP. The Fermi level is set at zero energy.

Table 6.4. Comparison of o-MoRuP, h-MoRuP, o-ZrRuP, h-ZrRuP.

Crystal	o-MoRuP	h-MoRuP	o-ZrRuP	h-ZrRuP
T_c (K)	15.5	>20 (pred.)	3.8	13.
Band width (eV)	6.9	7.2	6.3	6.7
N(E_F)	0.46	0.53	0.33	0.44
		(state/eV/atom)		
B (GPa)	304.7	300.4	235.2	237.9
B'	4.03	4.06	3.84	2.95
TE (eV)/f.u.	−893.733	−893.547	−749.592	−749.598
Vol./f.u. (Å^3)	39.91	40.22	45.07	45.08
No. electrons/f.u.	19	19	17	17
		Q^* (electrons)		
Mo/Zr	5.751	5.769	3.330	3.301
Ru	8.275	8.232	8.529	8.502
P	4.974	4.964	5.143	5.235
		5.072		5.119

6.4 Other recent studies on metals and alloys

Advanced metallic alloys have important applications in energy-related science and technology. Example applications include steam turbines, boilers for coal conversion power plants, and materials for the entire steel-based industry. Recent trends demand advanced alloys that can withstand harsh high temperature, high pressure, and corrosive environments in order to increase the efficiency, reduce the cost, prolong the life, and meet the increased emissions standards that are required for a range of devices. In this section, we present recent results on the electronic structure of two classes of such metallic alloys obtained using the OLCAO method: The Mo-Si-B alloys and the MAX phase compounds. These new results have yet to be published. These materials potentially have improved mechanical properties that can extend current performance limits. A fundamental understanding of their electronic structure is the starting point.

6.4.1   Mo-Si-B alloys

Many metallic alloys, such as the Ni-based superalloys, have a long tradition of application in energy related industry such as coal conversion power plants. However, many of these alloys have reached their ultimate performance limits (Bewlay et al., 2003). A new generation of novel materials that possess outstanding physical properties and high temperature oxidation resistance is needed. Understandably, finding the right materials involves trade-offs in their different properties under different conditions. Equally important is the consideration of their processing cost for commercial viability. Alloys of refractory metals (Nb, Mo, W, etc.) have shown great promise in meeting the more stringent performance requirements due to their high melting temperature (Burk et al., 2009, Kruzic et al., 2005, Sakidja et al., 2008). In particular, the major crystalline phases ($MoSi_2$, Mo_3Si, Mo_2B, Mo_5Si_3 and Mo_5SiB_2) within the Mo-Si-B system have been studied by several groups both experimentally and theoretically. These materials have a relatively high melting temperature. There is evidence that $MoSi_2$ offers excellent oxidation resistance yet has relatively poor fracture toughness and is brittle. Mo_5Si_3, on the other hand, has poor

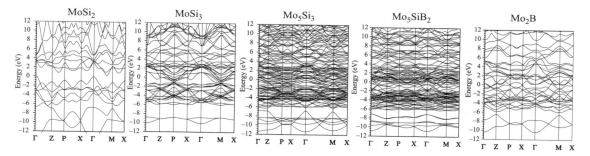

Fig. 6.9.
Band structures of the five crystalline phases in the Mo-Si-B phase diagram. The Fermi level is set at the zero energy.

Table 6.5. Calculated effective charge and DOS at the Fermi level $N(E_F)$ in the Mo-Si-B crystals. There are two Mo sites and two Si sites in both Mo_5Si_3 and Mo_5SiB_2.

Crystal	$MoSi_2$	Mo_3Si	Mo_5Si_3	Mo_5SiB_2	Mo_2B
			Mulliken effective charge Q^* (electrons)		
Mo	6.346	6.002	6.021, 6.048	6.026, 5.729	5.743
Si	3.827	3.994	3.946, 3.977	4.025, 3.516	–
B	–	–	–	3.516	3.515
			$N(E_F)$ (States/eV-cell)		
	–	3.90	11.0	9.6	4.2

oxidation resistance but better creep resistance at elevated temperature. The addition of B can significantly increase the oxidation resistance at temperatures ranging from 800°C to 1300°C. There is a trade-off between fracture toughness (high Mo content) and oxidation resistance (high B content). Also, the ductility of the alloy can be increased by spinel incorporation. In Mo_3Si, improvement of oxidation resistance can be achieved by adding Cr. All this indicates that a large set of parameters are at play in determining the optimum material in Mo-based alloys. To gain a fundamental understanding at the atomistic level in the Mo-Si-B system, the OLCAO method has been applied to calculate the electronic structure and bonding in these crystals. Figure 6.9 shows the calculated band structures of these five crystals. It can be seen that the results for the five crystals are very different due to their different crystal structures and the variations in the concentration of Mo, Si, and B among them. Four crystals are binary alloys and Mo_5SiB_2 is a ternary alloy. All are metals with a Fermi level except $MoSi_2$ which is a small gap semiconductor with an indirect band gap of 0.31 eV and a larger valence band (VB) width. Mo_5Si_3 and Mo_5SiB_2 have much more complex band structures and larger density of states at the Fermi level $N(E_F)$. The presence of B in Mo_5SiB_2 reduces $N(E_F)$ as compared to Mo_5Si_3. This correlates with the increase in the covalent character of the crystal and an increase in the bulk, shear, and Young's moduli.

The calculated Mulliken effective charges, Q^*, in the five crystals are summarized in Table 6.5. On average, Mo and B atoms gain charge whereas Si loses charge. Since the five crystals have different proportions of Si and B, the overall balance of the charge transfers and atomic composition is a critical aspect for their electronic bonding. Bond order calculations reveal a strong B-B bond in Mo_5SiB_2 while the B-B bond in Mo_2B is rather weak. Also important are the Mo-Mo bonds with short bonds and a large number of bonds or bond density.

6.4.2   MAX phases

An important class of inter-metallic alloys with a large number of applications are the so-called MAX phase compounds, or $M_{n+1}AX_n$ (M = a transition metal, A = Al, X = C or N). The MAX phases are layered transition metal carbides or nitrides with the rare combination of metal and ceramic properties. The MAX phase compounds were discovered in the 1970s (Nowotny, 1971) but remained relatively unnoticed until quite recently (Barsoum, 2000). Within the last ten years there has been a surge of interest in this class of materials both experimentally and theoretically (Lin et al., 2007, Wang and Zhou, 2009). Due to their unique structural arrangement, with alternating strong M-C bonds and relatively weaker M-A bonds in the crystallographic c-direction with both covalent and metallic characteristics (see Fig. 6.10), these thermodynamically stable alloys possess some exceptionally desirable properties such as damage resistance, oxidation resistance, excellent thermal and electric conductivity, machinability, and fully reversible dislocation-based deformation. Moreover, they are relatively cheap, lightweight, and have some outstanding

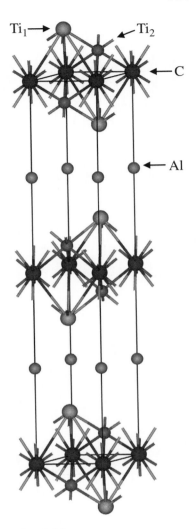

Fig. 6.10.
Crystal structure of hexagonal Ti_3AlC_2.

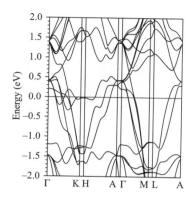

Fig. 6.11.
Band structure of Ti_3AlC_2 near the Fermi level.

high temperature properties that make them excellent candidate materials for use in many modern technologies.

The OLCAO method has been used to investigate the electronic structure and bonding in several MAX phases compounds. We present results for the most well-known phase, Ti_3AlC_2, as an example. The calculated band structure of Ti_3AlC_2 is shown in Fig. 6.11. The calculated total, atom-resolved and orbital resolved partial DOS is shown in Fig. 6.12.

The band structure is typical of many metals. The greatest concentration of states at the Fermi level in Ti_3AlC_2 is along the K-H line. From the total DOS for Ti_3AlC_2, it can be seen that the DOS at the Fermi level for Ti_3AlC_2 is at a minimum. This property is often used as a guide to the relative stability of the MAX phase compounds and other metallic alloys. The total DOS can be easily resolved into the atom-specific and orbital specific PDOS. In the case of Ti_3AlC_2, the DOS at the Fermi level in Ti_3AlC_2 is from the d-electrons of the Ti2 atom where the x^2-y^2 e_g states are the dominant contributors, but they are not associated with any major peak structures because all the large peaks are at higher energies. The DOS at the Fermi level for Ti1 and Ti2–$3d$ electrons are 0.554 and 2.288 respectively in units of states/[eV cell] with negligible contributions from Al and C.

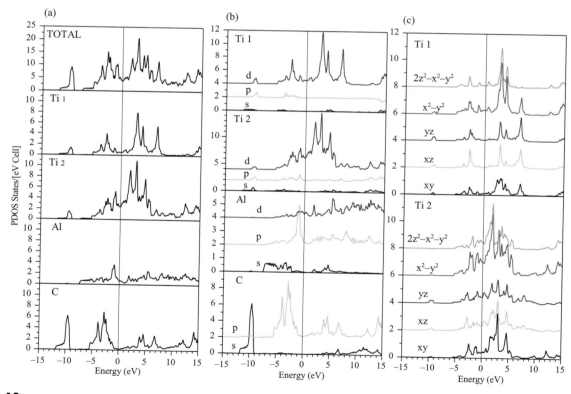

Fig. 6.12.
Calculated DOS and PDOS of Ti_3AlC_2: (a) total; (b) orbital resolved atomic type; (c) e_g and t_{2g} resolved d orbitals of Ti1 and Ti2.

The effective charge Q^* and bond order values in Ti_3AlC_2 are also calculated. The effective charges for Ti1, Ti2, Al, and C are respectively 3.39, 3.66, 2.97, and 4.66 elelctrons. Thus, the metallic elemets, especially Ti2, lose charge to C. We also found the bond order values for the Ti1-C, Ti2-C, and Ti2-Al bonds to be 0.199, 0.215, and 0.150 respectively, confirming the fact that the interlayer Ti-C form much stronger bonds than the Ti-Al and Ti-Ti bonds.

References

Anisimov, V. I., Zaanen, J., & Andersen, O. K. (1991), *Physical Review B*, 44, 943.

Arponen, J. & Pajanne, E. (1979), *Annals of Physics*, 121, 343–89.

Bardeen, J., Cooper, L. N., & Schrieffer, J. R. (1957a), *Physical Review*, 108, 1175.

Bardeen, J., Cooper, L. N., & Schrieffer, J. R. (1957b), *Physical Review*, 106, 162.

Barsoum, M. W. (2000), *Progress in Solid State Chemistry*, 28, 201–81.

Bednorz, J. G. & Müller, K. A. (1986), *Zeitschrift für Physik B Condensed Matter*, 64, 189–93.

Beno, M. A., Soderholm, L., Capone, D. W., et al. (1987), *Applied Physics Letters*, 51, 57–9.

Bewley, B., Jackson, M., Subramanian, P. & Zhao, J. (2003), *Metallurgical and Materials Transactions A*, 34, 2043–52.

Bharathi, A., Hao, L. Y., Sundar, C. S., et al. (1990a), Angular Correlation Studies on Yttrium Barium Copper Oxide ($Yba_2cu_3o_7$) Superconductor. *In:* Jean, Y. C., (ed.) *Proceedings of International Workshop on Positron and Positronium Chemistry* (Singapore:World Scientific), 512–17.

Bharathi, A., Sundar, C. S., Ching, W. Y., et al. (1990b), *Physical Review B*, 42, 10199.

Burk, S., Gorr, B., Trindade, V. B., Krupp, U., & Christ, H. J. (2009), *Corrosion Engineering, Science and Technology*, 44, 168–75.

Callaway, J. (1991), *Quantum Theory of Solids* (New York: Academic Press).

Ching, W. Y. & Callaway, J. (1973), *Phys. Rev. Lett.*, 30, 441–43.

Ching, W. Y. & Callaway, J. (1974), *Phys. Rev. B*, 9, 5115–21.

Ching, W. Y. & Callaway, J. (1975), *Phys. Rev. B*, 11, 1324–29.

Ching, W. Y. (1986), *Solid State Communications*, 57, 385–88.

Ching, W. Y. & Gu, Z. Q. (1987), *J. Appl. Phys.*, 61, 3718–20.

Ching, W. Y., Xu, Y., Zhao, G. L., Wong, K. W., & Zandiehnadem, F. (1987), *Phys. Rev. Lett.*, 59, 1333–6.

Ching, W. Y. & Gu, Z. (1988), *J. Appl. Phys.*, 63, 3716–18.

Ching, W. Y., Zhao, G. L., Xu, Y. N., & Wong, K. W. (1989a), *Mod. Phys. Lett. B*, 3, 263–9.

Ching, W. Y., Zhao, G. L., Xu, Y. N., & Wong, K. W. (1989b), Comparative Study of Band Structures of Tl-Ca-Ba-Cu-O and Bi-Ca-Sr-Cu-O Superconducting Systems. *In:* Baaquie, B. E., Chew, C. K., Lai, C. H., Oh, O. H., & Phua, K. K. (eds.) *Prog. High Temp. Supercond.* (Singapore: World Scientific).

Ching, W. Y., Xu, Y. N., Harmon, B. N., Ye, J., & Leung, T. C. (1990), *Phys. Rev. B*, 42, 4460–70.

Ching, W. Y., Zhao, G. L., Xu, Y. N., & Wong, K. W. (1991), *Phys. Rev. B*, 43, 6159–62.

Ching, W. Y. (1994), Local Density Calculation of Optical Properties of Insulators. *In:* Ellis, D. E. (ed.) *Electronic Density Functional Theory of Molecules, Clusters, and Solids* (Dordrecht, The Netherlands: Kluwer Academic Publishers).

Ching, W. Y., Huang, M.-Z., & Zhong, X.-F. (1994), *J. Appl. Phys.*, 76, 6047–9.

Ching, W. Y. & Huang, M.-Z. (1996), *J. Appl. Phys.*, 79, 4602–4.

Ching, W. Y., Huang, M.-Z., Hu, Z., & Yelon, W. B. (1997), *J. Appl. Phys.*, 81, 5618–20.

Ching, W. Y. & Xu, Y. N. (2000), *J. Magn. Magn. Mater.*, 209, 28–32.

Ching, W. Y., Gu, Z.-Q., & Xu, Y.-N. (2001), *J. Appl. Phys.*, 89, 6883–5.

Ching, W. Y., Xu, Y.-N., & Rulis, P. (2002), *Appl. Phys. Lett.*, 80, 2904–6.

Ching, W. Y., Xu, Y.-N., Ouyang, L., & Wong-Ng, W. (2003a), *J. Appl. Phys.*, 93, 8209–11.

Ching, W. Y., Xu, Y.-N., & Rulis, P. (2003b), *J. Appl. Phys.*, 93, 6885–7.

Fung, P. C. W., Lin, Z. C., Liu, Z. M., et al. (1990), *Solid State Communications*, 75, 211–16.

Gu, Z. Q. & Ching, W. Y. (1986), *Phys. Rev. B*, 33, 2868–71.

Gu, Z. Q. & Ching, W. Y. (1987a), *Phys. Rev. B*, 36, 8530–46.

Gu, Z. Q. & Ching, W. Y. (1987b), *J. Appl. Phys.*, 61, 3977–78.

Gu, Z. Q., Lai, W., Zhong, X., & Ching, W. Y. (1992), *Phys. Rev. B*, 46, 13874–80.

Gu, Z. Q., Lai, W., Zhong, X. F. & Ching, W. Y. (1993), *J. Appl. Phys.*, 73, 6928–30.

Gu, Z. Q., Xu, Y.-N. & Ching, W. Y. (2000), *J. Appl. Phys.*, 87, 4753–55.

Herbst, J. F., Croat, J. J., Pinkerton, F. E. & Yelon, W. B. (1984), *Physical Review B*, 29, 4176.

Huang, M.-Z. & Ching, W. Y. (1994), *J. Appl. Phys.*, 76, 7046–48.

Huang, M.-Z. & Ching, W. Y. (1995), *Phys. Rev. B*, 51, 3222–25.

Huang, M.-Z. & Ching, W. Y. (1996), *J. Appl. Phys.*, 79, 5545–47.

Huang, M.-Z., Ching, W. Y. & Gu, Z.-Q. (1997), *J. Appl. Phys.*, 81, 5112–14.

Hutchings, M. T. (1964), Point-Charge Calculations of Energy Levels of Magnetic Ions in Crystalline Electric Fields. *In:*Frederick, S. & David, T. (eds.) *Solid State Physics* (NewYork: Academic Press).

Jack, K. H. (1948), *Proceedings of the Royal Society of London. Series A. Mathematical and Physical Sciences*, 195, 34–40.

Jack, K. H. (1951), *Proceedings of the Royal Society of London. Series A. Mathematical and Physical Sciences*, 208, 216–24.

Kresin, V. Z., Morawitz, H., & Wolf, S. A. (1993), *Mechanisms of Conventional and High T_c Superconductivity* (New York: Oxford University Press).

Kruzic, J., Schneibel, J., & Ritchie, R. (2005), *Metallurgical and Materials Transactions A*, 36, 2393–402.

Lafon, E. E. & Lin, C. C. (1966), *Physical Review*, 152, 579.

Li, Y. P., Gu, Z. Q., & Ching, W. Y. (1985), *Phys. Rev. B*, 32, 8377–80.

Lin, Z., Li, M. & Zhou, Y. (2007), *J. Mater. Sci. Technol.*, 23, 145.

Maeda, H., Tanaka, Y., Fukutomi, M. & Asano, T. (1988), *Japanese Journal of Applied Physics*, 27, L209.

Mcmillian, W. L. (1968), *Physical Review*, 167, 331.

Mohn, P. & Wohlfarth, E. P. (1987), *Journal of Physics F: Metal Physics*, 17, 2421.

Nowotny, V. H. (1971), *Progress in Solid State Chemistry*, 5, 27–70.

Ouyang, L. & Ching, W. Y. (2001), *J. Am. Ceram. Soc.*, 84, 801–5.

Pickett, W. E., Freeman, A. J. & Koelling, D. D. (1981), *Physical Review B*, 23, 1266.

Pickett, W. E. (1989), *Reviews of Modern Physics*, 61, 433.

Rath, J., Wang, C. S., Tawil, R. A., & Callaway, J. (1973), *Physical Review B*, 8, 5139.

Sakidja, R., Perepezko, J. H., Kim, S., & Sekido, N. (2008), *Acta Materialia*, 56, 5223–44.

Schilling, A., Cantoni, M., Guo, J. D., & Ott, H. R. (1993), *Nature*, 363, 56.

Sheng, Z. Z. & Hermann, A. M. (1988), *Nature*, 332, 138.

Shirotani, I., Takaya, M., Kaneko, I., Sekine, C., & Yagi, T. (2000), *Solid State Communications*, 116, 683–6.

Singh, M., Wang, C. S., & Callaway, J. (1975), *Physical Review B*, 11, 287.

Stoner, E. C. (1938), *Proceedings of the Royal Society of London. Series A. Mathematical and Physical Sciences*, 165, 372–414.

Sundar, C. S., Bharathi, A., Ching, W. Y., et al. (1990), *Phys. Rev. B*, 42, 2193–99.

Sundar, C. S., Bharathi, A., Ching, W. Y., et al. (1991), *Phys. Rev. B*, 43, 13019–24.

Wang, C. S. & Callaway, J. (1974), *Physical Review B*, 9, 4897.

Wang, J. & Zhou, Y. (2009), *Annual Review of Materials Research*, 39, 415–43.

Wong-Ng, W., Ching, W. Y., Xu, Y.-N., Kaduk, J. A., Shirotani, I., & Swartzendruber, L. (2003), *Phys. Rev. B*, 67, 144523/1–144523/9.

Wong, K. W. & Ching, W. Y. (1989a), *Physica C*, 158, 1–14.

Wong, K. W. & Ching, W. Y. (1989b), *Physica C*, 158, 15–31.

Wong, K. W. & Ching, W. Y. (2004), *Physica C*, 416, 47–67.

Wu, M. K., Ashburn, J. R., Torng, C. J., et al. (1987), *Phys. Rev. Lett.*, 58, 908.

Xu, Y. N., Ching, W. Y., & Wong, K. W. (1988), *Phys. Rev. B*, 37, 9773–76.

Xu, Y. N., Ching, W. Y., & Wong, K. W. (1990), *Mater. Res. Soc. Symp. Proc.*, 169, 41–44.

Xu, Y. N., Gu, Z.-Q., & Ching, W. Y. (2000), *J. Appl. Phys.*, 87, 4867–69.

Xu, Y. N., Rulis, P., & Ching, W. Y. (2002), *J. Appl. Phys.*, 91, 7352–54.

Zhao, G. L., Xu, Y., Ching, W. Y., & Wong, K. W. (1987), *Phys. Rev. B*, 36, 7203–6.

Zhao, G. L., Ching, W. Y., & Wong, K. W. (1989), *J. Opt. Soc. Am. B*, 6, 505–12.

Zhong, X. F. & Ching, W. Y. (1988), *J. Appl. Phys.*, 64, 5574–76.

Zhong, X. F. & Ching, W. Y. (1989a), *Phys. Rev. B*, 40, 5292–95.

Zhong, X. F. & Ching, W. Y. (1989b), *Phys. Rev. B*, 39, 12018–26.

Zhong, X. F. & Ching, W. Y. (1990), *J. Appl. Phys.*, 67, 4768–70.

Zhong, X. F., Ching, W. Y. & Lai, W. (1991), *J. Appl. Phys.*, 70, 6146–48.

Zhong, X. F. & Ching, W. Y. (1993), *J. Appl. Phys.*, 73, 6925–27.

7

Application to Complex Crystals

In this chapter, we describe application of the OLCAO method to the very loosely defined subject of complex crystals. These materials could also be classified under the previous chapters as insulators or metals. The common theme for the materials in this chapter is that they all have relatively complex structures or they are closely related to those crystals that have complex structures in terms of the number and variety of atoms as well as their geometric configurations. Yet, they are neither non-crystalline materials nor are they crystals with defective structures that require complex structural modeling. These are considered separate issues and they will be addressed in Chapters 8 and 9 respectively. Further, the complex crystals that are related to biomolecular systems will also be separately discussed in Chapter 10.

7.1 Carbon-related systems

7.1.1 Bucky-ball (C_{60}) and alkali-doped C_{60} crystals

The discovery of the third crystalline form of carbon, C_{60} fullerenes, in 1985 (Kroto et al., 1985) was an exciting bit of news that stimulated much experimental and theoretical research. Even more surprising was the report that when intercalated with alkali elements, C_{60} crystals became superconducting with a T_c approaching 28 K (Rosseinsky et al., 1991). The OLCAO method was applied to investigate the electronic structure and optical properties of the C_{60} crystal (Ching et al., 1991). The calculated band structure showed a direct band gap of 1.34 eV at X and 1.87 eV at the zone center. The electron and hole effective masses at X were estimated to be 1.45 m_e and 1.17 m_e respectively. The calculated optical spectrum of f.c.c C_{60} was very rich in structure with various sharp peaks that were attributed to the unique combination of the molecular cluster of C_{60}, which gives rise to the localized bands, and the periodic f.c.c lattice which gives the critical points in the Brillouin zone. Five disconnected absorption bands were identified in the 1.4 eV to 7.0 eV range. It was also found that the direct transition at the X point is symmetry forbidden and that the optical transition threshold comes from the transition to the second CB similar to the case of the Cu_2O semiconductor. This led to the speculation that an exciton may form at X in f.c.c C_{60} as in Cu_2O. All the structures in the optical spectrum can be assigned to critical point transitions superimposed

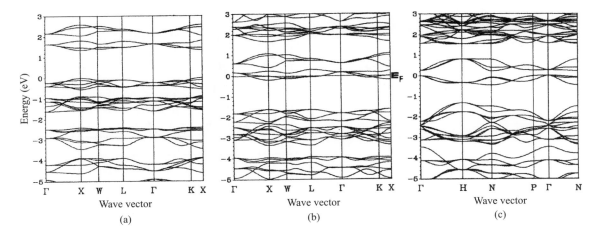

Fig. 7.1.
Band structure of: (a) f.c.c. C_{60}, (b) f.c.c. K_3C_{60}, and (c) b.c.c. K_6C_{60}. The zero energy is either at the top of the valence band or at the Fermi level.

on the band to band transitions. The calculated energy loss function was in excellent agreement with experiment.

The discovery of superconductivity in K and other alkali doped C_{60} crystals where T_c increases with pressure (Hebard et al., 1991, Kelty et al., 1991, Rosseinsky et al., 1991, Tanigaki et al., 1991) immediately led to extensive OLCAO calculations of the alkali doped C_{60} crystals. Figures 7.1, 7.2, and 7.3 show the comparison of the band structures, the imaginary parts of the dielectric functions, and the electron energy loss spectra of f.c.c C_{60}, f.c.c K_3C_{60} and b.c.c K_6C_{60} using experimentally determined crystal parameters (Huang et al., 1992a, Ching et al., 1991). Also calculated were the KC_{60} and K_2C_{60} crystals in order to see the evolution of the electronic structure as the amount of intercalated K varies (Xu et al., 1991). The main interest in these calculations were the properties related to superconductivity, especially the density of states at the Fermi level, the shape and structure of the Fermi surfaces, and the estimation of T_c based on BCS theory in comparison to the measured values.

Figure 7.1 shows that both f.c.c C_{60} and b.c.c K_6C_{60} have band gaps whereas K_3C_{60} is a metal with its Fermi level lying in a group of narrow bands. The calculated DOS at the Fermi level $N(E_F)$ of 17 states/[eV cell] was in excellent agreement with the experimental value of value of 20 states/[eV cell] (Tycko et al., 1991). Figure 7.2 shows that the optical absorption curve in the f.c.c C_{60} crystal has five disconnected bands in the 1.4 to 7.0 eV range. These disconnected bands started to merge when more K was added. Figure 7.3 compares the calculated electron energy loss functions with the measured spectra (Sohmen et al., 1992). They are in very good agreement.

The superconducting K_3C_{60} is metallic. Figure 7.4 shows the calculated Fermi surface in f.c.c. K_3C_{60}. There is a large hole pocket centered at Γ which is approximately spherical in the k_x-k_y plane and somewhat cylindrical in the

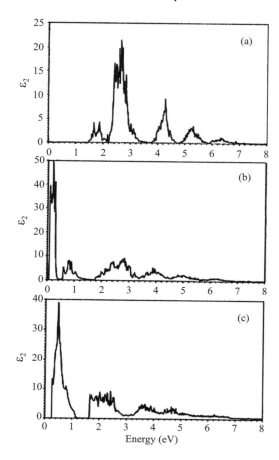

Fig. 7.2.

Calculated imaginary part of the dielectric function of: (a) f.c.c. C_{60}, (b) K_3C_{60}, and (c) K_6C_{60}.

k_x-k_z plane. The volume and the surface of this hole pocket were estimated to be $0.051 \, \text{Å}^{-3}$ and $0.90 \, \text{Å}^{-2}$ respectively.

In addition to K_3C_{60}, the OLCAO calculation was extended to other binary and pseudobinary alkali-doped C_{60} with three alkali atoms. They are RbK_2C_{60}, Rb_2KC_{60}, Rb_3C_{60}, Rb_2CsC_{60}, and Cs_3C_{60} (Huang et al., 1992b). With the exception of the hypothetical Cs_3C_{60}, there was an approximate linear relationship between the calculated values of $N(E_F)$ and the lattice constant with a slope of 14 states/[eV C_{60}] per Å. It is well known that in BCS superconductors, $N(E_F)$ is the most important parameter affecting T_c. Based on the calculated $N(E_F)$ values and the McMillian formula (McMillian, 1968) for T_c in BCS superconductors, a set of parameters necessary for the evaluation of T_c were identified that would yield the best agreement with the measured T_c values in the five alkali doped crystals. This same set of parameters was then used to evaluate T_c in K_3C_{60} crystal a different pressures. Experimental data indicated a steady decrease in T_c in alkali-doped C_{60} crystals under isotropic pressure (Huang et al., 1992b, Huang et al., 1993b, Xu et al., 1992). The result is illustrated in Fig. 7.5. It shows that optimized parameters do give

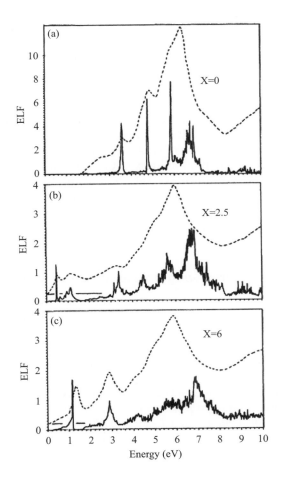

Fig. 7.3.
Calculated electron-energy-loss function of: (a) f.c.c. C_{60}, (b) K_3C_{60}, (c) K_6C_{60}. The dashed lines are the experimental data for K_xC_{60} with x = 0, 2.5, and 6 (see original reference for details).

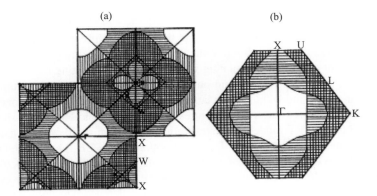

Fig. 7.4.
Calculated Fermi surfaces of K_3C_{60} in: (a) $K_z = 0$ plane, (b) $K_y = 0$ plane of the fcc BZ.

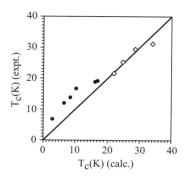

Fig. 7.5.
Calculated T_c vs. measured T_c in alkali doped C_{60}.

good agreement in T_c for RbK_2C_{60}, $Rb_2K\,C_{60}$, Rb_3C_{60}, and Rb_2CsC_{60}, but the T_c for K_3C_{60} under pressures from 0.0 to 2.8 GPa were underestimated, although the general trend in the pressure dependence is seen to be correct. The conclusion was that conventional BCS theory is valid for alkali-doped C_{60} superconductors but that the estimation of T_c requires more rigorous theory and calculations.

The results from the electronic structure calculation of f.c.c C_{60} and K_3C_{60} using the OLCAO method were used to construct the positron potential similar to that used in the high T_c superconductors discussed in the last chapter (Lou et al., 1992). The result was consistent with the experimental observation of a large difference in positron life times in C_{60} and K_3C_{60}. It was concluded that positrons may be distributed outside the molecule or inside the molecule depending on the amount of K doping. Both experimental and calculated results show strong evidence of high sensitivity of the positron density distribution in the K-doped C_{60} fullerenes.

7.1.2 Negative curvature graphitic carbon structures

The C_{60}-based fullerene crystals discussed above and the carbon nanotubes (CNT) to be discussed in the next section all have positive Gaussian curvatures when 5-membered rings are introduced in the two-dimensional sp^2 bonded graphitic structures with only 6-membered rings. More than a century ago, Schwarz suggested the possibility of periodic graphite-like carbon structures (PGCS) with positive Gaussian curvature and a minimal surface which contains 7- and/or 8-membered rings dubbed schwarzites (Schwarz, 1890). Several such PGCS models were proposed in the early 1990s including the Mackay and Terrones model (Mackay and Terrones, 1991), the Vanderbilt and Tersoff model (Vanderbilt and Tersoff, 1992), and several others by Lenosky et al. (Lenosky et al., 1992) and Adams et al. (O'Keeffe et al., 1992).

The OLCAO method was used to study the electronic structure of the Vanderbilt and Tersoff (VT) model (Ching et al., 1992) and 15 other PGCS models (Huang et al., 1993a). These periodic models have cubic structures in a simple cubic, b.c.c, or f.c.c lattice, 24 to 216 atoms in the unit cell, densities ranging from 0.051 to 0.111 C atom per Å^3, and they contain different numbers of 7- or 8-membered rings in addition to the regular 6-membered rings. Of these 18 models, eight of them are insulating with minimal gaps ranging from 0.17 eV to 2.96 eV, seven are metallic and one is a semi-metal. The band structure of the 168 atom VT model in an f.c.c lattice and that of a 24 atom "polybenzene" model with benzene-like ring structures in a simple cubic lattice are shown in Fig. 7.6. The VT model is a semiconductor with an indirect band gap of 0.48 eV. The top of the VB is at Γ and the bottom of the CB is at X with large differences in the hole effective mass ($-0.78\,m_e$ for heavy hole and $-0.26\,m_e$ for the light hole) and electron effective mass ($1.82\,m_e$). These values are very different from those of either graphite or the C_{60} fullerenes. The "polybeneze" model constructed by (O'Keeffe et al., 1992) can be considered as having four hexagonal benzene-like rings on the face of a tetrahedron. It has the largest

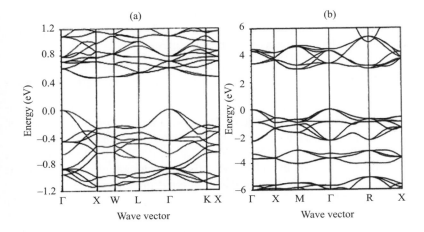

Fig. 7.6.
Calculated band structure of: (a) VT model and (b) Polybenzen model (see original references for details).

indirect band gap (2.96 eV) among all the 16 PGCS models calculated. The top of the VB is at Γ and the bottom of the CB is at M and the direct band gap at Γ is 3.39 eV. Careful analysis of the electronic structures of these 16 models failed to clearly reveal any correlations between structural properties and the controlling factors for the gap formation, although it was found that models containing only isolated 7- or 8-membered rings tended to be insulating.

The above studies on the complex three-dimensional schwarzite carbon structures with quasi-two-dimensional bonding was extended to a "perfect" amorphous graphite carbon model constructed by Townsend et al. (Townsend et al., 1992). The Townsend model has an f.c.c lattice with 1248 C atoms and contains random fractions of 5-, 6-, 7-, and 8-membered rings and no broken bonds and therefore has both positive and negative Gaussian curvatures. The bond length and bond angle distributions in this model have small deviations with an average bond length of 1.42 Å and an average bond angle of 119.81°. The Townsend model is very interesting because it provides a model of weak localization in a random system with two-dimensional bonding in a three-dimensional structure. The large periodicity of the model (a = 42.92Å) implies that the model represents a truly amorphous structure of infinite extension. Many actually believed that such structures exist in amorphous carbon films. The electronic structure and the transport properties of the Townsend model were studied using the OLCAO method (Huang and Ching, 1994). It is a metal with an $N(E_F)$ value of 0.098 states/[eV atom]. The DOS shows a dip at the Fermi level and the width of the occupied VB is 21.5 eV which is close to graphite and C_{60}. The very different properties of the Townsend model with respect to the VT model or other PGCS models indicates that the topological disorder introduced in the Townsend model has a profound effect on the overall electronic and optical properties of the system, notwithstanding that the local sp^2 bonding picture remains the same. Transport properties were calculated based on the Kubo–Greenwood formula (Greenwood, 1958). A zero temperature resistivity of 1160 $(\mu\Omega\text{-cm})^{-1}$ was obtained. The relatively high resistivity for a metallic structure is attributed to the existence of localized states at the Fermi level. The details of the transport property's calculations with the OLCAO method were discussed in Chapter 3 and will also be revisited in Chapter 8 under the subsection of metallic glasses.

7.2 Graphene, graphite, and carbon nanotubes

Graphite, along with diamond and C_{60}, was considered to be one of only three forms of carbon that exist in nature. However, shortly after the discovery of C_{60}, it was discovered that another type of carbon exists in the form of cylindrical tubes called carbon nanotubes (CNTs) (Iijima, 1991). A CNT can be single-walled (SWCNT) or multi-walled (MWCNT). Even more surprising was the discovery of a single sheet of graphite, referred to as graphene (Novoselov et al., 2004), which has fascinating and intriguing properties associated with its two-dimensional structure. These two discoveries fueled two successive waves of intense research on their special properties, fundamental physics, and possible applications ranging from use in metal oxide semiconductors and field effect transistors (MOSFETs) to novel drug delivery systems. In this subsection, we briefly discuss the application of the OLCAO method to graphite, graphene, and carbon nanotubes focusing mostly on the SWCNT.

7.2.1 Graphene and graphite

Figure 7.7 (a) to (c) compares the band structure, the DOS, and the optical absorption in the form of imaginary part of the dielectric function (ε_2) of graphite and graphene calculated by using the OLCAO method with sufficiently large number of k-points. They both have a two-dimensional hexagonal lattice with 6-member rings and each carbon bonds to three other carbons. In graphite, there are very weak inter-layer interactions in the c-direction. Alkali atoms can be easily intercalated between these layers (Zabel and Solin, 1992). Graphene is a true two-dimensional system. Its electronic structure was calculated in exactly the same manner as in graphite except the c axis was set to be 30 Å to avoid any inter-layer interaction. Figure 7.7 (a) shows that both graphite and graphene are semi-metals with zero band gaps at the H point. The total DOS shown in Fig. 7.23 (b) show that they have the same occupied band width of 20.5 eV and peak positions. The only difference is that graphene has shaper peaks due to the difference in van Hove singularities which depends on the dimensionality. These differences are more pronounced in the empty conduction band region and this leads to a larger difference in the absorption spectra ε_2 shown in Fig. 7.23 (c) which are resolved in the planar (x-y) and the perpendicular (z-) directions. The z-direction peak at 11.8 eV is shifted down to 10.6 eV in graphite showing the effect of inter-layer interactions in the later. All other peaks in ε_2 (at 4.0 eV, 14.5 eV and between 22.5 to 28.5 eV) are from the planer directions. They are less affected by the dimensionality and are only slightly broadened in the case of graphite. It can also be observed that the broader peak in the x–y plane at 8.7 eV in graphene appears as a slope extended from the higher peak which can again be attributed to the weak interaction in the z-direction in graphite. These peak structures can be roughly

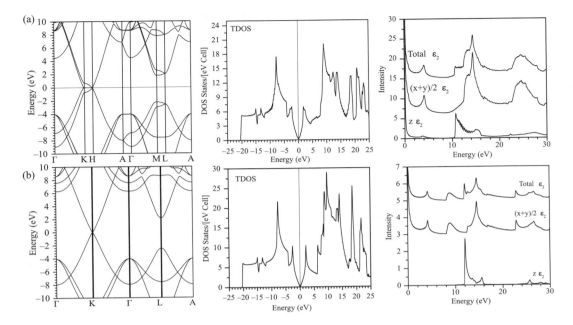

Fig. 7.7.
(a) Band, (b) Total DOS and (c) ε_2 of graphite (top) and graphene (bottom).

identified by the peak to peak transitions from the DOS diagram modulated by the matrix elements and be assigned to various transitions between different σ, π bonding states to σ^*, π^* anti-bonding states. It should also be pointed out that the magnitudes of the peaks in the graphene plot need to be scaled because of the assumption of the artificial 30 Å separation in the c-direction for the graphene layers to maintain the periodic boundary condition in the c-direction. The differences between the two reflect the weak inter-layer interactions in graphite. The degeneracy at the H point for graphene can be removed by applying strain, introducing defects, and so on which leads to all sorts of interesting phenomena including the non-locality of states near the Dirac point (Abanin et al., 2011) which opens the door for new technology based on the peculiar electronic properties of a pure two-dimensional lattice (Castro Neto et al., 2009).

7.2.2 Carbon nanotubes

The outstanding mechanical and electrical properties of CNTs make them ideal for a wide range of applications. However, for nano-electronics applications, widespread commercial and industrial use has not been realized due to the practical difficulty of sorting SWCNTs into monodispersed constituents and then placing them into predetermined arrangements (Zheng et al., 2003). The fundamental forces that govern SWCNT interactions are being studied in the hope of overcoming the sorting and placement barriers. The main motivation of our study on the electronic structure and optical properties of a large number of SWCNTs was to investigate their role in several long-range interactions (French et al., 2010), such as the van der Waals—London dispersion which will be further discussed in Chapter 12.

Each SWCNT is uniquely defined by its chirality vector [n,m], which denotes the geometric construction of the circumferential wrapping direction along a graphene sheet (Barros et al., 2006 Popov, 2004). Each SWCNT with a unique [n.m] has unique electronic structures (e.g., band gap, van Hove singularities (vHs), etc.). Their common composition and chemical bonding coupled with symmetry considerations makes it possible to categorize them by particular features. The three categories are armchair, zigzag, and chiral depending on the arrangements of the C atoms with respect to the axial direction of the SWCNT. They have been noted to have predictable trends (Charlier et al., 2007, Popov, 2004, Saito et al., 2005). The properties of SWCNTs change as a function of the chirality vector. The most well-known being the metallic versus semiconducting behavior that is dependent on whether $(n-m)/3$ is an integer or not. A SWCNT of infinite diameter would have the same electronic structure as graphene with strict planer sp^2 bonding. The bonding in the axial direction would resemble the in-plane bonding found in graphene and the bonding in the x-y plane would mimic the out-of-plane bonding found in graphene.

For a large but finite diameter SWCNT, the effects of periodicity would cause a set of discrete allowable states along a SWCNT's axial direction. Because the C-C bonds would still be nearly perfect sp^2 bonds, it is possible to derive a SWCNT's band structure by selecting the allowable states from that of graphene's band diagram. Such an approximation is called the "zone-folding" approximation. However, this is valid only for SWCNTs of diameter greater than about $10\,\text{Å}$. Below this threshold, the curvature of the SWCNT wall becomes sufficiently pronounced such that the bonds gain a significant sp^3 character and the zone folding scheme no longer works as well, especially for predicting the optical properties. Hence, *ab initio* calculations become necessary.

Table 7.1 lists 73 SWCNT whose electronic structure and optical properties were calculated using the OLCAO method. The calculations were carried out in a periodic cell by orienting the SWCNT with the axial direction (z direction) and by making the x and y periodicity large in comparison with the CNT's radius. The CNTs are classified as armchair, zigzag, and chiral according the [n,m] labels. Also listed is the chiral angle (in degrees), the radius of the tube, the number of atoms involved in each tube, N_A, and whether they are metals, semiconductors, or semi-metals. Many exciting studies on SWCNT have been conducted with different chiralities, but they were confined to properties associated with states lower than $6\,\text{eV}$ because they relied on the tight binding approximation, which artificially distorts the bands (Popov, 2004).

Figure 7.8 shows the calculated band structures, the DOS, and the imaginary dielectric function in the radial and axial directions for three SWCNTs: the armchair [24,24] which is a metal, the zigzag [30,0] which is a semi-metal and

Table 7.1. List of SWCNTs calculated using the OLCAO method and their characteristics.

n,m	r (Å)	Angle	Geometry	Lambin sp_2–sp_3	N_A
3,3	2.034	30.00	armchair	M	12
4,4	2.712	30.00	armchair	M	16
5,5	3.390	30.00	armchair	M	20
6,6	4.068	30.00	armchair	M	24
7,7	4.746	30.00	armchair	M	28
8,8	5.424	30.00	armchair	M	32
9,9	6.102	30.00	armchair	M	36
10,10	6.780	30.00	armchair	M	40
11,11	7.458	30.00	armchair	M	44
12,12	8.136	30.00	armchair	M	48
13,13	8.814	30.00	armchair	M	52
14,14	9.492	30.00	armchair	M	56
15,15	10.170	30.00	armchair	M	60
16,16	10.848	30.00	armchair	M	64
17,17	11.526	30.00	armchair	M	68
18,18	12.204	30.00	armchair	M	72
19,19	12.882	30.00	armchair	M	76
20,20	13.560	30.00	armchair	M	80
21,21	14.238	30.00	armchair	M	84
22,22	14.916	30.00	armchair	M	88
23,23	15.594	30.00	armchair	M	92
24,24	16.272	30.00	armchair	M	96
6,0	2.349	0.00	zigzag	SM	24
9,0	3.523	0.00	zigzag	SM	36
12,0	4.697	0.00	zigzag	SM	48
15,0	5.872	0.00	zigzag	SM	60
18,0	7.046	0.00	zigzag	SM	72
21,0	8.220	0.00	zigzag	SM	84
24,0	9.395	0.00	zigzag	SM	96
27,0	10.569	0.00	zigzag	SM	108
30,0	11.743	0.00	zigzag	SM	120
7,0	2.740	0.00	zigzag	SC	28
8,0	3.132	0.00	zigzag	SC	32
10,0	3.914	0.00	zigzag	SC	40
11,0	4.306	0.00	zigzag	SC	44
13,0	5.089	0.00	zigzag	SC	52
14,0	5.480	0.00	zigzag	SC	56
16,0	6.263	0.00	zigzag	SC	64
17,0	6.655	0.00	zigzag	SC	68
19,0	7.437	0.00	zigzag	SC	76
20,0	7.829	0.00	zigzag	SC	80
22,0	8.612	0.00	zigzag	SC	88
23,0	9.003	0.00	zigzag	SC	92
25,0	9.786	0.00	zigzag	SC	100
26,0	10.178	0.00	zigzag	SC	104
28,0	10.960	0.00	zigzag	SC	112
29,0	11.352	0.00	zigzag	SC	116
5,2	2.445	16.10	chiral	SM	52
6,3	3.107	19.11	chiral	SM	84
7,4	3.775	21.05	chiral	SM	124

(continued)

Table 7.1. Continued

n,m	r (Å)	Angle	Geometry	Lambin sp$_2$–sp$_3$	N$_A$
7,6	4.411	27.46	chiral	SM	508
8,2	3.588	10.89	chiral	SM	56
8,3	3.855	15.30	chiral	SM	388
8,4	4.143	19.11	chiral	SM	112
8,5	4.446	22.41	chiral	SM	172
8,6	4.762	25.28	chiral	SM	296
8,7	5.089	27.80	chiral	SM	676
9,3	4.234	13.90	chiral	SM	156
9,4	4.514	17.48	chiral	SM	532
9,5	4.810	20.63	chiral	SM	604
10,2	4.359	8.95	chiral	SM	248
10,4	4.889	16.10	chiral	SM	104
10,5	5.178	19.11	chiral	SM	140
11,2	4.746	8.21	chiral	SM	196
11,8	6.468	24.79	chiral	SM	364
4,2	2.071	19.11	chiral	SC	56
5,1	2.179	8.95	chiral	SC	124
6,1	2.567	7.59	chiral	SC	172
6,2	2.823	13.90	chiral	SC	104
6,4	3.413	23.41	chiral	SC	152
6,5	3.734	27.00	chiral	SC	364
7,5	4.087	24.50	chiral	SC	436
9,1	3.734	5.21	chiral	SC	364

the chiral [8,7] which is a semiconductor. All three have a large radius (16.27, 11.73, and 27.80 Å respectively) and in principle their electronic structure and optical properties should resemble that of graphene. The chiral [8,7] with 676 atoms in the unit cell is the largest SWCNT we have studied. The changes in the band structures from Γ to Z (axial direction) for these three different SWCNTs are very revealing. They have resulted in different DOS near the band gap or the Fermi level as well as variations in the peak positions and their magnitudes in the absorption curve. These results should be compared with that of graphene in Fig. 7.23 which should be the result for SWCNTs in the limit of infinite radius.

The OLCAO method has also been applied to a few MWCNT. They are [3,3]/[8,8], [4,4]/[9,9], [5–5]/[10,10], and [6,6]/[11,11] for the armchair class with a 3.39 Å radial separation, and [7,0]/[15,0], [8,0]/[16,0] for the zigzag class with a 3.13 Å radial separation. These double walled CNTs are concentric cylinders with a radial difference that is close to, but slight lower than, the inter-layer separation in crystalline graphite (2.62 Å). The calculated electronic structure and optical properties indicate that they can be reasonably approximated as the sums of the SWCNTs. This is consistent with the similarities in the electronic structure of graphite and grapheme shown in Fig. 7.23. The structures and hence the resulting electronic structure of real MWCNTs can be much more complicated and may not be in the form of concentric cylinders with a common axis.

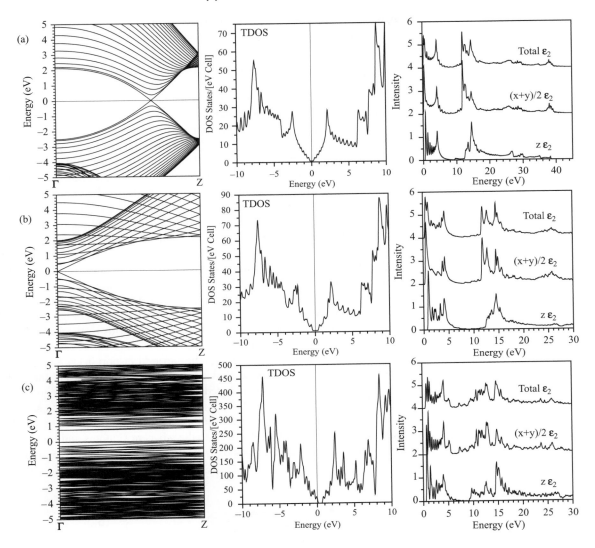

Fig. 7.8.
(a) Band, (b) Total DOS, and (c) ε_2 for three SWCNTs: armchair [24–24] (top), zigzag [30–0] (middle), and chiral [8–7] (bottom).

7.3 Polymeric crystals

Conjugated polymers with intervened polymer chains constitute a major materials class with its own unique structures and properties. Many of them can exist in complex crystalline forms which allows for the use of computational methods developed for crystalline solids. This is very beneficial because many electronic structure calculations on polymers use molecular scale quantum chemical methods and so by using solid state methods a complementary picture that is highly effective in dealing with issues such as longer ranged inter-chain interactions can be drawn. In this subsection, we describe the application of the OLCAO method to one such crystal called poly(di-n-hexylsilane) or polysilane.

Polysilanes are σ-conjugated high polymers with a silicon backbone and carbon-based side chains. They have some very remarkable linear and non-linear optical properties rooted mostly from the one-dimensional band structure that is associated with the Si backbone chain. The OLCAO method was used to study the electronic structure of one of the polysilane crystals, Poly(di-n-hexylsilane), (abbreviated as *Pdn6s*) whose crystal structure was accurately determined by Patnaik and Farmer (Patnaik and Farmer, 1992) using x-ray diffraction. The orthorhombic cell of *Pdn6s* (space group *Pna2*$_1$, Z = 2) contains 4 Si, 48 C, and 104 H atoms. The band structure of *Pdn6s* was calculated using the OLCAO method and is shown in Fig. 7.9 together with a sketch of its crystal structure and the Brillouin zone. A direct band gap of 3.5 eV was obtained which is consistent with the experimental value of about 4.0 eV. Figure 7.9 also shows the one-dimensional (1D) band from the Si backbone in the k_z direction (from Γ to Z) whereas the other bands are almost dispersionless. From the band curvatures of the HOMO (highest occupied molecular orbital) state (or the top of the VB) and the LUMO (lowest unoccupied molecular state) state (or the bottom of the CB), the effective mass components in the z-direction were estimated to be $-0.42m_e$ and $+0.19m_e$ respectively for hole in the VB and electron in the CB. These are typical effective mass values for semiconductors and this indicates that the carrier transport in crystalline *Pdn6s* will be along the Si backbone chain. From the partial DOS and the plot of the wave functions, it was concluded that the HOMO state is predominately derived from the Si-3p and C-2p orbitals whereas the LUMO state has a significant mixing from the Si-3d orbitals as well. Mulliken charge analysis indicated that Si loses 0.51 electrons to the adjacent C atoms and each C atom at the end of the side chain gains 0.70 electrons mostly at the expense of the H

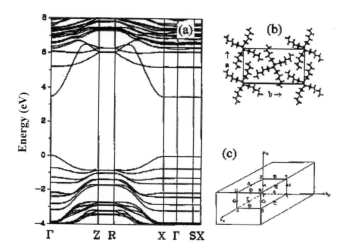

Fig. 7.9.
(a) Calculated band structure of Pdn6s crystal; (b) Projection of the PDn6s crystal on the *a-b* plane. (c) Brillouin zone of the crystal.

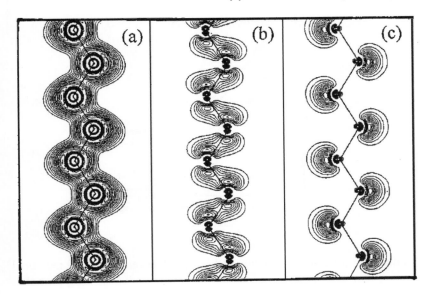

Fig. 7.10.
Distribution of the charge on the Si backbone in Pdn6s: (a) Total valence charge; (b) The HOMO state at Γ; and (c) The LUMO state at Γ (see original references for details).

atoms attached to it. So, on average, the Si backbone losses about one electron to the C side chains.

Figure 7.10 shows the valence charge distributions of the Si backbone and the HOMO and LUMO states in a plane containing the Si backbone. The strong covalent bonding character is evident and the lobe structure in the charge density plots indicates that the bonding is not strictly along the Si-Si chain direction but is strongly influenced by the side chains. The distribution of the LUMO state shows the anti-bonding character of a highly delocalized state with admixtures of Si-3s, Si-$3p_y$, and Si-$3d_{3z^2-r^2}$ orbitals.

Crystalline poly(di-n-hexylsilane) has some very interesting optical properties which have deep implications for the application of polymeric materials and polysilanes in particular. The structure of the temperature dependent VUV optical spectrum of Pdn6s is believed to be due to changes in the Si backbone conformation which, in turn, changes the electronic structure and long wave absorptions (French et al., 1992, Miller et al., 1985, Schellenberg et al., 1991). In the VUV spectrum shown in Fig. 7.11, the origin of the sharp peak at the absorption edge, labeled E1, is particularly interesting because it may be either an excitonic peak or it may be due to the one-dimensional nature of the band structure. To answer this question, accurate optical calculations based on the electronic structure were necessary. The OLCAO method was used to calculate the inter-band optical properties of the polysilane crystal (Ching et al., 1996). The calculated spectrum showed strong anisotropy in the direction parallel to the c-axis (the chain direction) and the plane perpendicular to the chain direction. The large optical absorption in the UV region observed in the experiment was reproduced and a sharp peak corresponding the experimentally observed $E_1{}'$ peak was also revealed. This peak has its major component in the chain direction and so it can be traced to the HOMO–LUMO transition in

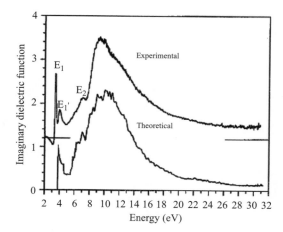

Fig. 7.11.

Comparison of the measured (upper) and calculated (lower) $\varepsilon_2(\omega)$ curves of Pdn6s crystal.

the one-dimensional-band structure of Fig. 7.9. The calculated and measured optical spectra are shown in Fig. 7.11. It was concluded that the sharp peak (E_1) was excitonic in nature because it was absent in the one electron calculation, whereas the second strong peak (E_1') had its origin in the one-dimensional nature of the band structure. The excellent agreement with the measured VUV data in Fig. 7.11 was achieved by applying an energy dependent scaling factor to account for the limited basis expansion in the OLCAO calculation that tends to shift the main peak toward higher energy values because of the limited variational freedom in the Bloch function obtained.

7.4 Organic crystals

In this section, we discuss the application of the OLCAO method to three types of organic crystals: (1) organic superconductors which are quasi two-dimensional charge transfer salts with complex crystal structures and low symmetry; (2) Fe(TCNE) crystals which belong to the class of organic-based magnets and which have tremendous application potential; and (3) the herap-athite crystal which has an interesting 150-year history and a crystal structure that has only very recently been resolved.

7.4.1 Organic superconductors

Organic superconductors, which may include the C_{60}-based supercon-ductors, are a special class of novel organic crystals with complex structures, fascinating properties, and the promise of future applications (Williams, 1992). The OLCAO method was used to study the electronic structure of two organic superconductors: κ-(BEDT-TTF)$_2$Cu(NCS)$_2$ and κ-(BEDT-TTF)$_2$Cu[N(CN)$_2$]Br (Ching et al., 1997, Kurmaev et al., 1999, Xu et al., 1995). They are both quasi two-dimensional charge transfer salts where BEDT-TTF stands for bis(ethylendithio) tetrathiafulvalenel and is abbre-viated as ET. Both are superconductors with transition temperatures of 10.4 K and 14.0 K respectively. The mechanism of organic superconductivity is still under debate but is believed to have a BCS ground state. κ-$(ET)_2Cu(NCS)_2$ has a monoclinic cell (space group P2$_1$, Z = 2) with 4 ET molecules and a total of 118 atoms, The κ-$(ET)_2Cu[N(CN)_2]Br$ crystal is almost twice as large. It has an orthorhombic cell (space group *Pnma*, Z = 4) and contains four ET molecules. These are fairly complex crystals with large numbers of atoms of different sizes and low symmetry. *Ab initio* electronic structure calculations of such crystals were a significant challenge at the time this work was being carried out. Previously, only semi-empirical methods were used, but these were not sufficient to give an accurate accounting of the electronic structures of organic superconductors. The crystal structure of κ-$(BEDT\text{-}TTF)_2Cu[N(CN)_2]Br$ is illustrated in Fig. 7.12.

Although different theories abound, the organic superconductors are likely to have a BCS type mechanism. Therefore the OLCAO electronic structure calculations have focused on the electron effective mass and Fermi velocity on the Fermi surface. This information is needed to interpret certain experimental measurements such as the Shubnikov–de Haas (SdH) effect, the de Haas–van Alphen experiment, and the two-dimensional angular correlation of positron annihilation measurements. An accurate Fermi surface mapping requires the solution of the secular equation at a large number of k points in the Brillion zone and a more accurate value of the density of states at the Fermi level $N(E_F)$. The effective mass over the Fermi surface was evaluated numerically from $m^* = (h^2/2\pi)dS(0)/dE$ where S is the closed Fermi surface. The calculated results for κ-$(ET)_2Cu(NCS)_2$ and κ-$(ET)_2Cu[N(CN)_2]Br$ crystals have some similarities but also some differences. Both crystals show an oval shaped hole pocket in the Fermi surface centered at the Γ and Z points of the BZ and marked van Hove singularity peaks slightly below the Fermi level. The calculated values of $N(E_F)$ are 12.8 and 27.4 states/[eV cell] for κ-$(ET)_2Cu(NCS)_2$ and κ-$(ET)_2Cu[N(CN)_2]Br$ respectively. On the basis of the number of ET molecules, the DOS at the Fermi level for the two would be 3.2 and 3.42 states/[eV ET]. This is consistent with the BCS interpretation that a higher value of $N(E_F)$ leads to a higher T$_c$. Figure 7.13 shows the calculated Fermi surfaces in the two crystals. Both crystals have a large enclosed oval-shaped hole surface. The Fermi surface for κ-$(ET)_2Cu[N(CN)_2]Br$ is more complicated than κ-$(ET)_2Cu(NCS)_2$ with four bands cutting through the Fermi level because it has twice as many ET molecules in the unit cell. For κ-$(ET)_2Cu(NCS)_2$, it has two orbits labeled as α and β shown in Fig. 7.13 (a). The effective masses on these two orbits were estimated to be $m^*_\alpha = 1.72m_e$ and $m^*_\beta = 3.05m_e$. These effective mass values are smaller than the measured values by a factor of two mainly because the mass enhancement factor due to electron–phonon and electron–electron interactions was neglected in the one electron band calculation. On the other hand, the ratio between m^*_α and m^*_β of 1.77 was actually in good agreement with the experimental value of 1.86. The effective mass for the hole orbit in κ-$(ET)_2Cu[N(CN)_2]Br$ was estimated to be only 0.28 m$_e$, much smaller than the value of 1.72m$_e$ in κ-$(ET)_2Cu(NCS)_2$. Experimental information on the Fermi surface properties of κ-$(ET)_2Cu[N(CN)_2]Br$ is much more difficult to obtain because of the

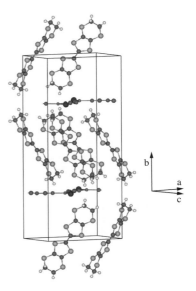

Fig. 7.12.
Crystal structure of κ-$(ET)_2Cu[N(CN)_2]Br$. The ET molecules are shown as dimers. The X molecule is in the plane perpendicular to the b axis.

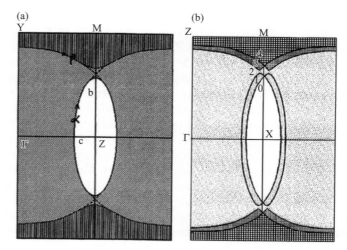

Fig. 7.13.
Calculated Fermi surface of: (a)
κ-(BEDT-TTF)$_2$Cu(NCS)$_2$, (b)
κ-(ET)$_2$Cu[N(CN)$_2$]Br (see original
references for details).

increased complexity in the crystal structure and its Fermi surface topology
(see Fig. 7.13 (b)). A plot of the electron charge density at the Fermi level
showed that electrons on the Fermi surface were localized mostly to the C=C
double bond in the ET molecule with non-equivalent weights due to the dimer-
ization of the ET stack. This was in complete agreement with the conclusion
from the ^{13}C NMR measurements on κ-(ET)$_2$Cu[N(CN)$_2$]Br.

It was pointed out that there is great similarity in the DOS near the Fermi
level between crystalline κ-(ET)$_2$Cu(NCS)$_2$ and the previously described crystalline K$_3$C$_{60}$. This is shown in Fig. 7.14. Conspicuously, there exists a positive
correlation between T$_c$, $N(E_F)$, the effective mass, and the proximity to the
van Hove singularities below the Fermi level in these two superconductors.
This pointed to the conclusion that the superconductivity in both crystals was
of BCS type in origin. Mulliken charge analysis indicated that on average the
ET molecule donates 0.45 electrons to Cu[NCS]$_2$. Analysis of the state wave
functions revealed that electron states at the Fermi level were derived from C
and S atoms in the ET molecule.

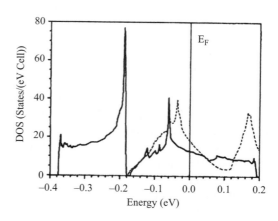

Fig. 7.14.
Calculated density of states of
k-(BEDT-TTF)$_2$Cu(NCS)$_2$ (solid line)
and K$_3$C$_{60}$ (dashed line).

The calculated DOS and partial DOS for different atoms and orbital components for the two organic superconductors were compared with data from x-ray fluorescence spectroscopy (Kurmaev et al., 1999). The measured Cu L_3, S $L_{2,3}$, C-K_α, N-K_α x-ray emission spectra (XES) near Cu 2p, S 2p, C 1s and N 1s thresholds were found to be in good agreement with the Cu (3d + 4s), S (3s + 3d), O 2p and N 2p partial DOS in the occupied valence band.

7.4.2 Fe-TCNE

Organic-based crystals with a general formula of $M^{II}[TCNE]_x \cdot zS$ represent a family of magnetic materials with interesting physical phenomena and potential uses in magnetoelectronic applications that may have performance improvements compared to their inorganic counterparts (Jain et al., 2007, Miller, 2007). In $M^{II}[TCNE]_x \cdot zS$, M is a transition metal (V, Mn, Fe, Co, Ni) and TCNE stands for tetracyanoethylene; S represent CH_2Cl_2. These organic compounds have magnetic ordering temperatures ranging from 44 K (M = Co, Ni) to 400 K [10] for M = V (Manriquez et al., 1991). It is generally believed that the magnetic ordering occurs via strong antiferromagnetic (AFM) exchange between the transition metal 3d and $[TCNE]^{\bullet -}$ π^* anion-radical unpaired spins (Gîrţu et al., 2000). Strong on-site Coulomb repulsion within $[TCNE]^{\bullet -}$ π^* and anti-ferromagnetic coupling between 3d electrons in M^{II} and $[TCNE]^{\bullet -}$ π^* play a critical role in their magnetic ordering. It is therefore of great importance that the electronic and magnetic structures of this class of compounds be investigated in detail both experimentally and theoretically. Although most of the study focused on the V-based organic magnets, only $[Fe^{II}(TCNE)(NCMe)_2][Fe^{III}Cl_4]$ or simply Fe-TCNE has a well determined crystal structure (Her, 2007). It has an orthorhombic cell containing two crystallographically non-equivalent types of Fe (Fe^{II} and Fe^{III}), seven types of C, four types of N, three types of Cl, and six types of H for a total of 112 atoms in the unit cell (space group *Pnam*, Z = 4) (see Fig. 7.15). The crystal shows a very complicated layered structure with two different forms of bent ($Fe^{II}(TCNE)^-$) layers interconnected by $[C_4(CN)_8]^{2-}$ ligands. The Fe^{II} bonds to six N in forming a canted octahedron relative to each other while the (Fe^{III})Cl_4 groups act like isolated molecules. The CH_2Cl_2 are solvent molecules with considerable disorder in the crystal and there is a large linear channel along the c-axis. Recent spin-polarized photoelectron spectroscopy (PES) shows large spin polarization for Fe-TCNE of 23% at the Fermi level (Caruso et al., 2009).

The spin polarized OLCAO method has been applied to investigate the electronic structure of F_E-TCNE crystal based on the LSDA approximation. The calculation used a full basis set consisting of atomic orbitals of Fe ([Ar] core plus 3d,4s,4p,5s,5p,4d), N (1s,2s,2p,3s,3p), C (1s,2s,2p,3s,3p) and H (1s,2s,2p). To achieve high accuracy, 90 *k* points in the irreducible portion of the Brillouin zone were used. The calculated spin-polarized band structure and DOS are shown in Fig. 7.16 (a) and 7.16 (b). It shows that Fe-TCNE is nearly a half-metal with a Fermi level close to the majority spin band from Fe from Fe^{II} and a small gap in the minority spin band. This is in line with the

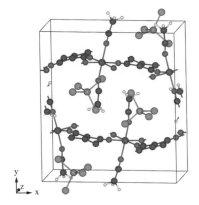

Fig. 7.15.
Sketch of the crystal structure of Fe-TCNE (see original references for details).

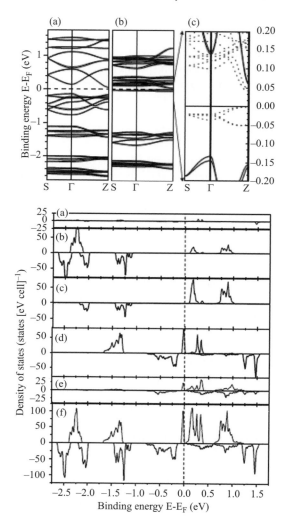

Fig. 7.16.
Calculated electronic structure of
Fe-TCNE. Upper panel: spin polarized
band structure: (a) spin minority, (b) spin
majority bands, (c) enlargement near the
Fermi edge along two high symmetry
directions. (- - -) is spin majority and
solid (−) spin majority. Lower panel:
total and partial DOS: (a) MeCN, (b) Cl,
(c) Fe^{III}, (d) Fe^{II}, and (e) TCNE. (f)
Total (see original reference for details).

experimental spin-polarized PES observation showing high spin polarization
at the Fermi level because the calculation is at zero temperature and thermal
effects are expected to reduce the 100% polarization for a half-metal electronic
structure. The calculation also indicated that F^{II}, Fe^{III}, and to a lesser extent
the N and C atoms, are antiferromagnetically coupled in this crystal.

These calculations were preliminary because strong intra-atomic correlations were not included. It is desirable to use the LSDA + U approach to
repeat the calculation similarly to the calculation for the YIG crystal described
in Chapter 6.

7.4.3   Herapathite crystal

Herapathite (HPT) is a complex organic crystal with a remarkable linear dichroism that has intrigued scientists for more than 150 years since its discovery in 1852 (Herapath, 1852). The strong linear dichroism in HPT and related structures has found numerous applications due to its polarizing properties. The electronic structure necessary for the explanation of its properties was unknown because its complex crystal structure was not fully resolved until recently (Kahr et al., 2009). HPT has a formula unit of $4QH_2^{2+}.C_2H_4O_2.3SO_4^{2-}.2I_3^-.6H_2O$ in an orthorhombic cell (space group $P22_12_1$, No. 18, $Z = 4$) where $Q = C_{20}H_{24}N_2O_2$ (quinine). Six water molecules are distributed over eight possible sites and the acetic acid molecule ($C_2H_4O_2$) is distributed over two possible sites. The crystal structure is shown in Fig. 7.17. The triiodide chains run along a clathrate channel formed by quinine molecules.

The calculation of crystalline HPT with 998 atoms in the unit cell is a great challenge in computational chemistry and physics. The electronic structure and linear optical properties of HPT were calculated using the OLCAO method (Liang et al., 2009). Figure 7.18 shows the calculated total DOS and the PDOS of group components: triiodide, quinine, sulfate, acetic acid, and water. The calculated HOMO–LUMO gap is in the range from 0.40 eV to 0.78 eV depending on the distribution of water and acetic acid molecules in the crystal. Within the range of -1.2 eV to $+2.0$ eV close to the HOMO–LUMO gap, the only states involved are those associated with I_3^- and quinine. Figure 7.19 shows the calculated $\varepsilon_2(\hbar\omega)$ up to 6 eV. The inset shows the averaged spectrum up to 40 eV in the far ultraviolet region. There is a huge doubled peak (marked as A and B) at 1.76 eV and 1.56 eV in the direction parallel to the b-axis of the crystal whereas absorptions in the perpendicular directions are negligible. The ratio of amplitudes for absorptions (or linear dichroism) in the orthogonal directions is almost 350! While anisotropic absorption in HPT crystal was described by its discoverer, the *ab initio* calculation provides the basis for this property in terms of its electronic structure. The broad peak at 15 eV originates from transitions in the quinine, sulfate, and solvent molecules.

The origin of the giant optical anisotropy in HPT was explained on the basis of detailed analysis of its electronic structure. Three factors should be considered for its proper interpretation: first, there are considerable interactions between the two I_3^- ions which breaks the symmetry so that a simple interpretation based on MOs of I_3^- with $C_{\infty h}$ symmetry is not adequate. Second, the two I_3^- ions are not strictly linear and their axial direction deviates from the z-direction (b-axis) by $\pm 12°$. Third, the I_3^- ions interact only weakly with the quinine molecules with some mixing from non-I components. By identifying all 22 molecular orbitals (MOs) (44 electrons) in the 2 I_3^- ions and the components of their wave functions, it was concluded that the strong transition peaks A and B in Fig. 7.19 originate from the two σ-like states at -0.39 and -0.18 eV with transition energies of 1.76 eV and 1.56 eV respectively. These two states, which give rise to the double peak in Fig. 7.19, originate from the splitting of the σ-like MO when the two I_3^- ions interact. They are dominated by the $5p_z$ orbitals with some mixing from $5p_y$ because the axial directions of

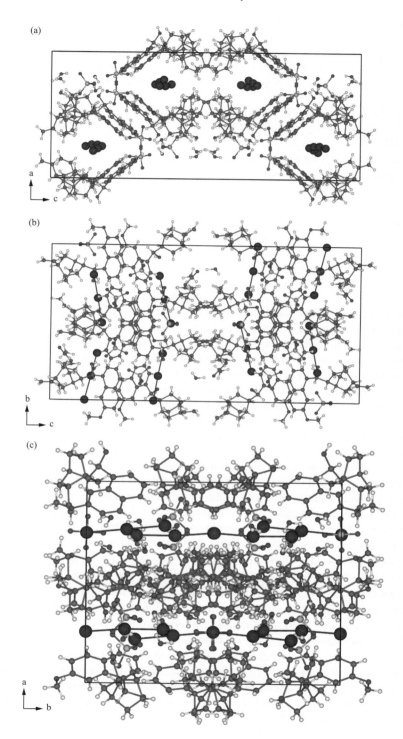

Fig. 7.17.
Crystal structure of HPT viewed in (a)
[010], (b) [100], and (c) [001] directions
(see original reference for details).

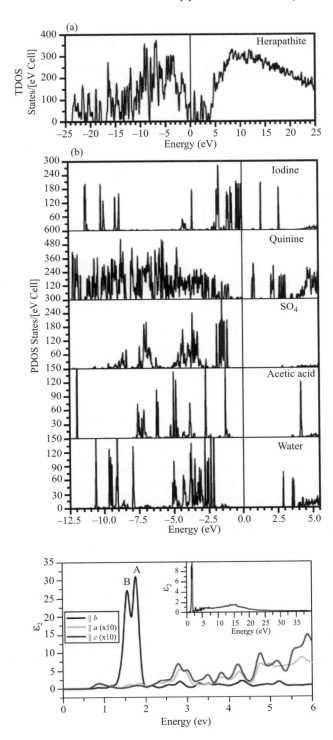

Fig. 7.18.
Calculated total (a) and partial (b) partial DOS of different groups in HPT crystal (triiodide, quinine, sulfate, acetic acid, and water).

Fig. 7.19.
Calculated $\varepsilon_2(h\omega)$ of HPT for components parallel and perpendicular to b axis. Inset: averaged spectrum up to 40 eV (see original reference for details).

the two I_3^- ions are not the same as the Cartesian z-direction in the calculation. The mixing allows for maximum overlap in the axial direction of I_3^- between the initial and final σ-like states resulting in large linear dichroism.

The effective charges Q^* on each ion in HPT were also calculated. For the six crystallographically non-equivalent I_3^- ions, the charges are −0.322, −0.071, −0.487, −0.382, −0.060 and −0.431 electrons for the six iodine atoms (I1, I2, I3, I4, I5, I6) respectively so that the ionic description of I_3^- for the tri-iodide chain (I1-I2-I3 or I4-I5-I6) is quite accurate. The central I atoms (I2 and I5) carry less charge which is typical for the 3-center 4-electron (3c, 4e) bonding in I_3^- (Landrum et al., 1997). However, the inter-I_3^- interactions in the chain are not negligible so we may view the I atoms in HPT as forming an infinite one-dimensional chain of I_3^- units running along the crystallographic b-axis.

7.5 Bioceramic crystals

Most bioceramics are phosphates. They are generally marked with very complex crystal structures. Non-stoichiometry, defect formation, and impurity contamination are the norm rather than exceptions. In this subsection we describe the application of the OLCAO method to two main groups of bioceramic crystals, the apatites and the tri-calcium phosphates.

7.5.1 Calcium apatite crystals

Fluorapatite (FAP) and hydroxyapatite (HAP) are two important bioceramic crystals belonging to the apatite family (MacConnell, 1973). HAP is known to be the primary constituent of the mineral portion of bones and teeth and, as such, plays a prominent role as a paradigm bioceramic material. FAP is known mainly as a mineral crystal used as a laser host material. Since F content in HAP is an important issue in biological processes related to surface absorption and diffusion, they are often studied together. There has also been a tremendous interest in the structures and properties of various proteins absorbed on HAP surfaces and their implications for various bioactivities (Hench, 1998, Koutsopoulos and Dalas, 2001).

The apatite crystal has a fairly complex structure with two formula units of $Ca_5(PO_4)_3X$, (X = F or OH) in the hexagonal cell (space group $P6_3/m$). For HAP, there are 44 atoms in the primitive cell with half occupation of the OH sites along the c-axis. There were some controversies concerning the orientation of the OH ions in HAP which are not totally settled, but the difference in the electronic structure due to OH orientation appears to be negligible. On the other hand, there is no such ambiguity in the position for the F ion in FAP. The crystal structure of HAP is illustrated in Fig. 7.20. There are two Ca sites in HAP and FAP, Ca1 and Ca2. The four Ca1 ions are usually labeled as columnar Ca because they form single atomic columns perpendicular to the basal plane. The six Ca2 ions are called the axial Ca because they create a channel defined by two groups of three Ca2 atoms with each forming a plane perpendicular to the c-axis. The two groups are slightly rotated with respect to each other

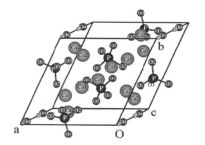

Fig. 7.20.
Sketch of the crystal structure of hydroxyapatite.

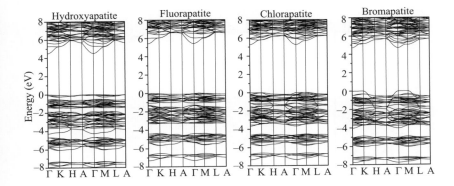

Fig. 7.21.
Calculated band structures for four apatite crystals HAP, FAP, ClAP, BrAP.

and the channel ions sit on the c-axis in the center of the Ca2 triangles. Ca1 loosely bonds to nine O ions whereas Ca2 bonds to six O ions. The PO_4 group is basically a tightly bonded tetrahedral unit within the crystal. Natural apatites have extensive substitutions for alkali, transition metal, and rare earth ions at both Ca sites.

The OLCAO method has been applied to calculate the electronic structures and optical properties of the four apatite crystals $Ca_{10}(PO_4)_6X_2$ (X = F, Cl, Br, and OH) (Rulis et al., 2004). Figure 7.21 shows the calculated band structures of these four apatite crystals. They are all wide band gap insulators with gap values ranging from 4.51 eV in HAP to 5.47 eV in FAP, and all have a very flat top to the valence band. The upper valence band show four peaks derived from O-2p orbitals while the lower valence band has multiple peaks. The HAP distinguishes itself from other apatite crystals by the presence of a sharp peak at the top of the VB originating from the OH ion. The conduction band states are dominated by the Ca ions. Mulliken effective charge and bond order calculations confirmed the strong covalent bonding in the PO_4 unit and weak ionic bonding between Ca and O. On average, Ca lost about 1.2 electrons to the surrounding ions. The PO_4 unit can be viewed as a negative ion having on average approximately an extra two electrons, thus the ionic description of -2 for PO_4 is different from the formal charge description of $(PO_4)^{-3}$. The calculated optical absorption spectra for HAP and FAP crystals are quite similar.

The OLCAO calculations on the HAP and FAP crystals were later extended to include (001) surface formation and XANES spectra. These will be discussed in Chapter 9.

7.5.2 α- and β-tricalcium phosphate

Tricalcium phosphate $(Ca_3(PO_4)_2)$ or TCP belongs to a large family of bioceramics that has shown many advantages in biomedical applications in clinical studies. TCP can accelerate the broken bone recovery process and facilitate growth of new bone tissue because of its biocompatibility with proteins that favor *in vivo* protein adsorption and cell adhesion. For that reason, TCP is the primary material for the development of the new generation of biomaterials and composite materials for special applications.

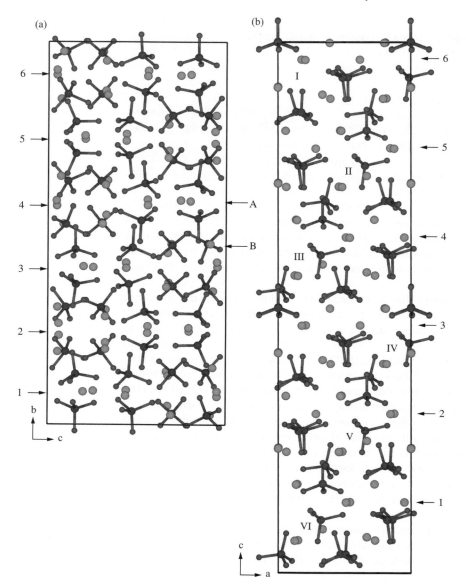

Fig. 7.22.
Crystal structure of: (a) α− and (b) β-TCP (see original reference for details).

There are two known phases of TCP crystals: α-TCP and β-TCP, both have extremely complicated structures containing several hundred atoms in the unit cell. Pure synthetic TCP crystals are not easy to grow because they often contain impurities such as Mg, Zn, and Si and are commonly porous. The bulk α-TCP crystal has a monoclinic cell containing 312 atoms per unit cell or 24 formula units of $(Ca_3(PO_4)_2)$ (space group $P2_1/a$). α-TCP can be viewed

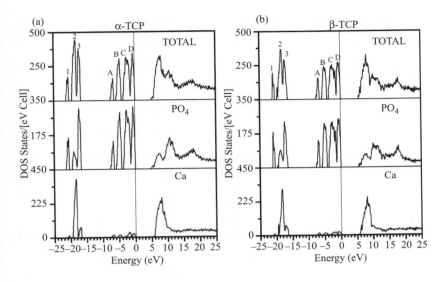

Fig. 7.23.
Calculated TDOS and PDOS of Ca and PO_4 group in α-TCP and β-TCP.

as consisting of two types of ion columns aligned along the (010) direction (see Fig. 7.22 (a)). Column A contains only Ca^{2+} ions whereas column B contains both Ca^{2+} and PO_4^{3-} ions. β-TCP (see Fig. 7.23 (b)) is reported to have a hexagonal cell (space group of $R3c$) with 273 atoms per unit cell or 21 formula units of $Ca_3(PO_4)_2$. In β-TCP, six Ca sites are half-occupied, implying that there should be three Ca vacancies in the unit cell. For electronic structure calculation, we modeled the structure of β-TCP by distributing the three Ca vacancies over the six half-occupied sites. The one with the lowest total energy was chosen as the most representative structure of β-TCP for the subsequent electronic structure studies.

The electronic structure and bonding for both α-TCP and β-TCP crystals were calculated using the OLCAO method. Figure 7.23 shows the calculated total and partial DOS for these two crystals. The gross features in the DOS are very similar to those of the other bioceramic crystals discussed above because they have similar structural units and composition. Both crystals are large gap insulators with direct band gap of 4.89 eV for α-TCP and 5.25 eV for β-TCP. Effective charge calculations show that on average, β-TCP has slightly less charge transfer per Ca than α-TCP.

The (010) surface model for α-TCP and (001) surface for β-TCP were also studied. They will be discussed in Chapter 9. Also studied were the mechanical properties of both TCP crystals using the VASP package within the plane wave pseudopotential method.

References

Abanin, D. A., Morozov, S. V., Ponomarenko, L. A., et al. (2011), *Science*, 332, 328–30.
Barros, E. B., Jorio, A., Samsonidze, G. G., et al. (2006), *Physics Reports*, 431, 261–302.
Caruso, A. N., Pokhodnya, K. I., Shum, et al. (2009), *Phys. Rev. B*, 79, 195202.

Castro Neto, A. H., Guinea, F., Peres, N. M. R., Novoselov, K. S. & Geim, A. K. (2009), *Reviews of Modern Physics*, 81, 109.

Charlier, J.-C., Blase, X. & Roche, S. (2007), *Reviews of Modern Physics*, 79, 677.

Ching, W. Y., Huang, M. Z., Xu, Y. N., Harter, W. G., Chan, F. T. (1991), *Phys. Rev. Lett.*, 67, 2045–48.

Ching, W. Y., Huang, M. Z. & Xu, Y. N. (1992), *Phys. Rev. B*, 46, 9910–12.

Ching, W. Y., Xu, Y.-N. & French, R. H. (1996), *Phys. Rev. B*, 54, 13546–50.

Ching, W. Y., Xu, Y.-N., Jean, Y. C. & Lou, Y. (1997), *Phys. Rev. B*, 55, 2780–83.

French, R. H., Meth, J. S., Thorne, J. R. G., Hochstrasser, R. M. & Miller, R. D. (1992), *Synthetic Metals*, 50, 499–508.

French, R. H., Parsegian, V. A., Podgornik, R., et al. (2010), *Reviews of Modern Physics*, 82, 1887.

Gîrţu, M. A., Wynn, C. M., Zhang, J., Miller, J. S. & Epstein, A. J. (2000), *Physical Review B*, 61, 492.

Greenwood, D. A. (1958), *Proceedings of the Physical Society*, 71, 585.

Hebard, A. F., Rosseinsky, M. J., Haddon, R. C., Murphy, D. W., & Glarum, S. H. (1991), *Nature*, 350, 600.

Hench, L. L. (1998), *Journal of the American Ceramic Society*, 81, 1705–28.

Her, J.-H., Stephens, P. W., Pokhodnya, K. I., Bonner, M., & Miller, J. S. (2007), *Angewandte Chemie International Edition*, 46, 1521–24.

Herapath, W. B. (1852), *Philos. Mag.*, 3, 161.

Huang, M. Z., Xu, Y. N., & Ching, W. Y. (1992a), *J. Chem. Phys.*, 96, 1648–50.

Huang, M. Z., Xu, Y. N., & Ching, W. Y. (1992b), *Phys. Rev. B*, 46, 6572–77.

Huang, M. Z., Ching, W. Y., & Lenosky, T. (1993a), *Phys. Rev. B*, 47, 1593–606.

Huang, M. Z., Xu, Y. N., & Ching, W. Y. (1993b), *Phys. Rev. B*, 47, 8249–59.

Huang, M. Z. & Ching, W. Y. (1994), *Physical Review B*, 49, 4987.

Iijima, S. (1991), *Nature*, 354, 56.

Jain, R., Kabir, K., Gilroy, J. B., Mitchell, K. a. R., & Wong, K.-C. (2007), *Nature*, 445, 291.

Kahr, B., Freudenthal, J., Phillips, S., & Kaminsky, W. (2009), *Science (Washington, DC, United States)*, 324, 1407.

Kelty, S. P., Chen, C.-C., & Lieber, C. M. (1991), *Nature*, 352, 223.

Koutsopoulos, S. & Dalas, E. (2001), *Langmuir*, 17, 1074–9.

Kroto, H. W., Heath, J. R., O'brien, S. C., Curl, R. F., & Smalley, R. E. (1985), *Nature*, 318, 162.

Kurmaev, E. Z., Shamin, S. N., Xu, Y. N., et al. (1999), *Phys. Rev. B*, 60, 13169–74.

Landrum, G. A., Goldberg, N., & Hoffmann, R. (1997), *Journal of the Chemical Society, Dalton Transactions Inorganic Chemistry*, 3605–13.

Lenosky, T., Gonze, X., Teter, M., & Elser, V. (1992), *Nature*, 355, 333.

Liang, L., Rulis, P., Kahr, B., & Ching, W. Y. (2009), *Phys. Rev. B*, 80, 235132.

Liang, L. Rulis, P., & Ching W. Y. (2010), *Acta Biomaterialia*, 6, 3763–71.

Lou, Y., Lu, X., Dai, G. H., Ching, W. Y., et al. (1992), *Phys. Rev. B*, 46, 2644–7.

Macconnell, D. (1973), *Apatite: Its Crystal Chemistry, Minerology, Utilization, and Geological Occurrences* (New York: Springer-Verlag).

Mackay, A. L. & Terrones, H. (1991), *Nature*, 352, 762.

Manriquez, J. M., Yee, G. T., Mclean, R. S., Epstein, A. J., & Miller, J. S. (1991), *Science*, 252, 1415–17.

Mcmillian, W. L. (1968), *Physical Review*, 167, 331.

Miller, J. S. (2007), *MRS Bulletin*, 32, 549.

Miller, R. D., Hofer, D., Rabolt, J., & Fickes, G. N. (1985), *Journal of the American Chemical Society*, 107, 2172–4.

Novoselov, K. S., Geim, A. K., Morozov, et al. (2004), *Science*, 306, 666–9.

O'keeffe, M., Adams, G. B., & Sankey, O. F. (1992), *Phys. Rev. Lett.*, 68, 2325.

Patnaik, S. S. & Farmer, B. L. (1992), *Polymer*, 33, 4443–50.

Popov, V. N. (2004), *Materials Science and Engineering: R: Reports*, 43, 61–102.

Rosseinsky, M. J., Ramirez, A. P., Glarum, S. H., et al. (1991), *Phys. Rev. Lett.*, 66, 2830.

Rulis, P., Ouyang, L., & Ching, W. Y. (2004), *Phys. Rev. B*, 70, 155104/1–155104/8.

Saito, R., Sato, K., Oyama, Y., Jiang, J., Samsonidze, G. G., Dresselhaus, G., & Dresselhaus, M. S. (2005), *Physical Review B*, 72, 153413.

Schellenberg, F. M., Byer, R. L., French, R. H., & Miller, R. D. (1991), *Physical Review B*, 43, 10008.

Schwarz, H. A. (1890), *Gesammelte Mathematische Abhandlugen* (Berlin: Springer).

E. Sohmen, Fink, J., & Krätschmer, W. (1992), *Europhys. Lett.*, 17, 51.

Tanigaki, K., Ebbesen, T. W., Saito, S., Mizuki, J., & Tsai, J. S. (1991), *Nature*, 352, 222.

Townsend, S. J., Lenosky, T. J., Muller, D. A., Nichols, C. S., & Elser, V. (1992), *Physical Review Letters*, 69, 921.

Tycko, R., Dabbagh, G., Rosseinsky, M. J., Murphy, D. W., Fleming, R. M., Ramirez, A. P., & Tully, J. C. (1991), *Science*, 253, 884–6.

Vanderbilt, D. & Tersoff, J. (1992), *Phys. Rev. Lett.*, 68, 511.

Williams, J. M. (1992), *Organic Superconductors (Including Fullerenes), Synthesis, Structure, Properties and Theory* (Eaglewood Cliffs, NJ: Prentice Hall).

Xu, Y. N., Huang, M. Z., & Ching, W. Y. (1991), *Phys. Rev. B*, 44, 13171–4.

Xu, Y. N., Huang, M. Z., & Ching, W. Y. (1992), *Phys. Rev. B*, 46, 4241–5.

Xu, Y. N., Ching, W. Y., Jean, Y. C., & Lou, Y. (1995), *Phys. Rev. B*, 52, 12946–50.

Zabel, H. & Solin, S. A. (eds.) (1992), *Graphite Intercalation Compounds Ii, Transport and Electronic Properties*, New York: Springer-Verlag.

Zheng, M., Jagota, A., Strano, M. S., et al. (2003), *Science*, 302, 1545–8.

8

Application to Non-Crystalline Solids and Liquids

This chapter is devoted to the application of the OLCAO method to non-crystalline systems and amorphous materials including frozen liquids. It is important to note that this type of system served as the original motivation for the OLCAO method. When the method was further developed to include self-consistency in the potential it was obvious that it could also be applied to complex crystalline systems with great ease. From a computational point of view, a large complex crystal is no different from a supercell model for non-crystalline solids. We will first briefly describe the early applications of the OLCAO method to amorphous semiconductors and insulating and metallic glasses before we move on to some more recent applications of current interest. The recent applications to non-crystalline materials require extensive structural modeling which is often accomplished by employing other *ab initio* methods. The last two sections on molten salts and concrete models are projects that are just now being initiated.

8.1 Amorphous Si and a-SiO$_2$

8.1.1 Amorphous Si and hydrogenated a-Si

Amorphous Si (a-Si) is a low cost photovoltaic material used in solar panels that generated tremendous research interest from the early 1970s that continues today. The first application of the OLCAO method to amorphous Si (a-Si) was with a 61-atom periodic continuous random tetrahedral network (CRTN) Henderson model (Ching and Lin, 1975, Ching et al., 1976). In these early days, diagonalization of a matrix equation of dimension 244×244 when only four orbitals ($3s, 3p_x, 3p_y, 3p_z$) per Si atom were used was considered to be a big accomplishment. The use of periodic boundary conditions in the model avoided artificial surface effects which are always present when finite clusters of atoms are used to represent amorphous materials. In spite of the crude nature of the potential, the truncated basis set used, and the fact that solutions were obtained at only four k points, the calculated DOS was actually in decent agreement with x-ray photoemission data. After the implementation of the orthogonalization scheme, the calculation was repeated and significant improvement was observed (Ching et al., 1976). Similar calculations were then applied to another periodic CRTN model, the 54 atom Guttman model,

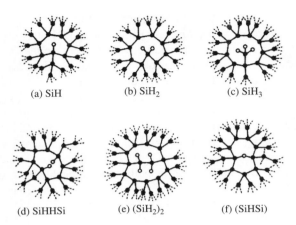

(a) SiH   (b) SiH$_2$ (c) SiH$_3$

(d) SiHHSi   (e) (SiH$_2$)$_2$ (f) (SiHSi)

Fig. 8.1.
Two-dimensional sketches (not to scale) of structural models for a-Si:H with six different H-bonding configurations as indicated.

which has smaller bond length and bond angle distortions (Ching et al., 1977). Interestingly, the CRTN models used in these studies were smaller than many of the complex crystals that were also being studied at this time. These works led to many subsequent OLCAO calculations on non-crystalline solids continuing on up to the present with the primary difference being that the periodic models used to mimic the infinite solid are now much larger and more realistic, exceeding thousands of atoms in some cases.

The calculation of the electronic structure of amorphous Si was extended to models of hydrogenated amorphous Si (Ching et al., 1979, Ching et al., 1980) and models containing defects such as O and F (Ching, 1980a, Ching, 1980b). It is well known that real amorphous Si contains many internal defects and broken bonds which are usually pacified by the incorporation of H to improve solar cell efficiency. To this end, a cluster type of model was hand-built and relaxed using a Keating potential instead of periodic CRTN models (Ching et al., 1979, Ching et al., 1980). Specific bonding configurations were built into the central portion of the cluster as illustrated in Fig. 8.1. These include the monohydride Si-H, dihydride SiH$_2$, trihydride SiH$_3$, saturated broken bond model (SiHHSi), polymeric model (SiH$_2$)$_2$, and bridging model (SiHSi). The calculated valence band local DOS of the H atoms in the broken bond model agreed well with photoemission experiments, which indicated that this was the most likely structure for H incorporation in hydrogenated a-Si. Because the cluster model was used, the LCAO method instead of the OLCAO method was adopted. Here, the outer shell of the cluster atoms contained only core states in the basis and no valence states as a cushion to minimize the surface effects. The cluster model was considered to be much less desirable than the periodic CRTN model but the inclusion of the defect configurations made it much more difficult to maintain the periodic boundary condition because of the directional aspect of the Si-Si covalent bonding.

8.1.2 Amorphous SiO$_2$ and a-SiO$_x$ glasses

The study of the CRTN model of amorphous Si naturally leads to the study of the amorphous SiO$_2$ glass (a-SiO$_2$). By inserting the bridging O atoms between all Si-Si pairs in the periodic model of a-Si, scaling the density to that of the glass, and relaxing the structure using a simple Keating type potential, an ideal periodic continuous random network (CRN) model for a-SiO$_2$ was created which circumvented the difficulty of maintaining the periodicity of the model for a system with directional bonding (Ching, 1982a). The model has no Si-Si bonds or dangling Si-O bonds and represents the ideal glass network structure most suitable for fundamental studies. The OLCAO method was applied to this 162-atom a-SiO$_2$ model generated from the Guttmann model (Ching, 1981, Ching, 1982b). The results show that the calculated DOS is very similar to the crystalline quartz (α-SiO$_2$) DOS and it is in very good agreement with experiment. This model is sufficiently large to enable us to evaluate the localization index (LI) based on the wave functions of the electronic states. A surprising finding was that while the states near the valence band edges were highly localized, as expected, those at the unoccupied CB edge were not localized, in contrast to what was believed at the time. This model was then utilized to study the electronic and vibrational structures of compressed a-SiO$_2$ for up to 20% volume reduction (Murray and Ching, 1989). In spite of the rather crude model for force calculations, the results were in qualitative agreement with experiments.

The above strategy of inserting O atoms between the Si-Si bonds in the a-Si model was used to create models for a-SiO$_x$ ($x = 0.5, 1.0, 1.5$) which were then used to model the Si-SiO$_2$ interface because it was a problem of great interest at the time (Ching, 1982a). Two types of a-SiO$_x$ models were constructed: the random bond model and the random mixture model. In the random bond model, the Si-O and Si-Si bonds in a-SiO$_x$ were randomly dispersed. In the random mixture model, the Si-O bonds and Si-Si bonds were each restricted to small clusters with no broken bond between them. Both types of models were then relaxed using the Keating potential as was done for a-Si and a-SiO$_2$. The electronic structure of both the random bond model and the random mixture model were calculated and the DOS was investigated in great detail according to the different local bonding configurations (Ching, 1982c). Through comparison with experimentally measured photoemission data, it was concluded that the random bond model was more appropriate for describing the structure of a-SiO$_x$ at the interface of SiO$_2$ and Si.

The above CRN model for a-SiO$_2$ was further developed into a much larger model of 1296 atoms and relaxed by using several different and more accurate potentials (Huang and Ching, 1996, Huang et al., 1999). The result was a nearly ideal topologically disordered CRN model with no broken bonds, small distortions in the Si-O bond length and O-Si-O bond angles, and a distribution of Si-O-Si bridging angles in close agreement with the experiment. This model is illustrated in Fig. 8.2. With increased computational power, the electronic structure and also the vibrational spectrum of this model were studied in great detail. It was shown that the low energy vibrational modes involve floppy motions of more than 10 atoms which have implications for the two-level tunneling model in glasses (Huang et al., 1999). By that time, the self-consistent version of the OLCAO method was fully developed and it could be applied to large systems. Figure 8.3 shows the calculated DOS of this

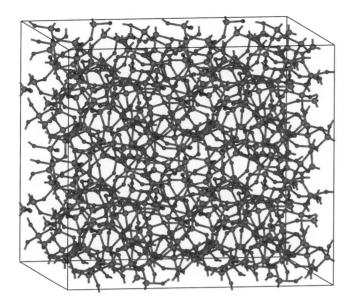

Fig. 8.2.
An amorphous SiO_2 glass model with 1296 atoms.

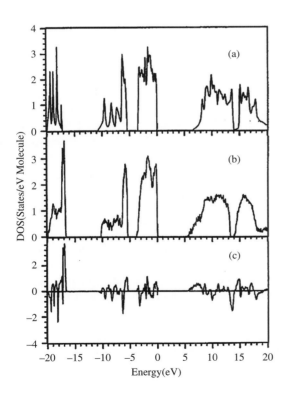

Fig. 8.3.
Calculated total DOS of: (a) α-quartz, (b) a-SiO_2, (c) difference between (b) and (a).

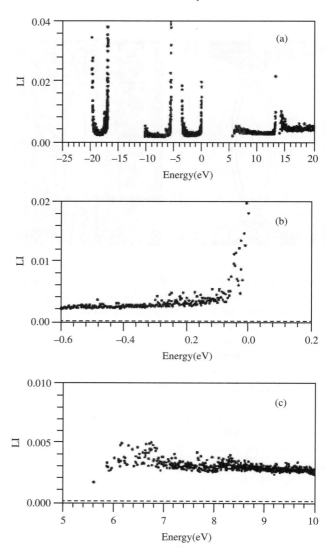

Fig. 8.4.
Localization index of electron states in
a-SiO$_2$. (a) For the entire range of states;
(b) for states at the top of the VB; (c) for
states at the edge of the CB. The dashed
horizontal line corresponds to $1/N =$
0.00013.

model in comparison with of that of α-quartz. From the calculated localization
index, the mobility edge at the VB edge was estimated to be less than 0.06
eV, much smaller than the mobility edge obtained from the smaller model. It
was also shown that there were absolutely no localized states at the CB edge
(see Fig. 8.4).

8.1.3　Other glassy systems

The early studies of a-SiO$_2$ glass using periodic models were soon extended to
studies of more complicated alkali-doped glasses such as $(Na_2O)_x(SiO_2)_{1-x}$
(Murray and Ching 1987, Murray et al., 1987) and sodium silicate glasses

with the addition of CaO (Veal et al., 1982). These were fairly small models, but the ionic bonds with the alkali ions provided the necessary flexibility in the bonding structure so that the periodic boundary conditions were easy to attain. Again, the DOS, the PDOS, the effective charges, and the localization index were calculated for these alkali-silicate models. Although these early non-self-consistent calculations used a rather crude overlapping of atomic potential model, the results were quite useful for providing an interpretation of experimental data. These studies showed the promise of the method and directly led to the far more accurate studies on complex models that are made by the OLCAO method today.

It was natural to question if a similar near-perfect CRN model could be created for another strongly covalently bonded glass such as amorphous Si_3N_4 (a-Si_3N_4). It turned out that the rigid requirements of maintaining tetrahedral bonding for Si and three-fold planer bonding for N while imposing periodic boundary conditions were too stringent. In spite of different trials and strategies (Ouyang and Ching, 1996), such a near-perfect CRN model for a-Si_3N_4 was not realized. Models generated by modifications from the crystalline phases of Si_3N_4 retain some of the crystalline order. This implies that a perfect CRN model for Si_3N_4 may not be possible and that it is more reasonable to expect the existence of some defective structures with non-ideal local bonding configurations in such materials. This experience later motivated us to use a molecular dynamics approach to obtain the initial models for disordered or non-crystalline structures and then later refine them into viable models for electronic structure calculations.

8.2 Metallic glasses

Metallic glasses constitute another class of non-crystalline solids with many applications. In contrast to the oxide glasses, the history of metallic glasses is relatively short and their structure and electronic properties are less well studied. Because metallic bonds are non-directional, it is far easier to constructed large periodic models for metallic glasses than it is to do so for a-SiO_2 or a-Si_3N_4. This point was quickly realized in the early 1980s when various a-SiO_2 and a-SiO_x models were investigated as described above. Modeling metallic glasses requires different strategies and puts a different focus on their electronic states. In this section, we describe several such applications to a few metallic glass systems.

8.2.1 Cu_xZr_{1-x} metallic glass

The first application of the OLCAO method to metallic glasses (MGs) was on the Cu-Zr system. This is one of the most common MGs with extensively reported experimental data but few theoretical calculations. The initial calculation used a small cell of only 39 atoms for the Zr-rich glass $Cu_{13}Zr_{26}$ (Jaswal et al., 1982) and 100 atoms for the Cu rich glass $Cu_{65}Zr_{35}$ (Jaswal and Ching, 1982). The sizes of the periodic cells were determined by the measured mass density of the samples. The models were then relaxed using the Monte Carlo technique (Metropolis et al., 1953) using a Lenard–Jones type of potential applicable to the random packing of hard spheres model for metallic glasses. The radial distribution function of the model was checked against experimental data and found to be satisfactory. A simple superposition of atomic potential model was used in these calculations. Nevertheless, the calculated DOS was in good agreement with experimental x-ray photoemission data and it showed large differences to the crystalline phase of the same composition. The focus of the study was primarily on the validity of the model. This was judged by the radial distribution function and the short range order coefficient, η_{AB}, in the binary MG as defined by Cargill and Spaepan (Cargill III and Spaepen, 1981), the shape of the calculated DOS, and its value at the Fermi level, E_F. The calculation was extended to Cu_xZr_{1-x} with a larger model of 90 atoms in the periodic cell (Ching et al., 1984). These early calculations on Cu_xZr_{1-x} MGs were very crude by today's standards. Very soon the calculations were extended to other MGs with larger models and more refined potentials.

8.2.2   Other metallic glasses

Apart from Cu_xZr_{1-x}, $Ni_{1-x}P_x$ is another MG that was intensely studied at this time. Specifically, electronic structure calculations of 100 atom models of $Ni_{75}P_{25}$ (Ching, 1985) and $Ni_{1-x}P_x$ (x = 0.15, 0.20, 0.25) (Ching, 1986) were performed. Mulliken charge analysis indicated that on average, P atoms gain 0.4 to 0.8 electrons from Ni for different x values ranging from 0.25 to 0.15. The high resistivity in $Ni_{1-x}P_x$ may be related to the existence of relatively localized states at the Fermi level and the calculated values of the DOS at E_F which were in good agreement with experimental values.

The magnetic properties of amorphous metals in amorphous Fe (a-Fe) (Xu et al., 1991, Zhong and Ching, 1994) and a-$Fe_{80}B_{20}$ (Ching and Xu, 1991) were investigated using the spin-polarized version of the OLCAO method with a much large model of 200 atoms. The model construction and relaxation were similar to that done for the other MGs but the spin-polarized calculation made it far more challenging. Figure 8.5 shows the spin-polarized DOS for the 200 atom model of a-Fe. The majority spin band has a peak at -0.5 eV while the minority spin band has a peak at 1.5 eV. The DOS at the Fermi level $N(E_F)$ for the minority spin band is three times as large as for the majority spin band. Figure 8.6 (top) shows a histogram distribution of the magnetic moments of the 200 Fe atoms in the a-Fe model. They range from 1.68 μ_B to 3.56 μ_B with the maximum at 2.25 μ_B and an average value of 2.46 μ_B. This is larger than the value of 2.15 μ_B calculated for the bcc crystalline Fe. The increased spin-magnetic moment, M_s, for Fe atoms in a-Fe is related to the larger exchange splitting in a-Fe as shown in Fig. 8.5.

In addition to the M_s, the orbital moments for a-Fe in the 200-atom model were also calculated from the spin-polarized wave function (Zhong and Ching, 1994). A spin-orbit coupling term was added to the Hamiltonian as described in Chapter 4. The distribution of the calculated orbital moments of

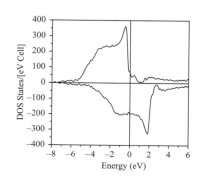

Fig. 8.5.

Calculated DOS for 200 atom model of a-Fe: Upper (lower) panel is for the spin-up (spin-down) states.

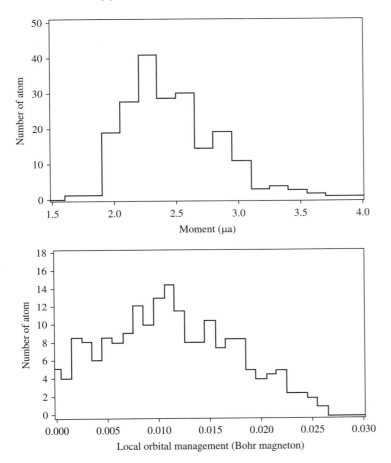

Fig. 8.6.
Distribution of spin moments M_S (top) and local orbital moments (bottom) in the 200 a-Fe model.

the Fe atoms is shown in Fig. 8.6 (bottom). They are widely distributed ranging from 0.0 to 0.026 μ_B with an average value of only 0.01 μ_B, much smaller than the 0.09 μ_B calculated for crystalline Fe. This low value for the average orbital moment in a-Fe indicates a strong quenching effect in the amorphous case and implies that the local spin moments may not have any preferred direction resulting in a random orientation of local moments in the metallic glass; a fact that has great implications for the ground state of a spin glass.

The spin-polarized OLCAO method was also applied to the a-$Fe_{80}B_{20}$ glass that was briefly discussed in Section 6.1.2 in Chapter 6. The average moment on each Fe atom was only 1.67 μ_B, smaller than that in a-Fe. The B atoms have opposite moments with an average value of -0.21 μ_B. This calculated value for the Fe moment is in excellent agreement with the measured value as illustrated in Fig. 8.7 which also includes the calculated Fe moments of crystalline Fe, FeB, Fe_2B, and Fe_3B discussed in Chapter 6 under Section 6.1.2.

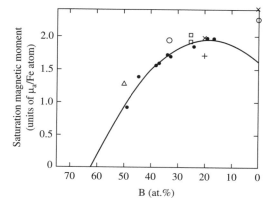

Fig. 8.7.

Average magnetic moment M_s per Fe atom in amorphous Fe-B films compared to the calculated values (see cited reference for details).

8.2.3 Transport properties in metallic glasses

The transport property of carriers in non-crystalline solids is an interesting and important subject (Mott and Davis, 1979). In metallic glasses, there are several intriguing phenomena such as negative temperature coefficients, the Mooij correlation for resistivity or thermopower, resistivity anomalies, the sign of the Hall coefficient, and so on that have attracted theorists wishing to provide an explanation (Zhao and Ching, 1989). The electronic structure of non-crystalline materials plays an important role in such explanations because many theories require specific parameters such as DOS at the Fermi level, the number of charge carries per atom, and elastic scattering time, which are not easy to obtain. In many cases, the choice of these parameters is influenced by the desire to obtain the anticipated results. It is therefore highly desirable to have accurate electronic structures for non-crystalline solids that can provide reliable parameters or at least set the boundaries for these parameters based on rigorous quantum mechanical solutions.

The electronic structure and transport properties of several metallic glasses were studied by the OLCAO method in the late 1980s using supercell models as described above. Three MG systems with very different characteristic and experimental observations were chosen for transport properties calculations (Ching et al., 1990, Zhao and Ching, 1989, Zhao et al., 1990): (1) a-Ni, a single component metastable transition metal MG; (2) a-$Mg_{1-x}Zn_x$, a free electron-like MG; and (3) a-$Cu_{1-x}Zr_x$, a strong scattering MG system with high resistivity. The procedures for such calculations were described in the method section in Chapter 4. The central quantity to be calculated was the conductivity function σ_E (Eq. 3.31) of the MG from which the dc conductivity due to elastic scattering of electrons near the Fermi level could be evaluated (Eq. 3.32). One can even go a step further to evaluate the thermopower S(T) which is sensitive to the curvature of σ_E near the Fermi level. The calculated transport properties of these three different classes of MGs were in surprisingly good agreement with limited experimental data in spite of the rather small supercell used to represent the amorphous structure (Zhao et al., 1990). Figure 8.8 shows the calculated σ_E and the average electron mobility $|D(E)|^2_{av}$

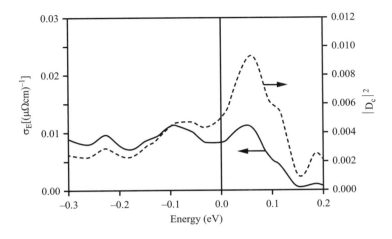

Fig. 8.8.
Conductivity function σ(E) (left-hand scale) and average mobility D(E) (right-hand scale) for *a*-Ni calculated with 100-atom model.

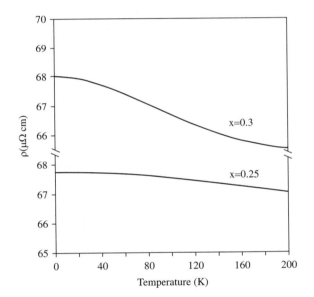

Fig. 8.9.
Calculated ρ(T) for *a*-Mg$_{70}$Zn$_{30}$ and *a*-Mg$_{75}$Zn$_{25}$.

approximately defined as $\langle \sigma_E \rangle = |D(E)|^2_{av}[N(E)]^2$ where N(E) is the DOS near the Fermi level in a-Ni. The Fermi level lies at a shallow minimum in σ_E and the calculated resistivity of 110($\mu\Omega$cm^{-1}) at 200 K was close to the reported experimental value (Zhao et al., 1990). Figure 8.9 shows the calculated temperature dependent resistivity in a-Mg$_{70}$Zn$_{30}$ and a-Mg$_{75}$Zn$_{25}$. The calculated resistivity values for a-Mg$_{70}$Zn$_{30}$ and a-Mg$_{75}$Zn$_{25}$ at 2K were 68 and 67.8($\mu\Omega$cm) respectively and were in line with experimental data. The results for the strong scattering a-Cu$_{1-x}$Zr$_x$ MG were equally encouraging. The calculated resistivity values of 197($\mu\Omega$cm) and 192 ($\mu\Omega$ cm) for a-Cu$_{60}$Zr$_{40}$ and a-Cu$_{50}$Zr$_{50}$ respectively were in reasonable agreement with the measured values in the range of 180–250 ($\mu\Omega$ cm). More importantly, the resistivity data

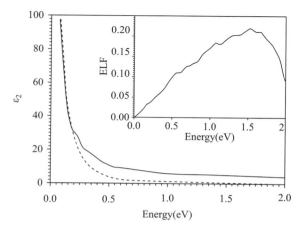

Fig. 8.10.

Calculated imaginary dielectric function for the graphitic carbon model. The dashed line is the Drude model. Inset shows the energy-loss function obtained from the complex dielectric function.

shows a negative gradient in a-$Cu_{1-x}Zr_x$ in agreement with observation (Zhao et al., 1990).

The OLCAO method was applied to study the transport properties of another non-crystalline solid using the same procedures, the Townsend model of amorphous graphitic carbon (Huang and Ching, 1994) which was briefly discussed in Section 7.1.2. The Townsend model with 1160 atoms is entirely different from the MG models discussed above. The system contains purely s and p electrons. The calculation showed that there was no minimum at E = 0 in σ_E and a positive temperature coefficient was obtained. The calculations also showed a positive thermopower at low temperature and high resistivity values of 1160 ($\mu\Omega$ cm) and 1235 ($\mu\Omega$ cm) at 0 K and 95 K respectively. Because the Townsend model is a hypothetical graphite network structure that may or may not exist in nature, there are no experimental data to compare.

An important aspect of supercell calculations for the linear optical properties in the Townsend model was that it enabled us to obtain the absorption curve at very low transition energies because there was no distinction between interband transitions and intra-band transitions. This is shown in Fig. 8.10. By fitting to the Drude expression at low energy, it is possible to estimate the relaxation time τ for the electron–electron scattering in a non-crystalline solid model. For the Townsend model, it was estimated that the Fermi velocity and electron mean free path are 2.61×10^4 m/s and 0.63 Å respectively.

8.2.4 Recent efforts on metallic glasses

As alluded to earlier, metallic glasses comprise a special material class with fascinating properties and increasing numbers of applications. From a fundamental point of view, the existence of both short range order (SRO) and intermediated range order (IRO) makes this class rather unique among the non-crystalline solids (Sheng et al., 2006). Recent work even reveals the existence of long range topological order in metallic glasses (Zeng et al., 2011). There have been many recent efforts to investigate the nature of the IRO and the fundamental role it plays in governing the material properties, especially the glass forming ability and the mechanical properties (Liu et al., 2010). It is generally believed that the formation of shear bands in metallic glasses can be at least partially explained by special atomic arrangements present in some forms of IRO (Peng et al., 2011). Most current research efforts appear to focus primarily on the classification of the geometric arrangements of the atoms involved in the IRO. There are many different and quite advanced models of edge sharing, corner sharing, and free volume enclosing clusters of atoms that are characterized by minute but identifiable features in the radial distribution functions. While such efforts do provide useful information, any material properties should ultimately be related to inter-atomic interactions based on quantum mechanics. This can be evidenced by the fact that a small addition of a new element in a multi-component metallic glass can alter its glass forming ability and properties significantly. In this respect, first-principles electronic structure calculations on well characterized metallic glass models are very important.

The OLCAO method is well suited to investigate the electronic structure of metallic glasses at a large scale due to its flexibility and efficiency. Because large models of up to several thousand atoms can be handled, the presence of different types of intermediate range order found in metallic glasses with different compositions and atomic species can be accommodated. Figure 8.11 shows a model of a $Cu_{0.5}Zr_{0.5}$ metallic glass with 512 atoms in a cubic cell with the calculated partial DOS beside it. The model was initially generated

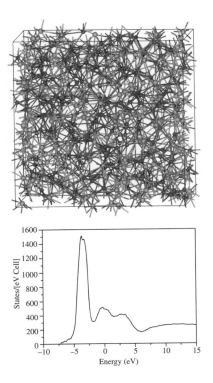

Fig. 8.11.
A 512-atom model of amorphous $Cu_{0.5}Zr_{0.5}$, along with its DOS.

by classical molecular dynamics and then fully relaxed using VASP. The electronic structure of this model is currently under study using the OLCAO method with a sufficiently large basis set. From the calculated partial density of states, the effective charges on each atom or group of atoms, and patterns of variation in the bond order values in relation to geometrical analysis of the atomic arrangements in the model, a deeper understanding of the fundamental properties can be ensured.

8.3 Intergranular glassy films

One of the most common defective features in structural ceramics, such as the silicon nitride and silicon carbide families, is the existence of thin intergranular glassy films (IGFs) between polycrystalline grains that result from the liquid phase sintering at high temperature. These IGFs have thicknesses of roughly 1–2 nm and are interconnected at the glassy triple junctions. The glass phase contains different proportions of Si, O, and N ions with far more complicated local bonding structures than are present in the crystalline part. The addition of rare earth elements from the sintering aids can drastically alter the structure and properties of the IGFs. Because the IGFs control the overall mechanical properties of the structural ceramics and play a critical role in device applications, there has been great interest in understanding the structure and properties of IGFs both experimentally and theoretically.

We consider IGFs to be a type of complex microstructure in a ceramic material and as such to belong to a non-crystalline system that should be covered in this chapter. Like the insulating glasses or metallic glasses discussed above, electronic structure calculations on IGFs require large supercells. Unlike the pure glassy models, the construction of periodic supercells containing IGFs is far more challenging. In this chapter we describe the application of the OLCAO method to three different IGF models in β-Si_3N_4 depending on the orientation of the crystal planes. These three models have the glass phase sandwiched between two basal planes (0001), two prismatic planes (10–10) and one basal and one prismatic plane of the hexagonal β-Si_3N_4 crystal. Although the mechanical and elastic properties of the IGF models are of great importance and were the prime motivation for the study, we focus in this chapter only on the results related to the electronic structure and bonding in the IGF models.

8.3.1 The basal model

The construction of the initial IGF model with basal planes started with the use of classical molecular dynamics (MD) with multi-atom potentials (Xu et al., 2005). Analysis of the pair distribution function for atoms in the IGF region showed the first peak at 1.63 Å and the second smaller peak at 1.73 Å corresponding to Si-O and Si-N pairs in the glassy region. This was consistent with experimental findings. This initial model was then further relaxed using VASP (Vienna *ab initio* simulation package) to obtain a more refined structure. The final periodic model had a dimension of 22.62 Å $\times$ 13.04 Å $\times$ 29.87 Å and

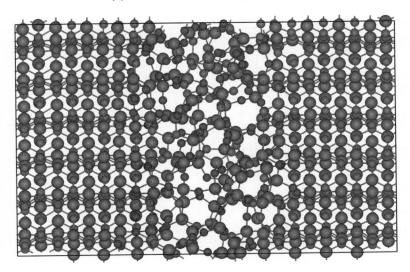

Fig. 8.12.
Sketch of the 798-atom basal model of
IGF in β-Si₃N₄.

798 atoms (322 Si, 336 N, and 140 O) and will be referred to as the basal-798 model. The basal-798 model has an IGF thickness of about 10 Å and is oriented perpendicular to the z-direction (see Fig. 8.12). A special characteristic of the basal-798 model is that the two inter-phase boundaries between the crystalline region and the glassy region are not well defined and sharp. This is partly due to the difficulty of constructing the initial model using MD while maintaining the periodic boundary conditions and ensuring structural stability. As a result, there are defective atoms within the crystalline layers close to the phase boundary and the layers even contain a few O ions. Within this less precisely defined IGF region, there are 76 Si, 112 O, and 50 N atoms corresponding to an effective $N/(N + O)$ ratio of 31% consistent with what is found in real samples.

The basal-798 model is of a reasonable size so that the OLCAO method can be effectively applied to investigate its electronic structure and bonding (Xu et al., 2005). The calculated total DOS (TDOS) and partial DOS (PDOS) for atoms in the bulk crystal region and the IGF region are shown in Fig. 8.13. Also shown are the TDOS and PDOS near the VB and CB edges. The basal-798 model is an insulator with a sizable LDA gap of about 3 eV. The PDOS of the atoms in the bulk region are the same as in β-Si₃N₄ and those in the IGF region resemble those of crystalline Si₂N₂O. The main difference is a wider O 2s band reflecting the glassy nature of the IGF. The defective structures at the inter-phase boundaries induce some defective states near the band edge (Fig. 8.13 (b)). A shallow acceptor level 0.1 eV above the VB edge can be traced to an N atom at the interface which was only two-fold bonded with short Si-N bond lengths. There are also at least three defect-related states at the bottom of CB which are associated with a three-fold bonded atom in the near-bulk region and another under-bonded Si atom in the IGF region connected to one N and two O ions. All these indicate that the OLCAO method was able to identify the atomic origin of particular states in a microstructure such as an IGF.

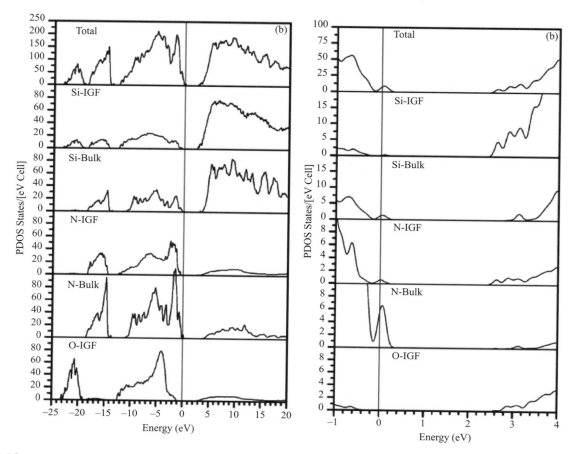

Fig. 8.13.
Calculated DOS and PDOS of the basal model of Fig. 8.12 (see original reference for details).

To investigate the effect of rare earth doping in IGFs, the consequence of Y doping in the basal-798 model was studied. Experimental evidence suggested that the rare earth ions are most likely to reside near the crystal/IGF boundaries. Several models were built in which 16 Si atoms (8 on each side of the IGF) were replaced by 16 Y atoms with simultaneous substitutions of 16 N by 16 O to maintain charge neutrality yielding $N/(N + O)$ and $Y/(N + O)$ ratios of 0.22 and 0.10 in the IGF region. The models were again fully relaxed using VASP and the one with the lowest energy was chosen as a representative Y-doped IGF model. The electronic structure and the mechanical properties of this Y-doped model were studied (Chen et al., 2005). The results indicate that Y-doping significantly enhances the mechanical properties of the IGF. More interestingly, it shows that the electrostatic potential difference across the film is increased by Y doping (see Fig. 8.14). Such calculations provide useful insights on issues related to the space charge model in ceramics with microstructures.

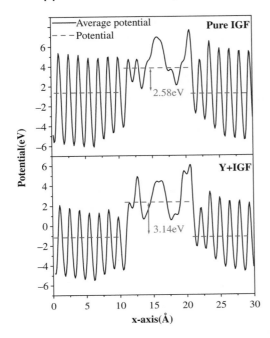

Fig. 8.14.
Electrostatic potential across the basal IGF at zero strain for (a) pure and (b) Y-doped IGF (see original reference for details).

8.3.2   The prismatic model

The second IGF model studied was the prismatic model. Experimental high resolution TEM pictures show that most of the needle-like polycrystalline grains in β-Si$_3$N$_4$ grow more easily in the (001) direction meaning that it is easier to form IGFs with the prismatic planes. A new protocol based on a multi-step simulated annealing technique was used to build the prismatic model which contains many fewer defective sites at the IGF/crystal boundary than the basal-798 model and which has a well defined inter-phase boundary (Ching et al., 2010). This new 907-atom model (hereafter referred to as the prismatic-907 model) has a dimension of 14.533 Å × 15.225 Å × 47.420 Å and an IGF region that is about 16.4 Å wide (see Fig. 8.15). The boundary between the crystal part and the IGF part was defined as the line between the top Si layer and the N layer closest to the IGF films. Using this reasonable criterion, the IGF portion contains 72 Si atoms, 32 N atoms, and 124 O atoms with a N/(N + O) ratio of 0.21. Within the IGF, the majority of the Si atoms are four-fold bonded to either 4 O (Si-O_4) or 3 O and 1 N (Si-NO$_3$). There are other four-fold bonded Si atoms in the form of Si-N$_2$O$_2$, Si-N$_3$O, and Si-N$_4$ as well as a few under-bonded Si (Si-NO$_2$, Si-N$_2$O) and one over-bonded Si (Si-O$_5$). For

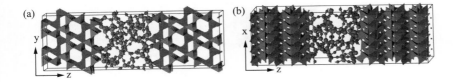

Fig. 8.15.
Sketch of the 907-atom prismatic model of IGF in β-Si$_3$N$_4$ in two orientations.

the anions, there are 7 N atoms that are two-fold bonded, only 2 O atoms that are three-fold bonded to Si, and one 1 O with a dangling bond. A remarkable feature is that there are no bonds between anions (N-N, N-O, or O-O bonds) so that the structure in the IGF region resembles that of a continuous random network model with defective features.

The electronic structure and bonding of the prismatic-907 model was studied using the OLCAO method similar to the case for the basal-798 model (Ching et al., 2010). Fully self-consistent calculation of a non-crystalline model of

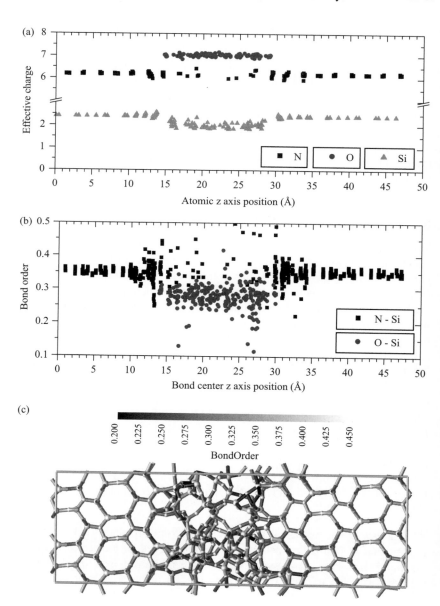

Fig. 8.16.

Calculated electronic structure of the prismatic IGF model: (a) Mulliken effective charges Q^* for Si, N, and O along the z axis; (b) bond order values along the z axis and averaged over the $x - y$ plane. (c) Same as (b) but illustrated in a color scheme (see cited reference for details).

this size is a considerable challenge. The calculated total and partial DOS show the presence of occupied defect-induced states at the top of the VB and at the bottom of the CB. These defect states can be traced to the presence of the under- and over-coordinated atoms in the IGF region. The LDA band gap ($\sim$ 4.0eV) is larger than that of the basal-798 model and close to that of crystalline β-Si_3N_4. The Mulliken effective charges Q^* on all atoms in the prismatic-907 model and the bond orders between all pairs were calculated and are shown in Fig. 8.16. The average Q^* values for Si and N in the bulk crystalline region are 2.41 and 6.19 electrons respectively, whereas in the IGF region, they are 2.02 and 6.18 electrons respectively. The effective charge for the O in the IGF is 7.03 electrons. These data indicate that the IGF is more ionic than the bulk crystalline part. The variations of Q^* within the IGF can be fairly large showing the disordered nature of the glassy region. Figure 8.16 also shows the calculated BO values which reflect the bond strengths for all the nearest-neighbor bonded atomic pairs. Within the IGF, very strong and very weak bonds exist depending on bond lengths and bond angles. The Si-O bonds are generally weaker than the Si-N bonds. The existence of very weak bonds has implications for the mechanical properties and the fracture behavior of the IGF under tensile strain discussed in (Ching et al., 2009) and (Ching et al., 2010).

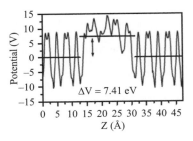

The electrostatic potential difference in the prismatic-907 model was also investigated using the self-consistent potential obtained from the OLCAO calculation evaluated on a dense $50 \times 50 \times 300$ mesh. Averaged over the x-y plane, the electrostatic potential along the z-axis shows that the average potential in the IGF region is higher than in the bulk crystalline region by 7.41 eV (see Fig. 8.17). This is much higher than that for the basal-798 model shown in Fig. 8.14. This clearly shows that the electrostatic potential difference between the IGF and the crystal depends on both the crystalline orientation as well as the chemical composition of the atoms in the IGF.

Fig. 8.17.
Calculated electrostatic potential of the prismatic IGF model along the z axis and averaged over the x–y plane. The aero potential is set at the averaged potential in the bulk region. The horizontal line indicates the averaged value in the IGF region.

Another remarkable application of the OLCAO method for IGF studies is the calculation of the spectroscopic properties of the IGF models. Figure 8.18 shows the calculated directionally resolved real and imaginary parts of the dielectric function of the prismatic-907 model in comparison with that of crystalline β-Si_3N_4. For the IGF model, it has a sharp peak at 9.0 eV in the xx component. The same sharp peak is present in the zz component of the β-Si_3N_4 crystal along the c-axis of the hexagonal cell. This is the same direction as the z axis of the prismatic-907 model. So the strong anisotropy in the IGF model is not due to the presence of the IGF layer. It is related to the crystalline orientation of the bulk β-Si_3N_4. Similar conclusions were reached in the calculated longitudinal sound velocities in the prismatic-907 model (Ching et al., 2010).

In additional to the linear optical absorptions in the prismatic model, the XANES/ELNES spectra of all atoms in the IGF region (Si–K, Si–L, N–K, and O–K edges) were calculated. This gave a sufficiently large sample of spectra for statistical analysis (Rulis and Ching, 2011) and will be discussed in Chapter 11.

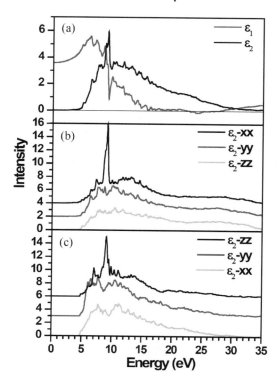

Fig. 8.18.

Calculated optical dielectric function of the prismatic IGF model and crystalline β-Si$_3$N$_4$: (a) real $\varepsilon_1(\hbar\omega)$ and imaginary $\varepsilon_2(\hbar\omega)$ parts for the IGF model; (b) $\varepsilon_2(\hbar\omega)$ of the IGF model resolved in the three direction (x, y, z). (c) $\varepsilon_2(\hbar\omega)$ of crystalline β-Si$_3$N$_4$ resolved in the three directions (x, y, z).

8.3.3   Prismatic-basal model (Yoshiya model)

Both the basal and the prismatic IGF models suffer the same drawback. They have the same crystallographic orientation on both sides of the IGF because of the need to maintain the periodic boundary condition. High resolution TEM images clearly show that the polycrystalline grains on each side of the IGF have different orientations. Yoshiya et al. built a large IGF model using classical MD techniques (Yoshiya et al., 2002) which has the basal plane on one side of the IGF and the prismatic orientation on the other side (hereafter referred to as the Yoshiya model). To maintain the periodic boundary condition, two oppositely oriented IGFs needed to be built into the model which made it about four times as large as the basal or the prismatic models discussed above. The Yoshiya model is shown in Fig. 8.19. It has a dimension of 23.022 Å × 23.028 Å × 82.359 Å, and contains a total of 3864 atoms. The IGF regions (two of them) are approximately 18.0 Å in width with 313 Si, 244 N, and 504 O atoms and a N(N + O) ratio of 0.33. The dangling bonds present in this model are minimal and it represents a more realistic IGF model in polycrystalline silicon nitride.

With an IGF model that large, it is not possible at this moment to relax it with *ab initio* codes or to do a self-consistent calculation with the OLCAO method due to limitations on the computational facility. However, we took the next best possible approach to study its electronic structure and bonding. In the crystalline part of the model, we used the site-decomposed Si and N potentials

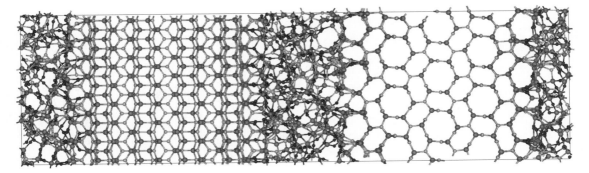

Fig. 8.19.
Sketch of the Yoshiya basal-prismatic IGF model.

from self-consistent calculations of β-Si$_3$N$_4$. In the IGF region, we used the Si, N, and O potentials from the orthorhombic Si$_2$N$_2$O crystal calculation. This strategy has been used in the early studies of the insulating and metallic glasses discussed in Sections 8.1 and 8.2. The results obtained from using this simplified potential are still highly reliable because in the OLCAO method, all the potential and charge densities are expressed in the form of atom-centered functions (see Chapter 3). The inter-atomic interactions due to local variations in the structure are fully reflected in the calculation which would not be the case if the approximation to the crystalline results were at the level of interaction parameters. For the basis set, we used an augmented minimal basis set that included the 3s and 3p orbitals of O and N and the 3d orbitals of Si. This is again a very reasonable approximation since we are mostly interested in the effective charge and bond order values and the electron states close to the gap region. Even with this approximation, the secular equation to be solved for the Yoshiya model, after core orthogonalization, is 38832 × 38832. Its full diagonalization is truly a formidable task. On the other hand, it also demonstrates the flexibility and the efficiency of the OLCAO method showing that it is capable of addressing some of the most complex material structures.

Figure 8.20 shows the calculated DOS of the Yoshiya model and the scattered plot of the calculated effective charges across the z-axis. The model shows an insulating gap in the range of 3–4 eV with defect-induced states at both the CB and the VB edges. The distribution of the effective charges shows a range of scattered values for atoms in the IGF and there is a gradual reduction in the range of variations from the IGF to the center of the bulk region. These results may change slightly if the Yoshiya model is further relaxed by *ab initio* methods but the essential features of the result are expected to remain the same.

The calculation of the electronic structure and bonding of the Yoshiya model is probably the largest model that the OLCAO method has been applied to so far. Obviously, it points to the direction of further applications to other similar or even larger IGF models with different characterizations and with various amounts of rare earth doping. Such calculations were dreams in the past and are now within reach. Using a similar strategy, the OLCAO method can be

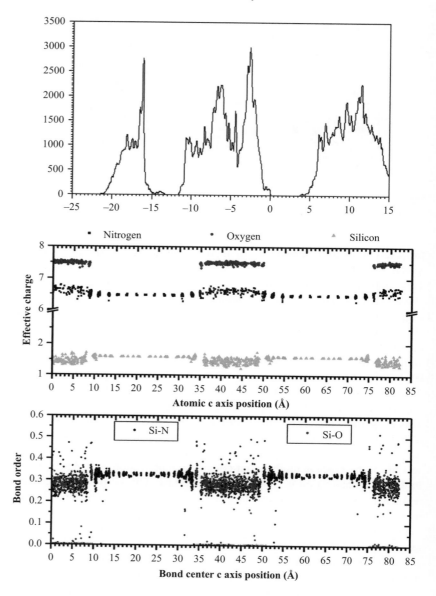

Fig. 8.20.
Calculated DOS (upper), effective
charges Q* (middle), and BO values
(lower) of the Yoshiya model.

made to study nano-particles and nano-structures in large models of several
thousand atoms and beyond. This is sufficiently large to make the structures
realistic for such systems.

8.4 Model of bulk water

A large portion of the condensed matter on the surface of the Earth is in the
form of liquids and liquids represent an important component of condensed

matter theory. A frozen liquid can be considered as an amorphous solid very much like how glasses are considered to be solids that have condensed from the liquid state upon cooling. The electronic structure of liquids is a subject of great importance and can be quite different from that of solids. In this section, we briefly describe the application of the OLCAO method to a model of water which is arguably the most well studied liquid of all time.

Water is one of the most common and intriguing substances on Earth yet it is far from being fully understood. The existence of a network structure in water is a fundamental issue that has not been fully resolved. At issue is the nature of the hydrogen bonds (HBs) which determine the local configurations of the network structure through intermolecular interactions (Lyakhov and Mazo, 2002). Several conflicting models regarding the network characteristics of water have been proposed based on the ways that experimental neutron scattering and x-ray absorption near edge structure (XANES) spectral data were interpreted. The main point of contention is whether the molecules in water form a network structure primarily composed of tetrahedrally hydrogen bonded H_2O or of one-dimensional filamentous structures.

A sufficiently large model for bulk water was constructed using a combination of different simulation methods involving both classical and *ab initio* MD and simulated annealing followed by accurate relaxation using VASP (Liang et al., 2011). The multi-step procedure produced a final model ($a = 21.6641$Å, density: 1gm/cc) for supercooled bulk water at ambient conditions. It contained 340 water molecules and very small variations in the O-H bond length and H-O-H bond angle such that the standard deviations from the mean were 1.00 ± 0.006Å and $106.32° \pm 2.528°$ respectively. The calculated radial distribution functions (RDF) for the O-H, H-H, and O-O pairs were in good agreement with data from neutron diffraction experiments. A ball and stick picture of the final relaxed water model is shown in Fig. 8.21.

The electronic structure of the 340-atom bulk water model was calculated using the OLCAO method. Figure 8.22 shows the calculated DOS and PDOS. The model has a HOMO–LUMO gap of 4.51 eV which may be slightly underestimated because of the LDA approximation. The element resolved partial DOS shows the strongly covalent nature of the bonding between O and H within the H_2O molecule. It is characterized by a sharp peak at -1.15 eV from the O lone pair and two other peaks at -6.64 eV and -18.68 eV. The bond order values between all pairs of atoms in the model were calculated using the Mulliken scheme as described in Chapter 4 (Eq. 4.5).

In liquid water, HBs (represented as O ... H) are much weaker than the O-H covalent bonds within the molecule yet they are sufficiently strong to dictate the intermolecular interaction. There are several different criteria that are commonly used to define HBs including using the local minimum in the potential curve as the upper limit; a geometric cutoff based on different sets of atomic configuration parameters, the near-neighbor hydrogen and oxygen criterion, the use of bond order (BO) values, and so on. The BO value characterizes the overall bond strength between two atoms and is far more precise than other criteria based purely on geometric parameters because it is derived from the quantum mechanical wave functions. Figure 8.23 displays the BO distribution

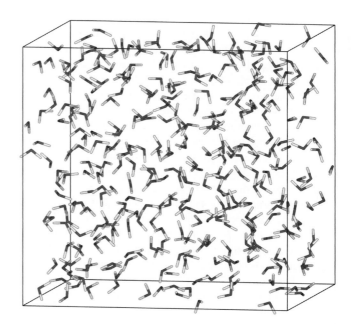

Fig. 8.21.

The bulk water model with 340 H_2O molecules.

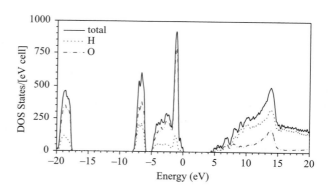

Fig. 8.22.

Total and partial DOS for the bulk water model.

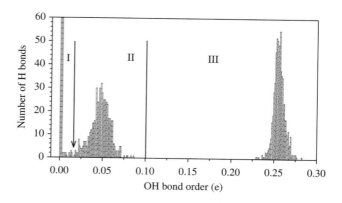

Fig. 8.23.

Distribution of H bond in the water model (see cited reference for details).

for (O, H) pairs with separations of less than 3.5 Å. Clearly, the bonds in the lower portion with BOs less than 0.1 are considered to be the HBs. The bonds in the upper portion, with BOs greater than 0.225 having a narrow peak centered at 0.27, are those of the covalent bonds within the H_2O molecules. According to this distribution, the HBs are the O ... H pairs with BO values in the range of 0.015 to 0.1 centered approximately at 0.05. The lower end of the distribution constitutes the weak HBs and the high end of the distribution represents the stronger HBs.

According to the calculated BO values for all the atoms in the model, the 340 water molecules can be divided into four groups with 2, 3, 4, and 5 HBs in which both the donor and acceptor hydrogen bonds are counted. The group with 4 HBs (2 donors and 2 acceptors) is the fully-bonded group which accounts for 85% of all the HBs. The four HBs in these molecules all tend to have similar HB strengths. Other groups with 2 (1 donor and 1 acceptor or 2 donors and no acceptors, 3.2%) or 3 (2 donors and 1 acceptor or 2 acceptors and 1 donor, 10.3%) HBs have a much smaller presence and they are usually identified as broken-bonded. The smallest group (1.5%) with 5 HBs (2 donors and 3 acceptors) is called the over-bonded group, and it has relatively weak HBs. The average HB number per molecule is 3.85 and this clearly supports the tetrahedral HB network model in bulk water. Figure 8.24 shows a scattered plot of the four types of HBs in relation to the distances ($r_{OO'}$) and angles ($\angle O'OH$, denoted as α) that characterize the HBs' separation and degree of alignment. It clearly shows that they are narrowly confined in the range of 2.6–2.8 Å for $r_{OO'}$ and 1°–15° for α respectively. This is a much more stringent criterion than previously suggested (Wernet et al., 2004).

The above *ab initio* calculation of a model of bulk water using the OLCAO method shows the effective extension of the method to liquid systems. By using a more rigorous definition of hydrogen bonds based on the calculation of bond order values, the network structure of water appears to be in favor of the tetrahedrally coordinated model with an average of 3.85 HBs per H_2O molecule. Furthermore, the XANES spectra (O–K edges) for *all* 340 oxygen atoms in this model were calculated and grouped according the number of HBs. They were shown to be consistent with the experimental data (Myneni et al., 2002, Wernet et al., 2004). This will be further discussed in Chapter 11. The work presented in this subsection clearly demonstrates that the criterion based on BO values to define the weak hydrogen bonding is accurate and efficient, and could be applied to more complex liquids, cement hydrates, biomolecular systems, and so on.

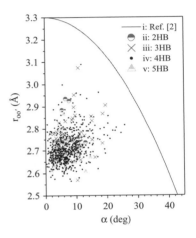

Fig. 8.24.
Scattered plots of different groups of H-bond (see cited reference for details).

8.5 Models for molten salts: NaCl and KCl

As another example of the application of the OLCAO method to liquids, we briefly discuss the calculation of the electronic structure and dielectric functions of two molten salts NaCl and KCl. The original motivation for this study was to evaluate the dielectric functions of the molten salts such that, with a specific mixing of the composition, they could act as a medium for stable suspension of certain nano-particles. This is achieved when the attractive and the repulsive parts of the Van der Waals forces (VdW) are balanced. Such a possibility has already been experimentally demonstrated in a gold particle over silica in the fluid medium of bromobenzen (Munday et al., 2009). This is of course a rather bold idea and is worth testing. The first step for such an estimation is to build supercell models of the molten salts of sufficient size in the same manner as for the water model discussed above. To this end, MD simulations of a 216 particle model were constructed with the density constrained to be the same as in the real salts (Private communication from V. B. Somani). Next the models were relaxed using VASP to obtain a stable structure. The result was a representative model of a snapshot of an instantaneous configuration of the frozen liquid. The OLCAO method was then applied to these models to obtain the electronic structure and the optical properties. Such models for molten salts are not difficult to build. One can start with a random mixture of Na (K) and Cl in equal proportion on the crystalline FCC lattice (NaCl structure). MD simulations at high temperature can then be carried out for sufficiently long simulation times. Such constructions have less stringent requirements because molten salts do not have directional bonds as was the case for the metallic glass. The RDF of the model must be compared against an experimentally measured one if it exists. A snapshot configuration is then taken and fully relaxed using

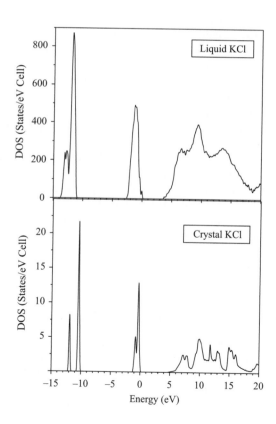

Fig. 8.25.

Comparison of DOS of liquid and crystalline KCl.

VASP. The OLCAO method is then used to calculate the electronic structure and dielectric function. Ideally, results should be averaged over several models or snapshots.

Figure 8.25 shows the calculated total DOS of a 216 atom model of molten KCl and compares it with the DOS of crystalline KCl. It is clear that the liquid phase is a broadened version of the crystalline phase with a wider band width and smaller band gap. The two peaks at -10.5 eV and -11.5 eV correspond to states from K-3p and Cl-4s. These two peaks merge in the molten KCl but their origin can still be clearly identified. There is a reduction of the band gap of about 1.5 eV.

Figure 8.26 compares the calculated imaginary part of the complex dielectric function, $\varepsilon_2(\hbar\omega)$, of crystalline KCl and molten KCl up to 30 eV. It can be seen that there are significant differences between the two, not just in the positions of the peaks but also in their relative intensities especially at energies near the edge onset. Similar calculations for molten NaCl show large differences in the structures of $\varepsilon_2(\hbar\omega)$ compared to that of molten KCl. The

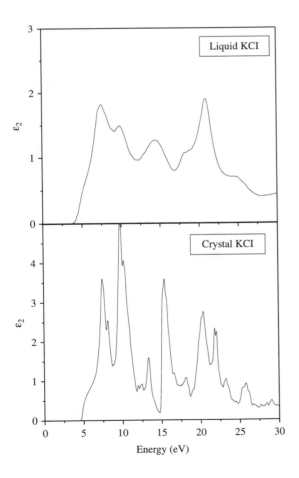

Fig. 8.26.

Comparison of imaginary dielectric function of liquid and crystalline KCl.

imaginary parts of the dielectric functions are used to evaluate the Hamaker coefficient of the medium. The VdW forces between particles in the molten salt medium with specific geometries can be evaluated using the Lifshitz theory (Parsegian, 2005). It is not clear to what extent the difference in $\varepsilon_2(\hbar\omega)$ will affect the sign and magnitude of the VdW forces.

The above results for molten salts are illustrative examples. In more rigorous calculations, larger models and a configurational averages over several models is desirable. Extension to other liquids in search of the right medium for the estimation of the VdW force is possible. These will be further discussed in Chapter 12.

8.6 Models for concrete

Concrete is made of cement hydrates or calcium silicate-hydrates (CSH) and it is perhaps the most utilized infrastructural material in the world. It is widely used as a building material in an industry with significant impact on energy and environmental issues. Enhanced life cycle performance for concrete can have far reaching consequences for sustainability and energy utilization. It is imperative to understand the nanoscale structure and properties of CSH to be able to efficiently improve the performance and durability of concrete materials. Although CSH has been researched for a long time by material scientists and engineers, its atomic structures are still contested and largely unknown. Past studies have indicated that CSH in cement paste occurs as a combination of closely related mineral crystallites that are arranged into a supercrystal structure. More recent theoretical studies by Pellenq et al. (Pellenq et al., 2009) seem to suggest that the CSH structure is far more complex and disordered but that it still retains short range ordering of SiO_4 tetrahedral units. They have proposed a cement-hydrate model with a Ca/Si ratio of 1.7 and a physical density of 2.6 gram/cc, consistent with real cement samples. The model has 672 atoms with a composition of $(CaO)_{1.65}(SiO_2)(H_2O)_{1.75}$ in a triclinic cell and periodic boundary conditions. The model is not crystalline but is instead amorphous as in a frozen liquid. Figure 8.27 shows a schematic diagram of this model illustrating the exceptionally complex nature of this material.

The Pellenq model was meticulously constructed using a combination of atomistic simulation methods including energy minimization using the General Utility Lattice Program (GULP) with classical pair potentials (Gale, 1997) and the Grand Canonical Monte Carlo method (Nicholson and Parsonage, 1982) by inserting the right amount of water molecules between crystalline-like planes of the rare mineral tobermorite. The calculated radial distribution function of this model was in good agreement with neutron scattering data and a calculation of its mechanical strength revealed a flexible response to both stretching and compression that was in line with cement materials. It is believed to be the most realistic model for CSH. However, the electronic structure calculation of this model has not been attempted so far. The OLCAO method would be ideal to investigate the electronic structure and bonding in the Pellenq model and other similar models. In particular, the role of hydrogen bonding in cements

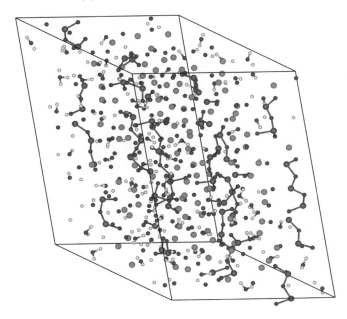

Fig. 8.27.
The sketch of the Pelleng model for C-S-H (see cited reference for details).

has seldom been discussed. We already showed in Section 8.4 that hydrogen bonding plays a critical role in the network structure of bulk water. It certainly will affect the cohesion and mechanical properties of cements. Additional modeling and *ab initio* calculations beyond the Pelleng model with different sized voids and degrees of water content in the CSH will be highly desirable. Such calculations will yield much deeper understanding of CSH materials and could lead to improved cement materials for all of its variety of applications.

References

Cargill III, G. S. & Spaepen, F. (1981), *Journal of Non-Crystalline Solids*, 43, 91–97.

Chen, J., Ouyang, L., Rulis, P., Misra, A., & Ching, W. Y. (2005), *Phys. Rev. Lett.*, 95, 256103/1–256103/4.

Ching, W. Y. & Lin, C. C. (1975), *Phys. Rev. Lett.*, 34, 1223–26.

Ching, W. Y., Lin, C. C., & Huber, D. L. (1976), *Phys. Rev. B*, 14, 620.

Ching, W. Y., Lin, C. C., & Guttman, L. (1977), *Phys. Rev. B*, 16, 5488–98.

Ching, W. Y., Lam, D. J., & Lin, C. C. (1979), *Phys. Rev. Lett.*, 42, 805–8.

Ching, W. Y. (1980a), *J. Non-Cryst. Solids*, 35–36, 61–66.

Ching, W. Y. (1980b), *Phys. Rev. B*, 22, 2816–22.

Ching, W. Y., Lam, D. J., & Lin, C. C. (1980), *Phys. Rev. B*, 21, 2378–87.

Ching, W. Y. (1981), *Phys. Rev. Lett.*, 46, 607–10.

Ching, W. Y. (1982a), *Phys. Rev. B*, 26, 6610–21.

Ching, W. Y. (1982b), *Phys. Rev. B*, 26, 6622–32.

Ching, W. Y. (1982c), *Phys. Rev. B*, 26, 6633–42.

Ching, W. Y., Song, L. W., & Jaswal, S. S. (1984), *J. Non-Cryst. Solids*, 61–62, 1207–12.

Ching, W. Y. (1985), *J. Non-Cryst. Solids*, 75, 379–84.

Ching, W. Y. (1986), *Phys. Rev. B*, 34, 2080–87.

Ching, W. Y., Zhao, G. L., & He, Y. (1990), *Phys. Rev. B*, 42, 10878–86.

Ching, W. Y. & Xu, Y. N. (1991), *J. Appl. Phys.*, 70, 6305–7.

Ching, W. Y., Rulis, P., Ouyang, L., & Misra, A. (2009), *Applied Physics Letters*, 94, 051907–3.

Ching, W. Y., Rulis, P., Ouyang, L., Aryal, S., & Misra, A. (2010), *Physical Review B*, 81, 214120.

Gale, J. D. (1997), *Journal of the Chemical Society, Faraday Transactions*, 93, 629–37.

Huang, M.-Z. & Ching, W. Y. (1996), *Phys. Rev. B*, 54, 5299–308.

Huang, M.-Z., Ouyang, L., & Ching, W. Y. (1999), *Phys. Rev. B*, 59, 3540–50.

Huang, M. Z. & Ching, W. Y. (1994), *Physical Review B*, 49, 4987.

Jaswal, S. S. & Ching, W. Y. (1982), *Phys. Rev. B*, 26, 1064–66.

Jaswal, S. S., Ching, W. Y., Sellmyer, D. J., & Edwardson, P. (1982), *Solid State Commun.*, 42, 247–49.

Liang, L., Rulis, P., Ouyang, L., & Ching, W. Y. (2011), *Physical Review B*, 83, 024201.

Liu, X. J., Xu, Y., Hui, X., Lu, Z. P., Li, F., Chen, G. L., Lu, J., & Liu, C. T. (2010), *Phys. Rev. Lett.*, 105, 155501.

Lyakhov, G. A. & Mazo, D. M. (2002), *Europhys. Lett.*, 57, 396–401.

Metropolis, N., Rosenbluth, A. W., Rosenbluth, M. N., Teller, A. H., & Teller, E. (1953), *J. Chem. Phys.*, 21, 1087–92.

Mott, N. F. & Davis, E. A. (1979), *Electronic Process in Non-Crystalline Materials* (Oxford: Clarendon).

Munday, J. N., Capasso, F., & Parsegian, V. A. (2009), *Nature*, 457, 170–73.

Murray, R. A. & Ching, W. Y. (1987), *J. Non-Cryst. Solids*, 94, 144–59.

Murray, R. A., Song, L. W., & Ching, W. Y. (1987), *J. Non-Cryst. Solids*, 94, 133–43.

Murray, R. A. & Ching, W. Y. (1989), *Phys. Rev. B*, 39, 1320–31.

Myneni, S., Luo, Y., Näslund, L. Å., Cavalleri, M., et al. (2002), *J. Phys: Condens. Matter*, 14, L213.

Nicholson, D. & Parsonage, N. G. (1982), *Computer Simulation and Statistical Mechanics of Adsorption* (New York: Academic Press).

Ouyang, L. & Ching, W. Y. (1996), *Phys. Rev. B*, 54, R15594–97.

Parsegian, V. A. (2005), *Van Der Waals Forces: A Handbook for Biologists, Chemists, Engineers, and Physicists* (New York: Cambridge University Press).

Pellenq, R. J.-M., Kushima, A., Shahsavari, R., et al. (2009), *Proceedings of the National Academy of Sciences*, 106, 16102–7.

Peng, H. L., Li, M. Z., & Wang, W. H. (2011), *Phys. Rev. Lett.*, 106, 135503.

Rulis, P. & Ching, W. (2011), *Journal of Materials Science*, 46, 4191–98.

Sheng, H. W., Luo, W. K., Alamgir, F. M., Bai, J. M., & Ma, E. (2006), *Nature*, 439, 419–25.

Veal, B. W., Lam, D. J., Paulikas, A. P., & Ching, W. Y. (1982), *J. Non-Cryst. Solids*, 49, 309–20.

Wernet, P., Nordlund, D., Bergmann, U., et al. (2004), *Science*, 304, 995–99.

Xu, Y. N., He, Y. & Ching, W. Y. (1991), *J. Appl. Phys.*, 69, 5460–62.

Xu, Y. N., Rulis, P., & Ching, W. Y. (2005), *Phys. Rev. B*, 72, 113101/1–113101/4.

Yoshiya, M., Tatsumi, K., & Tanaka, I. (2002), *Journal of the American Ceramic Society*, 85, 109.

Zeng, Q., Sheng, H., Ding, Y., et al. (2011), *Science*, 332, 1404–6.

Zhao, G. L. & Ching, W. Y. (1989), *Phys. Rev. Lett.*, 62, 2511–14.

Zhao, G. L., He, Y., & Ching, W. Y. (1990), *Phys. Rev. B*, 42, 10887–98.

Zhong, X. F. & Ching, W. Y. (1994), *J. Appl. Phys.*, 75, 6834–36.

Application to Impurities, Defects, and Surfaces

9

Crystal imperfection is a pervasive issue because it profoundly affects the properties and applications of all types of materials. Localized defects, extended defects, impurities, and surfaces are topics that are intensely studied in all classes of materials. *Ab initio* calculations of materials with defects that use periodic boundary conditions must use sufficiently large supercells to minimize the defect–defect interactions and render the defect concentration consistent with or close to what is known to exist in real materials. In surface calculations, slab geometry configurations are usually used so that there will be two surfaces separated by a vacuum region that is sufficiently thick to make surface–surface interactions become negligible. The OLCAO method is ideally suited for such large calculations because of its efficiency. In this chapter, we discuss the application of the OLCAO method to several cases of vacancies, impurities, surfaces, and interfaces. Grain boundaries (GBs) in crystals are considered to be a kind of extended defect and are thus included in this chapter. The presence of defects modifies the local atomic scale structure of a perfect crystal and therefore geometry optimization or relaxation must be carried out before any electronic structure calculations can be done. Many of the recent OLCAO calculations on defects relied on structural optimizations carried out by other efficient *ab initio* methods such as VASP.

9.1 Isolated vacancies and substitutional impurities

9.1.1 Isolated vacancies

The most common crystal defect is the isolated vacancy. *Ab initio* calculation of isolated vacancies is not an easy task and it is typically more difficult to treat than substitutional impurities because the removal of an atom from the lattice tends to induce relatively larger lattice distortions. For that reason, semi-empirical methods using classical pair potentials were more common and efficient for explaining experimental observations. In addition, vacancies in ionic compounds are usually associated with other defects such as impurities or dopants because substitution of a charged atom may result in the simultaneous creation of other defects for charge compensation. This more complicated defect scenario will be discussed in a later section. However, from the beginnings of the OLCAO method, Si has been a favorite system to test various types of calculations and so Si was used for early calculations of isolated vacancies. Supercell calculations of Si with 16, 54, or 128 atoms in the diamond lattice with one atom removed were carried out using a semi-*ab initio* approach (Ching and Huang, 1986). This simple test indicated that a supercell of about 222 atoms was needed to minimize the defect–defect interaction between adjacent replicated supercells.

The first detailed calculation of a vacancy using the well-developed OLCAO method was the case of an isolated O vacancy (V_O) in sapphire (α-Al_2O_3) using a 120-atom supercell of the hexagonal lattice (Xu et al., 1997). This was a very important calculation because V_O always exists in most oxides and it greatly influences their transport properties when used for device applications. The minimum V_O-V_O separation was 9.535 Å due to periodic boundary conditions. The four nearest neighbor (NN) Al atoms and the 12 next nearest neighbor (NNN) O atoms were allowed to relax either towards or away from the vacancy site. The calculation showed that the NN Al atoms moved away from the V_O by about 16% and the NNN O atoms moved towards the vacancy by about 8% with a defect formation energy of 5.83 eV. Figure 9.1 shows the calculated DOS of the V_O model in α-Al_2O_3 compared to that of the defect free crystal. It shows a deep level F center (marked as S) which is doubly occupied. Defect-induced states near the CB edges (marked as P1, P2, and P3) and below the occupied O-2s band and the upper VB are clearly visible at −20.3 eV and −7.8 eV. Figure 9.2 shows a plot of the square of the wave function for the F center. Also shown is an isoelectronic charge surface for this state in the three dimensional supercell surrounding the O and Al ions. Figure 9.2 shows that the state S draws charge from the surrounding Al and O ions and in no way can be considered to be highly localized as one might have expected. There is even a trace of its charge near the supercell boundary indicating that the supercell used in the calculation may be barely sufficient. Such detailed information on the nature of defect states can only be obtained from *ab initio* calculations.

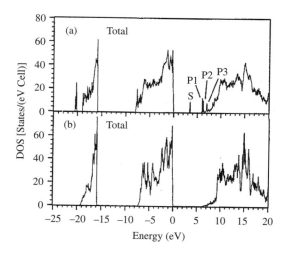

Fig. 9.1.

The total DOS of the supercell of α-Al_2O_3: (a) with a relaxed O vacancy; (b) perfect crystal without the O vacancy. Defect states in the gap are as marked.

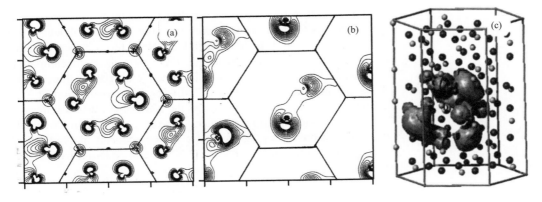

Fig. 9.2.
Contour plot of the square of the wave function of the S state in Fig. 9.1: (a) in a plane perpendicular to the c axis and containing the O ions; (b) in a plane above the vacancy and close to the two Al ions. (c) A three-dimensional plot of the isoelectronic surface.

The F center calculation described above was extended to an F^+ center in which the deep defect state S is occupied by only one electron. To maintain charge neutrality in the supercell, the total number of valence electrons was reduced by one with a simultaneous reduction of the background positive charge by one electron. The result showed that the state for F^+ or S^+ is lower than the S state by about 0.25 eV, which is consistent with optical experiments. Optical transitions from F and F^+ states to the unoccupied defect induced states were also calculated and they are in generally good agreement with experimental observations taking into consideration that the LDA approximation used in the calculation may have led to a reduced band gap and perhaps a shift in these unoccupied defect induced states.

9.1.2 Single impurities or dopants

The example of an isolated dopant in a crystal is exemplified by the case of Y substation in crystalline α-Al_2O_3. Y has a very low solubility in crystalline α-Al_2O_3 yet its presence and segregation to grain boundaries or other types of interfacial regions is an important topic in structural ceramics. It is well known that a small addition of Y influences both the properties and microstructures in α-Al_2O_3 and increases the adhesion of the oxide scales in all aluminum containing metals. This is the so-called Y-effect. Y in α-Al_2O_3 was regarded as a donor, despite being isoelectronic with Al. Thus, a detailed study of Y in alumina using *ab initio* techniques could be very revealing.

The *ab initio* OLCAO method was applied to a 120-atom supercell of α-Al_2O_3 in the hexagonal lattice similar to the V_O calculation presented above. An Al atom was replaced by a Y and the same relaxation scheme based on the total energy calculations was carried out (Ching et al., 1997). The calculated defect formation energy was fairly large at 4.79 eV, which may be an overestimate because of the crude nature of lattice relaxation, but it is consistent with the fact that Y has an extremely low solubility in alumina. It was found that the NN O atoms moved away from the Y site by about 8%, due mostly to the larger size of the Y ion, and the NNN Al ions moved toward Y by 5%. The calculated electronic structure in the form of site decomposed PDOS is shown in Fig. 9.3. It shows that the doped Y introduces three empty levels in the gap near the CB edge, thus justifying the classification of Y as a donor impurity. Orbital decomposition of the state shows that the lower one has an exclusive Y-4d ($3z^2$-r^2) component whereas the other two (doubly degenerate) originate from the xy, yz, zx, and x^2–y^2 components of Y-4d. Figure 9.3 also shows the strong interaction of the NN O and NNN Al with Y because their

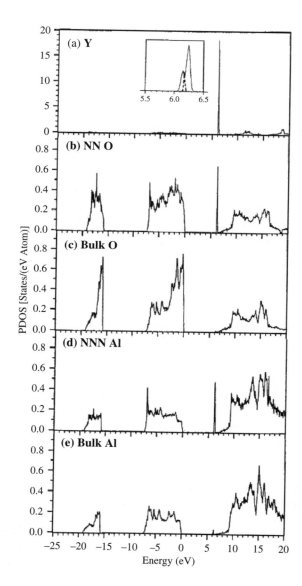

Fig. 9.3.

PDOS per atom for the supercell of α-Al$_2$O$_3$ with a substituted Y at Al site. (a) Y atom (the inset shows that the peak is due to three defect levels); (b) NN O atoms,(c) bulk O atom, (d) next NN Al atom, and (e) bulk Al atom.

PDOS deviate significantly from those of the bulk crystal. Mulliken effective charge and bond order calculations indicated that the Y-O pairs form stronger and more covalent bonds than the Al-O bonds do because of the participation of the Y-4d electron. This analysis could explain the strengthening aspect of the Y-effect in Y-containing alumina.

Hydroxyapatite (HAP) is an important bioceramic because of its relation to the mineral component of bones and other hard tissues in mammals. It is

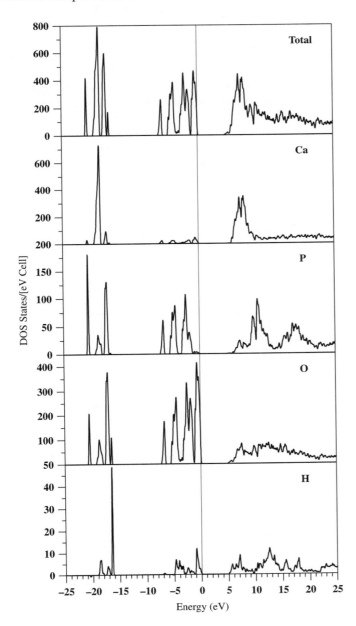

Fig. 9.4.
PDOS of Ca, P, O, H in the $2 \times 2 \times 2$ supercell of hydroxyapatite.

bioactive and is used as a bioceramic coating (i.e. in metallic implants) and in bone fillers. HAP readily absorbs impurities so that stoichiometric HAP is not common. The doping of metal ions into crystalline HAP is believed to be related to many health related issues. The change in the electronic structure and physical properties of impurity-doped HAP at the atomic level is thus a subject of considerable importance. Calculations on perfect crystals of HAP were discussed in Chapter 5. Here we show the application of the OLCAO method to the two most common metal impurities, Mg and Zn in HAP. Both metal elements are supposed to replace Ca, at either the Ca1 or Ca2 sites. These are very recent calculations so a large $2 \times 2 \times 2$ hexagonal supercell with 384 atoms was used. As mentioned above, accurate relaxation is necessary to obtain correct defect configurations and the models used in the present calculation where obtained by Matsunaga et al. (Matsunaga, 2008, Matsunaga, 2010). They have investigated the structure and formation energy of stoichiometric

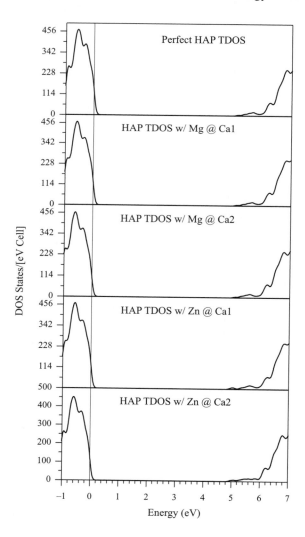

Fig. 9.5.
PDOS of Ca, P, O, H in the $2 \times 2 \times 2$ Mg-and Zn doped supercell of hydroxyapatite.

HAP with one impurity ion (Zn, Mg, Sr, Cu, Ni, Ba, or Cd) at either the Ca1 or Ca2 site in a 384-atom supercell, and they have also investigated a model with one vacancy at the Ca1 site with the addition of nearby protons. The OLCAO method has been applied to these VASP relaxed models to obtain information on the electronic structure and bonding of doped HAP. Figure 9.4 shows the calculated PDOS of Ca, P, O, and H in the 384-atom supercell of HAP. This is to be compared with the PDOS of Mg and Zn when they replace a Ca1 or Ca2 in the same energy range shown in Fig. 9.5. It can be seen that neither Mg nor Zn substitution at the Ca sites introduces impurity related defect levels in the gap or near the gap edges. From the PDOS spectra of Mg or Zn and in comparison with the PDOS of the perfect crystal, it is possible to see the interactions between Mg or Zn and the surrounding atoms and the differences between substitution at the Ca1 site or the Ca2 site. The occupied Mg (Zn) levels in the VB show that it contributes to all four major segments of the upper VB. In the OLCAO method, it is very easy to resolve the DOS into the PDOS of individual atoms or orbitals and thus show specific interactions between Mg (Zn) and its near neighbors. From the calculated total energies, it was found that both Mg and Zn prefer to replace Ca1 over Ca2 but the total energy differences are very small, 0.316 eV for Mg and 0.226 eV for Zn. This implies that, consistent with the almost identical PDOS in both cases, substitutions at both sites are possible.

These calculations clearly show the effectiveness of using the OLCAO method to analyze the defect structures of complex systems at the atomic level. Further discussion about the proton compensated Ca1 vacancy will be considered in Section 9.3 below.

9.2 Vacancies and impurities in $MgAl_2O_4$ (spinel)

We next discuss the issue of defects and its relation to the optical properties in a specific compound, $MgAl_2O_4$ or spinel. This is an example of how *ab initio* calculations can yield useful information to guide experimental work that leads to new or improved applications. It is well known that optical materials with high refractive indices are very important in certain applications related to specialized instrumentation. In particular, optical materials with a higher refractive index and with transmission at a specific wave length, say at 193 nm, are of considerable interest. Spinel is one such candidate material that was briefly discussed in Chapter 5. It has a large band gap, high mechanical strength, and a sufficiently high refractive index (1.72). However, site inversion and defect contamination issues are also common. To assess the electronic effect on the optical properties due to these defects is a problem of practical importance. The OLCAO method is ideally suited for such tasks and below we discuss this type of investigation using spinel as an example.

9.2.1 Strategy

All defect calculations in spinel begin with a supercell of sufficient size. We used a relatively small 56-atom cubic cell, denoted by $(Mg_8)[Al_{16}]O_{32}$, as the

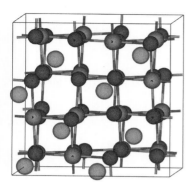

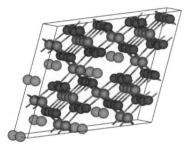

Fig. 9.6.

Sketch of the 56-atom cubic supercell
(left) and the next larger $2 \times 2 \times 2$
supercell of the primitive cell with 112
atoms (right) for $MgAl_2O_4$ spinel.

supercell for spinel. The band structure and optical properties of this supercell
were calculated first so that they could be used as a benchmark for comparison
with the defect calculations. Figure 9.6 illustrates the 56-atom cubic supercell
and the next larger $2 \times 2 \times 2$ supercell of the primitive cell with 112 atoms.
Figure 9.7 shows the calculated band structure, DOS, and optical absorption in
the form of the frequency dependent imaginary part of the dielectric function
$\epsilon_2(\hbar\omega)$ for the cubic supercell. The results are exactly the same as the results
obtained when using the primitive cell. Because all LDA calculations tend
to underestimate the band gap of insulators and because the locations of the
defect induced levels in the band gap are important for relating the results
to experimental data, we used a scaling factor f_g that will account, at least
partially, for the effect of the band gap under estimation. The value f_g is defined
as the ratio between the experimental band gap $E_g(exp)$ and the calculated band
gap E_g or $f_g = E_g(exp)/E_g$. In spinel, we used the value of $E_g(exp) = 7.8\,eV$
(Bortz and French, 1989) and the calculated value of $E_g = 5.55\,eV$ to obtain
$f_g = 1.41$ which is an approximate value since both $E_g(exp)$ and E_g are subject
to some uncertainty. All calculated defect levels (relative to the top of the
VB) were scaled with f_g in order to be comparable with transmissions at a
specific wave length of interest. So, it is then convenient to use $E_{scal} = E_{cal} \cdot f_g$
where E_{scal} and E_{cal} are the scaled and calculated energy levels respectively.
For example, if we were interested in $\lambda = 193\,nm$ (photon energy of $E_{scal} =$
$6.43\,eV$), which is the frequency at which water has a refractive index of 1.44
as the immersion fluid, then we would be looking for spectral features near
$E_{cal} = 4.56\,eV$. We further assumed that defects at low concentration do not
change the overall band gap and the established scaling factor f_g. To make a
direct connection to laboratory data, we paid special attention to the absorption
coefficient $\alpha = \varepsilon_2(\hbar\omega)/E_{scal}$ (in units of 1/cm) at $E_{scal}(\hbar\omega)$. The frequency-
dependent refractive index $n = \sqrt{\varepsilon_1(\hbar\omega)}$ can also be obtained in most
cases.

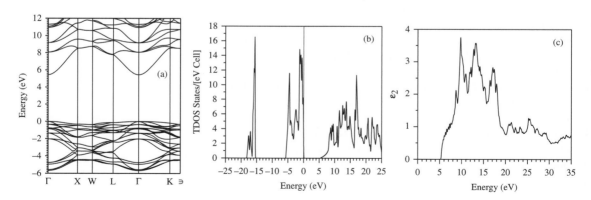

Fig. 9.7.
Calculated (a) band structure, (b) total density of states, and (c) imaginary part of the dielectric function of the 56-atom supercell of $MgAl_2O_4$.

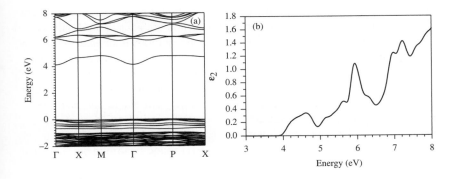

Fig. 9.8.
Calculated band structure (a) and $\varepsilon_2(\hbar\omega)$ (b) in spinel with one (Mg, Al) pair inversion.

9.2.2   Effect of inversion

The supercell calculations with one or more inversion pairs (Al, Mg) have also been investigated. The normal spinel in the 56 atom supercell can be written as $(Mg_8)[Al_{16}]O_{32}$ where () denotes the tetrahedral site and [] denotes the octahedral site, and a partially inversed spinel as $(Mg_{1-2\lambda}Al_{2\lambda})[Mg_{2\lambda} Al_{2-2\lambda}]O_4$, or $(Mg_{8-16\lambda} Al_{16\lambda})[Mg_{16\lambda} Al_{16-16\lambda}]O_{32}$. The inversion parameter λ designates the fraction of the octahedral sites occupied by Mg^{2+}, and calculations for the whole inversion range (λ from 0 to 0.5) were carried out. The number of inversion pairs, n, of (Mg and Al) depends on the inversion parameter and can take on values of $16\lambda = 1, 2, 3, 4, 5, 6, 7, 8$. For $n \geq 2$, there are many possibilities for multiple inversion sites and the final result should be the statistical average of many configurations. We attempted several configurations for each n and fully relaxed the structures using VASP. The one with the lowest total energy was chosen as the representative configuration for that λ and its electronic structure and optical absorption were calculated using the OLCAO method.

Figure 9.8 shows the calculated band structure and $\varepsilon_2(\hbar\omega)$ from 1 to 8 eV for the inverse spinel with one (Mg, Al) pair inversion. The results are almost the same as in Fig. 9.7 for the normal spinel. Similar results are also obtained for other inverse spinels of different λ. The average minimum gap is about 4.7 eV, well above the $E_{scal} = 4.5$ eV established for critical absorption at 193 nm. It is clear then that the main effect of inversion is at the top of the valence band (VB) which is fully occupied and does not affect the optical absorption at all, at least at the scaled critical energy of photons with wave length near 193 nm. The inverse spinel is essentially a disordered system in which many atomic configurations are possible (except for $n = 1$). The true behavior should be a statistical average of a large number of configurations especially for $n > 4$. This study was for a limited number of configurations. However, based on all of the results obtained, the indication is that they will not drastically change even with a larger sample set.

9.2.3   Effect of isolated vacancies

We next consider the effect of isolated vacancies at the O, Mg, and Al sites. Supercell calculations were carried out for V_O, V_{Mg}, and V_{Al} with the structures fully relaxed using VASP. The calculated band structure and $\varepsilon_2(\hbar\omega)$

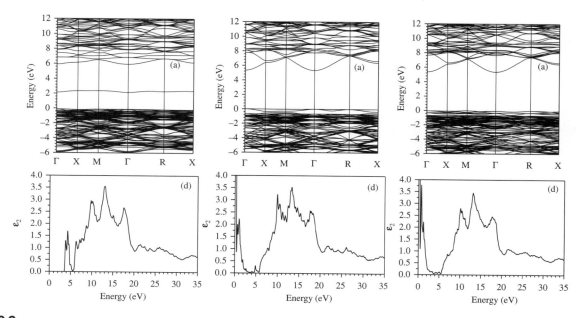

Fig. 9.9.
Calculated band structure (top) and $\varepsilon_2(\hbar\omega)$ (bottom) for V_O, (left), V_{Mg}, (middle) and V_{Al} (right) in $MgAl_2O_4$.

curves are shown in Fig. 9.9. The presence of V_O results in deep levels (occupied) in the gap and an increase in the band gap up to 6.03 eV. This impurity band is only 0.22 eV wide and is located at about 2.36 eV above the top of the VB. There is significant absorption at the impurity level after scaling the energy with $\varepsilon_2(E_{scal}) = 1.15$ due to transitions from this defect level to the empty CB states. In the case of V_{Mg}, the induced defect levels are at the top of the VB and are composed of the p-orbitals of nearby O ions. The top-most band is unoccupied (acceptor-like) to which optical transitions from lower occupied states are possible. Whereas there are significant absorptions at energies below 3.0 eV and above 5.0 eV, the absorption at $E_{scal} = 4.5$ eV is actually quite small ($\varepsilon_2(E_{scal}) = 0.04$). Thus V_{Mg} may not have an adverse effect on absorption at $\lambda = 193$ nm. The results for V_{Al} are qualitatively similar to V_{Mg} with defect levels at the top of the VB and the top-most band unoccupied. Optical transitions from occupied states to this empty defect state can result in absorptions in the energy range of interest. At E_{scal}, the ε_2 value is rather small ($\varepsilon_2(E_{scal}) = 0.07$) but slightly larger than in the V_{Mg} case. The adverse effect on absorption due to this defect at the frequency of interest is considered to be not severe.

It may be questioned whether or not the 56-atom supercell is sufficient for the above calculations. So, the calculation for V_O was repeated with the 112-atom supercell that is shown in Fig. 9.6 (b). Compared with the results obtained from the 56-atom supercell, it was found that the electronic structures are slightly different. The defect level is shifted to a slightly higher energy (2.70 eV from the top of the VB) and the defect band is narrower (0.14 eV). Because the defect concentration is a factor of 2 less than the 56-atom supercell, the optical absorption at E_{scal} due to V_O is reduced to 0.24. This calculation indicates that the size effect is important for optical absorption in crystals containing defects and impurities. Larger supercells should be used whenever possible to estimate absorptions at lower concentrations by extrapolation.

9.2.4   Effect of Fe substitution

Fe is a transition metal ion with a $3d^6 2s^2$ configuration in the neutral atom. When substituted at the tetrahedral Mg site (Fe1), it would be in a d^4 electronic configuration, or Fe^{2+} state. When substituted at the octahedral Al site (Fe2), it would be in the d^3 electronic configuration, or Fe^{3+} state. The 56 atom supercell was used to investigate the band structure and optical absorptions of Fe substitution at both sites. The structures were fully relaxed using VASP and spin-polarization was not considered. From the calculated total energies, it was found that Fe1 has a lower energy than Fe2 by 1 eV for the supercell, or less than 0.02 eV per atom. This energy is relatively small and will depend on the concentration. It is reasonable to assume that the occurrence of both Fe1 and Fe2 are possible. Figure 9.10 shows the calculated band structure and absorption curves for Fe substitutions at the tetrahedral and the octahedral sites. The crystal field of the spinel lattice splits the five 3d levels of Fe into a two-fold e_g state and a three-fold t_{2g} state. For Fe1, the occupied e_g band is lower than the partially occupied t_{2g} band and the band gap for Fe1 is 5.77 eV, slightly larger than in the perfect crystal. For Fe2, the empty e_g states are above the partially occupied t_{2g} states. Optical transitions in both cases could occur via several routes: (1) from the occupied VB states to the unoccupied portion of the Fe 3d states, (2) from the occupied to the unoccupied states

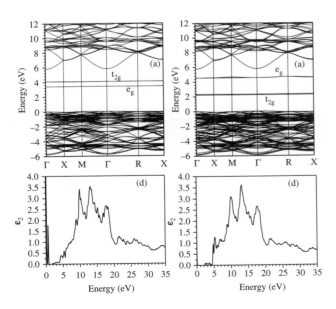

Fig. 9.10.
Calculated band structures (upper) and $\varepsilon_2(\hbar\omega)$ (lower) of Fe1 (left side) and Fe2 (right side) impurity in $MgAl_2O_4$.

within the Fe $3d$ manifold. This could result in large absorptions within the energy range of interest. We found that for Fe1 substitution, the absorption at E_{scal} of 193 nm is not as large as in the V_O case ($\varepsilon_2(E_{cal}) = 0.38$). For Fe2 substitution, the absorption is present in the gap region. At $E_{scal} = 193$ nm, the $\varepsilon_2(E_{scal})$ is 0.16, less than half of $\varepsilon_2(E_{scal})$ in Fe1. However, in the present case, there is a rapid increase in absorption at energies slightly above E_{scal}. These values depend, of course, on the size of the supercell used or on the defect concentration in the real sample. The use of the 56-atom supercell implies an exceptionally high defect concentration. Also not discussed here is the effect of intra-atomic correlation within the Fe 3d levels which could affect the absorption at the critical energy of interest.

The above studies about the optical absorption in crystalline $MgAl_2O_4$ under several defect and impurity based scenarios indicate that some scenarios may have adverse effects on the optical absorption at the specific frequency of interest. The most severe case is the oxygen vacancy (V_O) which is also the most likely defect to occur in spinel other than the Mg-Al pair inversion. On the other hand, the effects from V_{Mg} and V_{Al} are expected to be minimal. Unlike the vacancy defects, Fe impurity, and most likely other defects, have localized deep levels in the band gap which may or may not affect the optical absorption at the same frequency of interest because the transitions to and from localized levels will be different in different cases. In the present study, considerable absorptions occurred at E_{scal} due to multiple possible transition sources. This is expected to be greatly reduced if a much larger supercell corresponding to a lower impurity concentration is used in the calculation. The comparison between V_O calculations using 56-atom supercell and the 112-atom supercell suggests that the size effect is considerable. Optical absorption at a specific frequency calculated using smaller supercells is greatly over-estimated. The above calculations demonstrate the effectiveness of theoretical modeling using the OLCAO method to gain important insights on the influence of defects in promising optical crystals which may be difficult to obtain by other means.

9.3 Impurity vacancy complexes

The above discussion about single vacancies and isolated impurities pertains to the most ideal case of lattice imperfection. In real materials, a non-isovalent impurity substitution usually results in the creation of other charge compensating defects, especially in ionic crystals. Thus, the next level in the complexity of lattice defects is the impurity vacancy complex. Here we discuss a few such examples. The first example is the V^{2+} substitutional impurity at a Li site which is accompanied by the creation of a vacancy at another nearby Li site in crystalline LiF (Harrison et al., 1981). Similar calculations were also carried out with Cu^+ in LiF (Harrison and Lin, 1981). These calculations were done almost 30 years ago using the more primitive LCAO method but the basic techniques were similar to more recent calculations. Lattice relaxations were not included and more simplified potential functions were used. Nevertheless, these early calculations pioneered the realistic first-principles treatment of defect complexes. In the case of V^{2+} in LiF, it was found that the 3d to 4s and 3d to 4p transitions were generally in good agreement with experimental findings. The wave functions involved were not localized on the V^{2+} ion as would have been assumed by ligand-field theory (McClure, 1959, Simonetti and Mcclure, 1977). Such work opened up a new approach to treat crystalline defects.

The second defect complex example using the fully developed OLCAO method was carried out in 1999 for Cr^{3+} and Cr^{4+} doping in crystalline YAG ($Y_3Al_5O_{12}$) (Ching et al., 1999). The traditional approach of studying absorption and emission of transition metal ions in laser crystals is that of a free electron in a crystal field (Powell, 1998). The spectroscopic terms and the multiplets of the single ions are determined with the help of adjustable parameters using experimental inputs. Although this approach for theoretical analysis has enjoyed tremendous success in the study of solid state laser operations, it does have limitations. For example, in most cases, reliable experimental data are difficult to obtain and getting the correct parameters is a challenging endeavor. In the case of multiple ions involving co-doping, the approach of a single ion in a perfect crystal is no longer applicable. It is therefore tempting to try the alternate approach of *ab initio* calculation within the one electron theory. The OLCAO method was applied to study the cases of Cr^{3+} and Cr^{4+} impurity states in YAG using a 160-atom cubic cell of YAG (Yen et al., 1997). Depending on whether the Cr was substituting for an Al at the octahedral site or the tetrahedral site, the result was the creation of either the Cr^{3+} or Cr^{4+} ion. For the case of Cr^{4+}, it was also necessary to consider co-doping by replacing a nearby Y with a Ca ($Cr^{4+} + Ca^{2+}$). This could be easily accommodated in the 160-atom cell. The calculations were fully self-consistent so as to accurately account for the lattice impurity interaction. The resulting wave function enabled estimation of the oscillator strengths for transitions between various one electron impurity levels (Yen et al., 1997). Lattice relaxation around the Cr and Ca was not considered and spin-orbit coupling was ignored. In spite of these simplifications, meaningful insights were obtained from the calculation. Figure 9.11 shows the calculated energy levels for Cr^{3+}, Cr^{4+}, and $Cr^{4+} + Ca^{2+}$ (3 different configurations) in the band gap of YAG. Because all the defect levels for $Cr^{4+} + Ca^{2+}$ lie within the calculated gap (4.7 eV) of YAG, no adjustment was made to match the measured gap of about 6.5 eV in YAG to account for gap underestimation due to the LDA approximation. Figure 9.11 shows that an isolated Cr^{3+} introduces three defect levels (A,B,C) and Cr^{4+} introduces four (A,B,C,D). When co-doped with Ca^{2+}, all four defect states from Cr^{4+} moved lower in energy and their relative separations were independent of the Ca^{2+} location. The relative spacing between the defect levels are in agreement with the reported absorption and emission peaks for Cr^{4+} ions in YAG. This gave some credence to the one electron calculation of defect states in laser crystals. Based on the calculated transition strengths using defect wave functions, an alternative model for saturable absorption that involves the excited state absorption (ESA) (Hercher, 1967) to the unoccupied CB was proposed. This model was later partially validated by photoconductivity measurements (Brickeen and Ching, 2000).

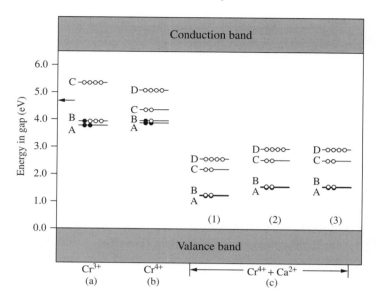

Fig. 9.11.
Calculated impurity levels of Cr^{3+}, Cr^{4+}, and three configurations of $(Cr^{4+} + Cr^{2+})$ co-doping in the gap of YAG. (See cited reference for details.)

    The above supercell calculation for defect complexes in laser crystals demonstrates that this will be a very promising approach to advance the field, if single ion multiplet calculations can be incorporated to give more accurate defect states in laser crystals (Ogasawara et al., 2000).

    The third example for multiple defects is the case of protonated Ca vacancy (V_{Ca}) in HAP. HAP crystals are generally characterized by their Ca/P ratio which is 1.67 in a perfect HAP crystal. Laboratory samples of HAP have Ca/P ratios ranging from 1.5 to 1.67 indicating that the samples are Ca deficient. Matsunaga et al. have constructed a thermodynamically stable defect model

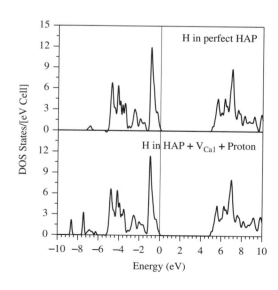

Fig. 9.12.
PDOS of H in the perfect (upper) and the V_{Ca1} + proton (lower) defect in hydroxyapatite.

containing a Ca vacancy at the Ca1 site and charge compensation by protonation (Matsunaga, 2010). The model was fully relaxed by using VASP on a large supercell. The OLCAO method was used to study the electronic structure and bonding of the model as was done for the single metal ion substitution discussed in Section 9.1.2. This enabled us to investigate the bonding near the vacancy site and the role of the introduced protons. The results showed that deviations in the electronic structure from the perfect crystal were almost negligible and that there were no defect states in the gap. Figure 9.12 compares the calculated DOS and the PDOS of H in the perfect crystal with the $V_{Ca1} + H$ model constructed by Matsunaga. The only difference is that the added protons introduced two H levels at -8.6 eV and -7.24 eV deep in the VB. Such bonding structures signify extreme stability and therefore provide a plausible explanation for the common occurrence of Ca deficiencies in HAP samples.

9.4 Grain boundary models

Grain boundaries (GBs) in crystals are generally considered to be some type of extended defects. Their structures are complicated but they are important because they tend to control the properties of the bulk materials (Sutton and Balluffi, 1995). *Ab initio* studies and simulations on GBs, especially in ceramic materials, have a relatively short history because large scale modeling is computationally demanding due to the complexity of GB structures and their low symmetry. Nevertheless, substantial progress has been made in recent years in the *ab initio* simulation of GBs of increasing complexity. Most of the recent investigations focused on GBs in $\alpha\text{-Al}_2O_3$ (Fabris and Elsässer, 2003, Marinopoulos and Elsässer, 2000). In fact, many consider the intergranular glassy films in polycrystalline ceramics discussed in Section 8.3 of the last chapter to be a more complex type of internal grain boundary. In this section, we consider several examples of the application of the OLCAO method to GB structures.

9.4.1 Grain boundaries in $\alpha\text{-Al}_2O_3$

One of the earliest calculations of GB models using the OLCAO method was for a near $\Sigma 11$ a-axis tilt GB in $\alpha\text{-Al}_2O_3$ (Mo et al., 1996a) based on a model constructed by Kenway (Kenway, 1994). The Kenway model is the closest model to the experimentally observed (by HRTEM) low energy near $\Sigma 11$ GB in alumina which has a 0.7° misorientation. The Kenway model with 72 Al and 108 O atoms is not periodic because it was constructed with a static lattice calculation using pair potentials. In order to assess the effect of the artificial surfaces from the Kenway model, a periodic model of bulk crystalline $\alpha\text{-Al}_2O_3$ with the same number of atoms and similar dimensions was constructed. Then OLCAO calculations were carried out on the systems with and without the surfaces. The calculation showed that the near $\Sigma 11$ GB did not introduce any gap states and that the defective states appeared near the O-2p derived VB and the Al-3s derived CB. Effective charge and bond order calculations showed an increase in the charge transfer from Al to O for atoms in the GB region due to the shorter Al-O bond lengths and lower coordination numbers of the atoms. The calculation of the electronic structure and bonding of the Kenway model was extended to include the optical properties which also showed that deviations from the bulk crystal results were minimal (Mo et al., 1996b).

More recently, the simulation of GBs in α-Al$_2$O$_3$ has been greatly extended. Here we describe the results for three such calculations on stoichiometric GBs that were prepared as either pure or Y doped models. They are Σ3, Σ37, and Σ31. These works further illustrate the importance of combining multiple simulation methods to obtain results that are relevant to experimental observations. The first example concerns a $(01–10)/[2–1–10]/180°$ ($\Sigma = 3$, or Σ3) GB (Chen et al., 2005a) model with 220-atoms (132 O, 88 Al). As with any GB model with periodic boundaries, the supercell model contains two oppositely oriented GBs. For the Y-doped case, four Y atoms replaced four Al atoms in the GB region or two Y ions per GB. The actual substitutional sites and the positions of the Y ions were determined by energy minimization. An important component of the study of the Σ3 GB was the inclusion of a theoretical tensile experiment to explore changes in the failure behavior of the system when Y was added to the GB. This study eventually led to similar tensile experiments on the IGF models that were briefly touched upon in Chapter 8. Figure 9.13 shows a comparison of the stress vs. strain curves obtained from theoretical tensile experiments performed on the perfect crystal, the undoped Σ3 GB, and the Y doped Σ3 GB. The stress and strain values at the failing point for the Y-doped GB are extended which indicates a strengthening of the GB due to Y doping. OLCAO calculations in the form of bond order, charge density plots, and PDOS analysis were used to interpret the theoretical experiment and to explain the mechanism for strengthening by Y doping.

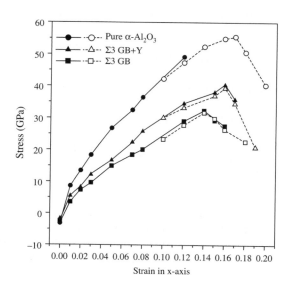

Fig. 9.13.
Stress vs. strain from theoretical tensile experiments for the three models: α-Al$_2$O$_3$ crystal, Σ3 GB + Y; Pure Σ3 GB. (See cited reference for details.)

The second example is the (01–18)/[04–41]/180° GB or the Σ37 GB in α-Al$_2$O$_3$. The supercell model contains 760 atoms (456 O, 304 Al). Molecular dynamic simulations on the undoped and Y-doped models were carried out (Chen et al., 2005b) using a classical Born–Myer–Hugging potential with additional angular terms (Blonski and Garofalini, 1997). The results showed different pre-melting and melting temperatures before and after Y doping. This was another demonstration of Y strengthening of the GB in α-Al$_2$O$_3$. Although the OLCAO method was not used in the analysis of this GB, we expect that a conclusion similar to that of the Σ3 GB calculation would be reached. The important message is to raise awareness of the possibility of combining MD simulations with *ab initio* OLCAO calculations as a viable approach for studying large and complicated GB structures. This has already been demonstrated in the case of the Yoshiya IGF model discussed in Section 8.3.3. Such a combination of methods could lead to improved fundamental understanding of temperature dependent properties and melting processes based on electronic structure and bonding calculations at instantaneous configurations.

The third example is a complicated Σ31 GB in α-Al$_2$O$_3$ (Buban et al., 2006). In this work, simulations were intimately coupled with experimental work using high resolution Z-contrast STEM imaging. The initial model was constructed using GULP for the static lattice simulation with a large supercell of 1240 atoms before being reduced to a periodic one with 700 atoms (two GBs). Finally, the model was fully relaxed using VASP and a grain boundary formation energy of 3.93 J/m^2 was obtained. The bonding configuration for atoms near the GB core was vividly observed in the Z-contrast STEM images (Buban et al., 2006) to be a 7-member ring of Al columns and this structure was also present in the model. The bonding structure for atoms in the vicinity of the GB core before and after the introduction of the Y ions was analyzed in terms of charge density maps and atomic coordinations. Figure 9.14 shows the local coordination of atoms involved in the 7-membered ring at the GB core. It was concluded that strengthening of the GB by Y replacement of Al at the center of the 7-member ring was the result of more and stronger Y-O bonds. The OLCAO method was used to verify the bond strengthening via the effective charge and bond order calculations as in the Σ3 GB case discussed above.

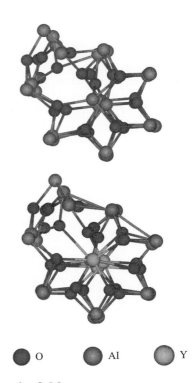

Fig. 9.14.
Local coordination of atoms involved in the 7-membered ring at the core of the Σ31 grain boundary in α-Al$_2$O$_3$. Left (right) without (with) Y.

9.4.2 Passive defects

There are some special types of defects in covalently bounded crystals such as Si or SiC. These defects introduce relatively small distortions in the bond lengths and bond angles, or introduce different types of covalent bonds that do not cause large perturbations in the overall electronic structure. They are difficult to detect experimentally and are generally called passive defects. In spite of their minimal impact on the electronic structure, they can still subtly affect the performance of devices in highly sensitive experiments. In this section, we describe two such passive defect models. Both models were used mainly for XANES/ELNES calculations using the supercell OLCAO method. Because of their passive nature, other types of measurements are

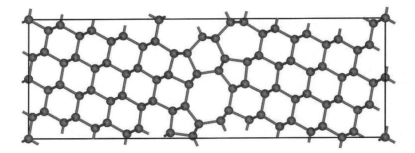

Fig. 9.15.
The 90-atom periodic model of the {113} extended planar defect in Si showing the 8-, 7-, 6-, and 5-membered rings.

not sufficiently sensitive to detect them. However, electronic structure and bonding is a prerequisite for spectral calculations and so we will discuss these two models here. The first model is the {113} extended planar defect in Si (Chen et al., 2002). It has 90 Si atoms with specially designed primitive lattice vectors (Kohno et al., 1998). This periodic model is shown in Fig. 9.15 and it contains 8-, 7-, 6-, and 5-membered rings. A bond order calculation showed very small deviations that were generally associated only with Si atoms in the 8-member or 5-member rings where the bond angle distortions tended to be larger than average. The overall electronic structure in the form of the atom-resolved PDOS showed very small difference and no gap states as expected. On the other hand, the differences in the XANES/ELNES spectra of individual atoms were more noticeable and this issue will be further discussed in Chapter 11 (Rulis et al., 2004). This model was used as a testbed for developing a spectral imaging technique based on calculated ELNES spectra.

The other passive defect model is a {122} $\Sigma = 9$ GB in β-SiC crystal. In this GB, so-called Si-Si or C-C "wrong bonds" are introduced due to the presence of 5-member rings. Two such periodic models with 64 atoms were constructed by Kohyama and co-workers (Kohyama, 1999, Kohyama, 2002, Kohyama and Tanaka, 2003). In the polar model, the two oppositely oriented GBs in the model have different wrong bonds, one with two Si-Si bonds and the other with two C-C bonds. In the non-polar model, each GB has one Si-Si wrong bond and one C-C wrong bond. These are illustrated in Figure 9.16. Because both the Si-Si and C-C "wrong bonds" are as strong as the Si-C bonds in β-SiC crystal, the disturbance to the overall electronic structure in the presence of these polar or non-polar GBs was expected to be small and to not introduce any deep levels into the band gap. The electronic structure and bonding in these two models were studied using the OLCAO method although the main focus was also on the ELNES/XANES spectra (Chen et al., 2002). It was concluded that the effect of the bond angle distortion was larger than the changes in bond lengths for both the polar and non-polar GBs. Such calculations will be very valuable for detecting subtle distortions that are introduced by the formation of different types of bonds under various dynamic configurational changes in real time. The OLCAO method, because of its efficiency, will be the ideal method to probe such variations.

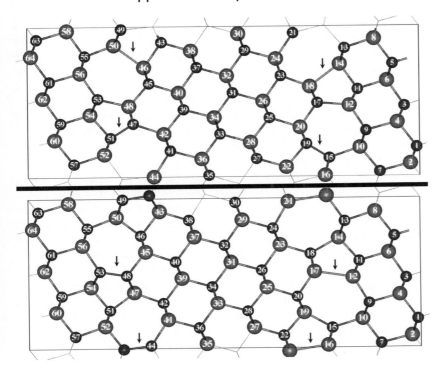

Fig. 9.16.
Sketch of the non-polar (upper) and polar (lower) {122} $\Sigma = 9$ grain boundary model in β-SiC. Arrows indicate wrong bonds.

9.4.3 Grain boundary in SrTiO$_3$

One of the early GB calculations, other than the Kenway model of a $\Sigma = 11$ GB in Al$_2$O$_3$, was the $\Sigma = 5$ {210} $< 001 >$ GB in SrTiO$_3$ (McGibbon et al., 1994). SrTiO$_3$ is a prominent functional oxide in which the presence of the GB greatly affects its electric properties (Imaeda et al., 2008, Shao et al., 2005). Two $\Sigma = 5$ GB models relaxed by a plane wave pseudopotential method were studied (Mo et al., 1999); one with 50 atoms, and the other 100 atoms and each containing two oppositely oriented GBs in the periodic cell. The 50 atom model was doubled in the c-direction to make the 100 atom model, and it was somewhat more realistic because of the presence of buckled Sr columns in the GB core. These were relatively small models when compared with more recent studies. The OLCAO method was applied to study the electronic structure and bonding in these two models and the results were compared to a parallel calculation on the bulk crystal. The calculated DOS of the GB model showed only minor differences in the lower part of the CB when compared to the bulk crystal. A bond order calculation showed that the Ti-O and Sr-O bonds in the GB region were weaker than those in the crystal. The key problem was to find out if the GBs in SrTiO$_3$ were charged or not. Extensive charge density analysis using the *ab initio* wave functions showed that there was no charge accumulation at the $\Sigma = 5$ GB and that there were no defect states introduced in or near the band gap. It is not clear if these small and stoichiometric models correspond to the real GB structures observed in experiments.

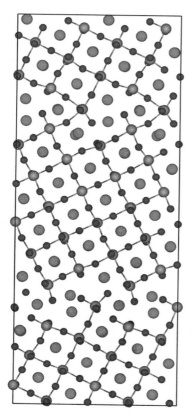

Fig. 9.17.

The sketch of the large 960 atoms Σ5 GB model in SrTiO₃. The linked atoms are O and Ti and the unlinked circles are Sr atoms.

In order to obtain a more detailed picture of the electronic structure and charge distribution for GBs in SrTiO₃, the above study on the $\Sigma = 5$ GB has been greatly expanded with a much larger supercell of 960 atoms shown in Figure 9.17. At this size, the interaction between the two GBs is negligible and much more reliable data can be obtained. The model was fully relaxed using VASP and its electronic structure calculated using the OLCAO method. Figure 9.18 shows the calculated total DOS for this model. The result is slightly different from the original model but is more accurate than that reported in (Mo et al., 1999). This model will be used to study the possible locations of O vacancy V_O at the GB core that has been discussed in (Kim et al., 2001). When V_O are introduced at different locations as suggested in (Kim et al., 2001), the calculated total energies show very small differences indicating that it is possible to have V_O at any of these locations. OLCAO calculations were carried out on these defect containing GB models. The results showed that a donor level is located near the bottom of the CB. Thus the O vacancy at the $\Sigma = 5$ GB in SrTiO₃ definitely will affect its electric properties profoundly. These models will also be used to calculate the O-K edge for O atoms at the GB and near the V_O, as a means to help identify the defective structure in real samples of SrTiO₃. This study also enables us to make a closer connection with the recent developments in aberration-corrected STEM experiments for identifying buried defects in SrTiO₃ (Krivanek et al., 2010, Zhang et al., 2003).

The examples given in this subsection on GB calculations indicate that the OLCAO method would be very effective in studying the electronic structure and spectroscopic properties of GBs and other complex microstructures. The IGF models discussed in the previous chapter can be considered as a special kind of complex GB. Such applications are not limited to oxides or nitrides, but can be used to study grain boundaries in metals and alloys as well.

9.5 Surfaces

Surface physics is one of the most active areas of materials research. A surface can be considered as the interface between a crystalline facet and the vacuum with the periodicity maintained only in the two-dimensional plane in the direction normal to the surface. The LCAO method has been used to solve surface problems since the early development of the method (Mednick and Lin, 1978).

Fig. 9.18.

Total DOS of the 960-atom grain boundary model in SrTiO₃.

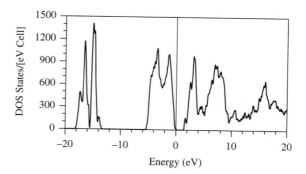

The applications of the modern OLCAO method to surfaces problems are comparatively rare with respect to studies on complex bulk materials. In this section, we describe two examples of surface calculations on two bioceramic crystals hydroxyapatite (HAP) and tri-calcium phosphate (TCP). The electronic structure and bonding of the bulk form of these materials has already been discussed in Chapter 5. The first application is to the O-terminated (001) surfaces of fluorapatite (FAP) and HAP. A supercell slab geometry was used in these calculations as is illustrated in Fig. 9.19. It was shown that in both FAP and HAP, the O-terminated (001) surface is more stable with calculated surface energies of 0.865 J/m^2 and 0.871 J/m^2 for FAP and HAP respectively. In FAP, the two surfaces are symmetric. In HAP, the orientation of the OH group along

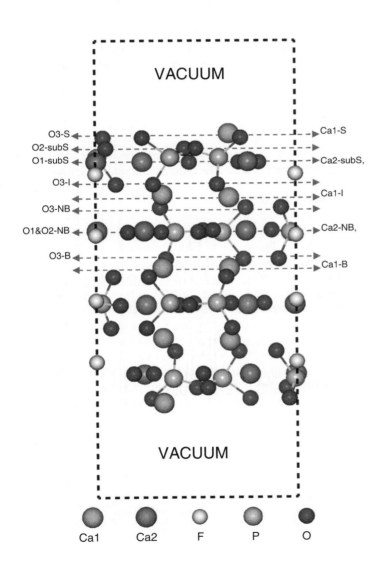

Fig. 9.19.
Surface model of FAP (001). Layers of atoms are labeled as explained in the cited reference.

the c-axis reduces the symmetry such that the top and bottom surfaces are no longer symmetric. The atoms near the surface and subsurface are significantly relaxed especially in the case of HAP. The largest relaxations occurred via the lateral movements of the O ions at the subsurface level. The electronic structures of the surface models in the form of layer-by-layer resolved partial density of states for all the atoms show systematic variation from the surface region towards the bulk region. The calculated Mulliken effective charge on each type of atom and the bond order values between cations (Ca, P) and anions (O, F) show different charge transfers and bond strength variations from the bulk crystal values. Electron charge density calculations show that the surfaces of both FAP and HAP crystals are mostly positively charged due to the presence of Ca ions at the surface. The total electron charge density $\rho(z)$ along the z axis (perpendicular to the surface and integrated over the x–y plane) and its deviation from neutral atomic charges at the same sites, $\delta\rho(z)$, are shown in Fig. 9.20. It can be seen that the charge distributions at the top and bottom surfaces in HAP (001) are asymmetric due to the orientation of the (OH) group. The lower surface, with the H in OH pointed in the negative z direction, has a slightly more positive surface. The positively charged surfaces have implications for the absorption of water and other organic molecules in an aqueous environment on the apatite surfaces which are an important part of its bioactivity.

The XANES/ELNES spectra of all the ions in these surface models were also studied (Rulis et al., 2007) and will be further discussed in Chapter 11. Marked differences can be seen between atoms of the same element at the surface, near the surface, and in the bulk.

The crystal structures of two phases of tri-calcium phosphates, α-TCP and β-TCP, and their electronic structures were described in Section 7.5. Here we specifically restrict our discussion to their surfaces in the same manner as for the FAP and HAP surfaces discussed above (Liang et al., 2010). TCP is one of the major bioceramics having many biomedical applications. It has been well documented that TCP can accelerate the broken bone recovery process and eventually be incorporated into new bone tissue. The bioactivity and biocompatibility of TCP are closely related to their surface chemistry and topography which have a strong influence on the *in vivo* protein adsorption, cell adhesion, and response from the host. The development of a new generation of biomaterials is aimed at enhancing desirable functions and properties, such as surface modifications that facilitate a particular protein adsorption or fabrication of hybrid composites with collagen or gelatin. The surface charge distribution in TCP crystals plays an important role in determining these functions.

The (010) surface of α-TCP and the (001) surface of β-TCP were studied by the OLCAO method using a slab geometry (Liang et al., 2010). The supercell models were obtained by cutting the bulk unit cell at 6 different positions along the $b(c)$ axis in α-TCP (β-TCP) and introducing a 15 Å wide vacuum region. The atomic positions of the surface models after relaxation are shown in Fig. 9.21. There were substantial movements of the atoms in the surface region because of the presence of the vacuum. However, in the bulk region interior to the surface, β-TCP had larger movements compared to α-TCP mainly because

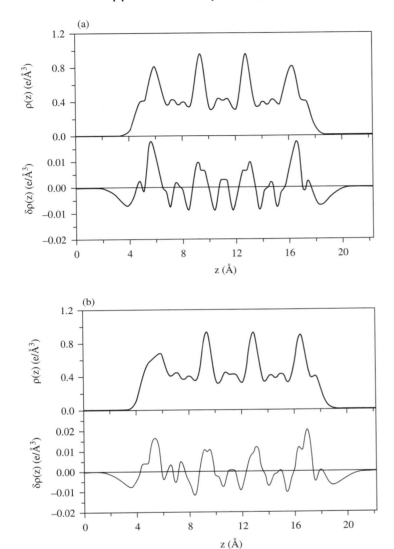

Fig. 9.20.
Electron charge density $\rho(r)$ and $\Delta\rho(r)$ across the bulk and surface in: (a) FAP (001) and (b) HAP (001).

of the presence of three Ca vacancy sites that are necessary to account for the half-occupancy of one of the Ca sites in β-TCP. The calculated surface energy values for α-TCP (010) and β-TCP (001) are 0.777 and 0.841 J/m^2, respectively. These values are very close to the surface energies of HAP and FAP (001) surfaces discussed above. The lower surface energy in α-TCP relative to β-TCP is related to the more even surface charge distribution in the α-TCP model. Surface related bands appear at the top of the VB and the bottom of the CB. They were identified from the wave functions as having large components from surface atoms. Also calculated were the effective charges and the bond order values for the two surface models. Ca atoms in α-TCP had

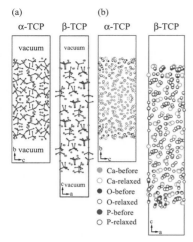

Fig. 9.21.
(a) Relaxed surface models of α-TCP (010) and β-TCP (001). (b) Atomic positions before and after the structural relaxation.

a larger charge transfer than in β-TCP similar to what we found in the bulk crystals. Also, a larger charge transfer occurred at the surface indicating that the material is more ionic when surfaces are present.

The total valence electron charge density distribution $\rho(r)$ and its deviation $\delta\rho(r)$ from the neutral atom configurations along the b (c) axis for α-TCP(β-TCP) are shown in Fig. 9.22. There are large peaks in the $\rho(b)$ and $\rho(c)$ plots corresponding to planes where the atoms are concentrated. Near the surfaces, the electronic charge distributions decay into the vacuum region. From the $\delta\rho(r)$, it is clear that the surfaces are losing some charge on average and are therefore positively charged. For α-TCP, the charge distributions at the top and bottom surfaces are symmetric as expected. For β-TCP, the two surfaces are asymmetric with the top one more positively charged than the lower one due to more concentrated Ca^{2+} ions on the top surface. Charge density maps also show that the surfaces of both TCP crystals are positively charged overall due to the presence of Ca ions near the surfaces.

9.6 Interfaces

Although the OLCAO method has been applied to grain boundary models and surface models discussed above, both can be classified as some kind of special interface. It has so far not been applied to interface models between bulk crystals. Solid–solid interfaces are important both in nanotechnology, coatings, and in physio or chemical absorptions on surfaces. A recent project has been to model the interface between Au and TiO_2. This system is interesting because of the observation of a preferential orientation for Au nano-particles on the $TiO_2(110)$ interface (Akita et al., 2008). Three types of models were constructed: stoichiometric, Ti-rich, and O-rich. Because of the large lattice mismatch, these models are not periodic in the direction perpendicular to the surface so that artificial surfaces are present. We can still focus on the interface parts of the model by taking appropriate measures to avoid the spurious effects

Fig. 9.22.
Electron charge density $\rho(r)$ (top) and $\Delta\rho(r)$ (lower) across: (a) b-axis of α-TCP and (b) c-axis of β-TCP.

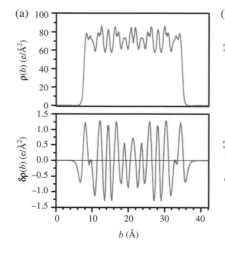
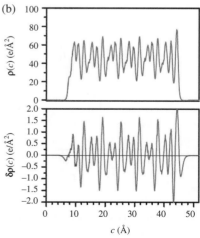

of the surfaces introduced. Fig. 9.23 shows a sketch for the stoichiometric model of the Au-TiO$_2$ interface.

We expect that the OLCAO method will play a very important role in the study of the electronic structure and bonding of different types of interfaces. These include nano-particles with coated surfaces and the protein–protein or protein–inorganic crystal interfaces. Also attractive is the investigation of protective layers on alloy surfaces to enhance their mechanical properties at extreme thermodynamic conditions of high temperature and pressure and in a corrosive environment. Some such OLCAO calculations are currently being planned.

References

Akita, T., Tanaka, K., Kohyama, M., & Haruta, M. (2008), *Surface and Interface Analysis*, 40, 1760–63.

Blonski, S. & Garofalini, S. H. (1997), *Journal of the American Ceramic Society*, 80, 1997–2004.

Bortz, M. L. & French, R. H. (1989), *Applied Physics Letters*, 55, 1955–57.

Brickeen, B. K. & Ching, W. Y. (2000), *J. Appl. Phys.*, 88, 3073–75.

Buban, J. P., Matsunaga, K., Chen, J., et al. (2006), *Science*, 311, 212–15.

Chen, J., Ouyang, L., & Ching, W. Y. (2005a), *Acta Materialia*, 53, 4111–20.

Chen, J., Xu, Y.-N., Rulis, P., Ouyang, L., & Ching, W. Y. (2005b), *Acta Materialia*, 53, 403–10.

Chen, Y., Mo, S.-D., Kohyama, M., Kohno, H., Takeda, S., & Ching, W. Y. (2002), *Mater. Trans.*, 43, 1430–34.

Ching, W. Y. & Huang, M. Z. (1986), *Solid State Commun.*, 57, 305–7.

Ching, W. Y., Xu, Y.-N., & Ruhle, M. (1997), *J. Am. Ceram. Soc.*, 80, 3199–204.

Ching, W. Y., Xu, Y.-N., & Brickeen, B. K. (1999), *Appl. Phys. Lett.*, 74, 3755–57.

Fabris, S. & Elsässer, C. (2003), *Acta Materialia*, 51, 71–86.

Harrison, J. G. & Lin, C. C. (1981), *Physical Review B*, 23, 3894.

Harrison, J. G., Lin, C. C., & Ching, W. Y. (1981), *Phys. Rev. B*, 24, 6060–73.

Hercher, M. (1967), *Appl. Opt.*, 6, 947.

Imaeda, M., Mizoguchi, T., Sato, Y., et al. (2008), *Physical Review B (Condensed Matter and Materials Physics)*, 78, 245320–12.

Kenway, P. R. (1994), *Journal of the American Ceramic Society*, 77, 349–55.

Kim, M., Duscher, G., Browning, N. D., et al. (2001), *Physical Review Letters*, 86, 4056.

Kohno, H., Mabuchi, T., Takeda, S., Kohyama, M., Terauchi, M. & Tanaka, M. (1998), *Physical Review B Condensed Matter and Materials Physics*, 58, 10338–42.

Kohyama, M. (1999), *Philosophical Magazine Letters*, 79, 659–72.

Kohyama, M. (2002), *Physical Review B*, 65, 184107.

Kohyama, M. & Tanaka, K. (2003), *Solid State Phenomena*, 93, 387.

Krivanek, O. L., Chisholm, M. F., Nicolosi, V., et al. (2010), *Nature*, 464, 571–74.

Liang, L., Rulis, P., & Ching, W. Y. (2010), *Acta Biomaterialia*, 6, 3763–71.

Marinopoulos, A. G. & Elsässer, C. (2000), *Acta Materialia*, 48, 4375–86.

Matsunaga, K. (2008), *Physical Review B*, 77, 104106.

Matsunaga, K. (2010), *Journal of the American Ceramic Society*, 93, 1–14.

Mcclure, D. S. (1959), Electronic Spectra of Molecules and Ions in Crystals Part Ii. Spectra of Ions in Crystals: Part Ii. Spectra of Ions in Crystals. *In:* Frederick, S. & David, T. (eds.) *Solid State Physics*(New York: Academic Press).

Fig. 9.23.
Sketch for a small stoichiometric model of Au–TiO$_2$ interface.

Mcgibbon, M. M., Browning, N. D., Chisholm, M. F., et al. (1994), *Science*, 266, 102–4.

Mednick, K. & Lin, C. C. (1978), *Physical Review B*, 17, 4807.

Mo, S.-D., Ching, W. Y., & French, R. H. (1996a), *J. Am. Ceram. Soc.*, 79, 627–33.

Mo, S.-D., Ching, W. Y., & French, R. H. (1996b), *J. Phys. D: Appl. Phys.*, 29, 1761–66.

Mo, S.-D., Ching, W. Y., Chisholm, M. F., & Duscher, G. (1999), *Phys. Rev. B*, 60, 2416–24.

Ogasawara, K., Ishii, T., Tanaka, I., & Adachi, H. (2000), *Physical Review B*, 61, 143.

Powell, R. C. (1998), *Physics of Solid-State Laser Materials* (New York: Springer Verlag).

Rulis, P., Ching, W. Y., & Kohyama, M. (2004), *Acta Mater.*, 52, 3009–18.

Rulis, P., Yao, H., Ouyang, L., & Ching, W. Y. (2007), *Phys. Rev. B*, 76, 245410/1–245410/15.

Shao, R., Chisholm, M. F., Duscher, G., & Bonnell, D. A. (2005), *Physical Review Letters*, 95, 197601.

Simonetti, J. & Mcclure, D. S. (1977), *Physical Review B*, 16, 3887.

Sutton, A. P. & Balluffi, R. W. (1995), *Interfaces in Crystalline Materials* (New York: Oxford University Press).

Xu, Y.-N., Gu, Z.-Q., Zhong, X.-F., & Ching, W. Y. (1997), *Phys. Rev. B*, 56, 7277–84.

Yen, H. L., Lou, Y., Xu, Y. N., Ching, W. Y., & Jean, Y. C. (1997), *Mater. Sci. Forum*, 255–257, 482–4.

Zhang, Z., Sigle, W., Phillipp, F., & Rühle, M. (2003), *Science*, 302, 846–9.

Application to Biomolecular Systems

10

Biomolecular systems are inarguably among the most complex of all known materials systems. An atomic and molecular approach is rapidly becoming the norm for fundamental research in the life sciences because of its ability to unlock many of the mysteries of biochemical reactivity and physiological function. *Ab initio* calculations on real biomolecular systems such as proteins, DNA/RNA chains, lipid bi-layers, colloidal virus particles, and so on that can extend up to the nanometer scale contain a minimum of several thousand atoms and are still a formidable task. In addition, the biomolecular systems that are most interesting and important are in an aqueous environment, which adds further complications. Most *ab initio* calculations on biomolecular systems focus on a small fragment of the structure or limit themselves to well-known structural subunits and leave larger scale calculations to molecular mechanics or molecular dynamics using well developed and highly accurate parameters such as in AMBER (Case et al., 2010). We believe that sooner or later, quantum mechanical treatment of large complex biomolecular systems of much larger size will be necessary for addressing questions such as hydrogen bonding, charge transfer and distribution, long range electrostatic interactions, and the effect of solvents. In this regard, the OLCAO method is ideally suited to making such advances because it is efficient for large systems and it is not restricted to any particular structures. In this chapter, we discuss several such examples. The first examples are the vitamin B_{12} cobalamins, followed by periodic models of both single and double stranded B-DNA. We then present some results on models of one of the most common and important proteins, collagen. We end up with a brief projection of other systems that may be of interest. Many of these calculations are currently in progress and the results presented here are preliminary.

10.1 Vitamin B_{12} cobalamins

Vitamin B_{12} (cobalamin) is an important molecule in medicine. It is capable of helping those suffering from pernicious anemia and it is an essential nutrient for mammals. No less than four Nobel prizes have been awarded in the past for work related to vitamin B_{12}. The structure of the B_{12} molecule is extremely complex but was finally resolved by Dorothy Hodgkin and co-workers after years of persistent x-ray diffraction experiments and with the assistance of emerging computing technology (Bonnett et al., 1955, Hodgkin et al., 1955). Once the structure was known, synthesis and analysis to predict chemical reactions became possible. In recent years, there has been an increased urge to understand the biological functions of the B_{12} enzymes at an atomic and electronic level. For example, why has nature chosen a Co-corrin unit in B_{12} for carrying out its particular biological function, instead of a Fe-porphyrin such as is present in many other important biomolecular systems?

The B_{12} cofactors belong to the general group of alkylcobalamins (RCbl) in which the vitamin B_{12} or the cyanocobalamin (CNCbl) is the most well-known one. The structure of alkylcobalamin consists of a cobalt corrinoid with a pendant nucleotide which occupies five of the six coordination sites of an octahedral Co(III). The sixth position is occupied by the R group with R = methyl (methylcobalamin or MeCbl), adenosyl (adenosylcobalamin or adoCbl), or a hydroxo group (hydroxocobalamin or OHCbl). This structure is illustrated in Fig. 10.1. CNCbl is biologically inactive whereas the other cobalamins are biologically active. All known reactions of B_{12}-dependent enzymes involve the making or breaking of the Co-C bond with the alkyl ligand (Banerjee, 1999). Of particular importance is the adoCbl which is a cofactor in several enzymatic reactions in which an H atom is interchanged with a functional alkyl group. Understanding the factors that influence the

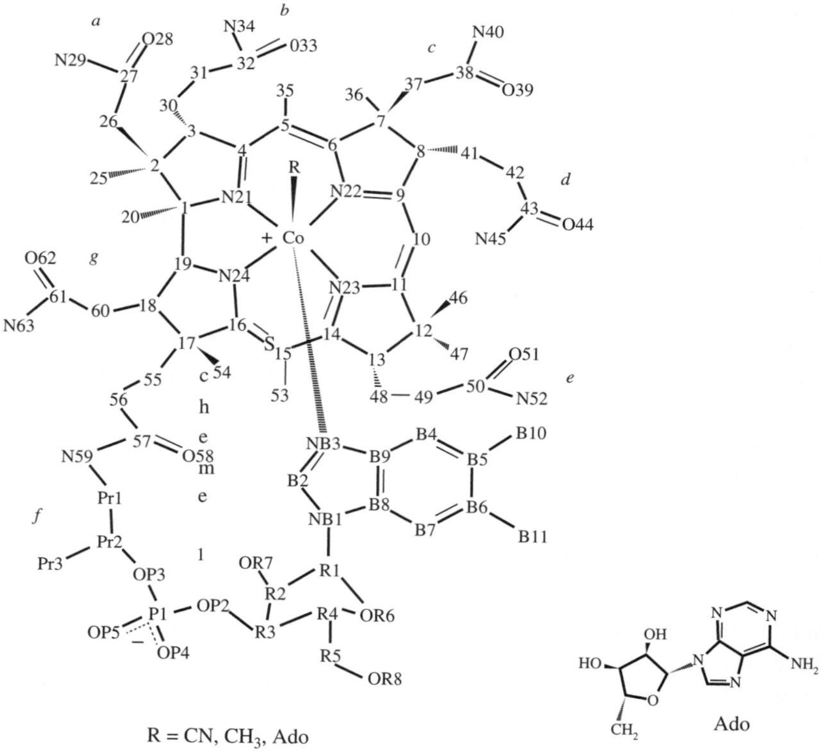

Fig. 10.1.
Sketch of structure of alkylcobalamin.

R = CN, CH₃, Ado

cleavage of the Co-C bond in B_{12} requires knowledge of the electronic structure and bonding in the alkylcobalamins. Early studies were based on simple
models that mimicked the B_{12} enzymes due to a lack of knowledge about
the precise crystal structures of the cobalamins at the atomic scale. This
changed when accurate high resolution structural data became available using
x-ray synchrotron radiation (Kratky et al., 1995, Randaccio et al., 2000). The
experimentally determined structure at high accuracy includes information
about the side chains (a–g in Fig. 10.1) which were always ignored in earlier
calculations of the B_{12} cofactors.

The OLCAO method was used to study the electronic structure and bonding
in CNCbl (Ouyang et al., 2003) using the experimentally determined crystal
structure (Randaccio et al., 2000) with all the side chains included. This was
the first time that the OLCAO method was applied to a complex biomolecule.
The focus of this study was on the PDOS as it was grouped according to
different structural units, the effective charge on each atom, and the bond order
values between pairs of atoms. The calculated PDOS in CNCbl is shown in
Fig. 10.2. The PDOS is broken down into seven different groups of atoms:
(1) Co atom, (2) C atom in CN; (3) N atom in CN, (4) sp^2 bonded C atoms
in the coring ring (CR), (5) sp^3 bonded C atoms in the CR, (6) N atoms in
the CR, and (7) the N atom in the benzimidazole that bonds to Co (labeled

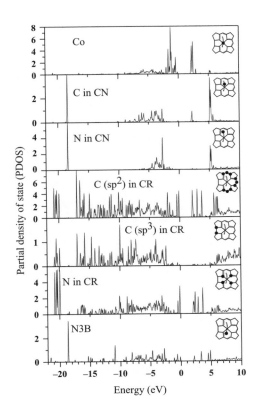

Fig. 10.2.
Calculated PDOS of CNCbl. The inset
shows the participating atom or atoms as
black dots.

as N3B). Figure 10.2 reveals some very important facts. First, the HOMO–LUMO gap of about 1.96 eV is determined by interactions between C-C and C-N atoms within the corrin ring. The Co states with large $3d$ components are actually slightly removed from the HOMO and LUMO states. The C-N bonding in CN is very strong as evidenced by the peak alignments at -18.6 eV and -2.2 eV. The sp^2 and sp^3 bonded C atoms in the CR have different PDOS which is expected, and there is a large difference between the PDOS of N in CR and N in N3B. The calculations also show states in the HOMO–LUMO gap arising from the charged PO_4 unit in the side chain f that contains the benzimidazole. The Mulliken effective charges on every atom and the bond order values between every pair of bonded atoms in CNCbl were calculated and schematically displayed in Fig. 10.3. Co loses about 0.3 electrons to the N due to charge transfer, and large BO values for C-C and C-N bonds in the CR are evident. The largest BO value is between C-N in the cyano group CN. The BO of 0.25 between Co and C in CN is larger than the BO between Co-N for N in the CR.

Similar calculations using the OLCAO method were carried out for the other cobalamins, namely, MeCbl, adoCbl, and OHCbl. In OHCbl, the Co is bonded to O in the hydroxyl group. Figure 10.4 shows the comparison of the PDOS of Co in these four cobalamins. The differences are quite obvious especially in the unoccupied states. The splitting of the t_{2g} orbitals from Co in OHCbl is larger

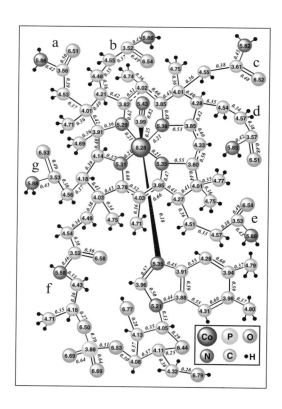

Fig. 10.3.

Calculated effective charges and bond order values in CNCbl.

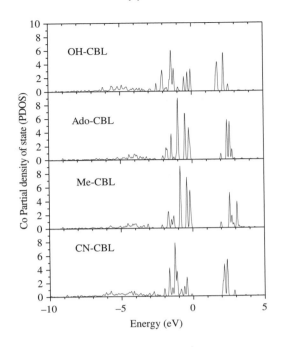

Fig. 10.4.
Calculated PDOS on OHCbl: Co, O in the 6th ligand, N in CN, NCR, N in benzimidazole.

Table 10.1. Q^* on Co and donor atoms and BO of coordination bonds.

	CNCbl[a]	MeCbl[b]	AdoCbl[c]	OHCbl
Q^*(Co)	8.28	8.29	8.29	8.18
Q^*(NB3)	5.35	5.34	5.36	5.34
Q^*(R,6th)	3.99(C)	4.81(C)	4.58(C)	6.81(O)
Q^*(N21-24)[d]	5.30	5.32	5.31	5.32
Q^*(N22-23)[d]	5.36	5.36	5.36	5.34
BO(Co-R,6th)	0.25	0.13	0.15	0.24
BO(Co-NB3)	0.18	0.16	0.15	0.18
BO(Co-Neq)[d]	0.21	0.23	0.23	0.20

than in the other cobalamins. Also, the LUMO states in OHCbl and CNCbl are significantly enhanced by the presence of the Co-$3d$ states. Table 10.1 lists the calculated Mulliken effective charges Q^* of Co and those atoms that bond to Co together with selected bond order values. It can be noticed that in all four cases Co is a cation. It loses about 0.10 electrons more in the OHCbl than in the other cobalamins whereas the Q^* of the N atoms in the corrin ring and that of N3B from benzimidazole are similar. From the bond order values between Co and the C in R or O in OH, we can see that bonding between Co and O in the 6th ligand is much stronger in CNCbl and OHCbl than in MeCbl or adoCbl, consistent with the fact that vitamin B$_{12}$ is biologically inactive while MeCbl and adoCbl are biologically active.

It is interesting to know if the above calculated electronic structure results can somehow be related to experimental data. Figure 10.5 shows a comparison of the measured optical absorption spectrum in CNCbl which is nearly 50 years old (Hill et al., 1964) with transitions in the Co-related electronic states from the calculation. The agreement is reasonable especially with respect to the three main absorption peaks α, β, and γ. Figure 10.6 compares the measured x-ray emission spectra (XES) with the properly broadened orbital-resolved PDOS of O, N, and Co in the occupied region. The agreements are very good which helps to verify the validity of the calculated results for cobalamins.

The results presented above on B_{12} are still considered to be preliminary and are mostly for demonstration purposes. Additional calculations with improved accuracy are contemplated in which the $[PO_4]^{-1}$ group is to be compensated by a Na ion. This could negate the spurious gap states related to the

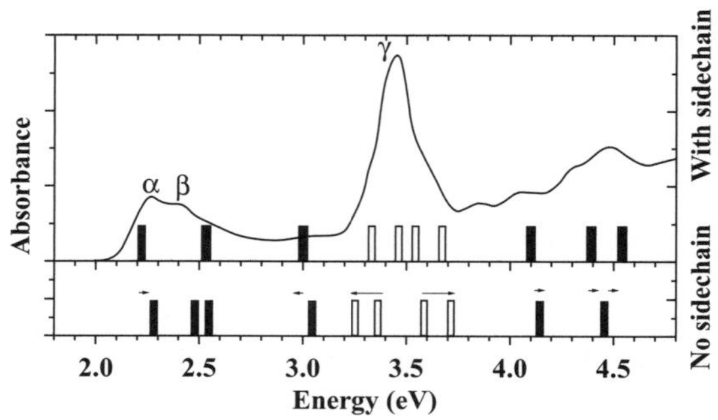

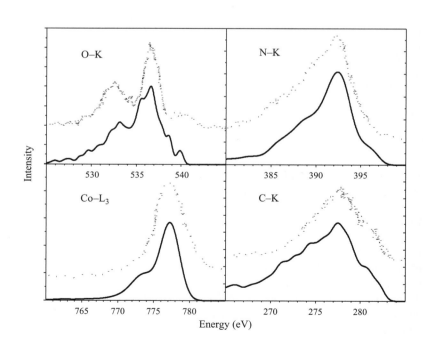

Fig. 10.5.
Optical absorption spectrum of CNCbl. The empty vertical segments represent transitions involving Co levels and the full vertical segments those of corrin.

Fig. 10.6.
Comparison of broadened orbital-resolved PDOS (solid lines) and measured soft-x-ray emission spectra (points) for O-K, Co-K, N-K, and C-K in CNCbl.

uncompensated charge group. Calculations could also be extended to the full cell containing the other molecules in the crystal with the possible addition of water molecules instead of just the isolated molecule. Such calculations can test the effect of intermolecular interactions and the effect of the solvents. Characterization of excited states of large biomolecules presents another serious challenge. Calculations on the XANES spectra of the Co edges will be discussed in Chapter 11.

10.2 b-DNA models

DNA or deoxyribonucleic acid is the fundamental building block of all known living organisms (Bloomfeild et al., 2000). DNA research is essential for understanding various subjects related to heredity, disease, medicine, agriculture, etc. In the area of nanotechnology, DNA is often considered simply to be a long polymer made of repeating nucleotides that can have special functional or structural applications such as the scaffolds for three-dimensional nanostructures.

The first accurate structural model of DNA was revealed by the famous work of Watson and Crick establishing it as a double helix (Watson and Crick, 1953). The phosphate-deoxyribose backbone consists of alternating phosphate (P) and sugar (S) groups that form a nucleotide chain with different base molecules attached to sugar. The sugar groups in DNA are in the form of a pentose with one O and five C atoms, and are joined by the phosphate groups that form the so-called phosphodiester bonds. The four DNA base molecules are adenine (A), thymine (T), guanine (G), and cytosine (C) (see Fig. 10.7). A and G are fused 5- and 6-membered heterocyclic rings called purines and C and T are 6-membered rings called pyrimidines. It is the sequence of these four base molecules along the backbone that encode the genetic information unique to a living organism. The unique feature of the DNA bases is the complementary base pairing across the double helix through hydrogen bonding that stabilizes the double helix (Kool, 2001). A bonds only to T and C only to G. The GC pair has three H bonds (1 H–N, 2 H–O) and the AT pair has two H bonds (1 H–N, 1 H–O). The H bonds are much weaker than the covalent bonds between atoms in the backbone but they are crucial for determining many of the physical properties of the molecule including the elasticity and strength between the two strands. Thus, double stranded DNA (dsDNA) with higher numbers of AT pairs have weaker interacting strands and are easier to separate into single stranded DNA (ssDNA) polymeric chains.

DNA exists in several conformations such as A-DNA, B-DNA, Z-DNA, and more, depending on the helical structure and other factors. However, the B-DNA is the most common one under the conditions of the living cell, and therefore it is often considered to be the most important. In this chapter, we restrict our discussion to the electronic structure and bonding of B-DNA models using the OLCAO method. Figure 10.8 shows a B-DNA model that is periodic in the axial direction (z-axis), with ten AT base pairs. Natural dsDNA molecules are very long, containing an estimated 3 billion base pairs, so that realistic quantum mechanical calculations are best realized with a periodic

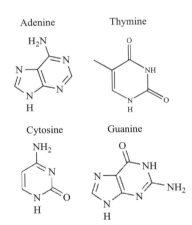

Fig. 10.7.
Chemical formulas of the four bases in DNA: A, T, C, and G.

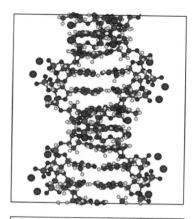

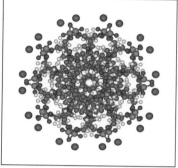

Fig. 10.8.

A periodic DNA model with 10 base pairs
and Na counter ions. The lower panel
shows the cross-section of the model.

model. The DNA in its relaxed state circles the axis of the double helix every
10.4 base pairs; therefore a periodicity of 10 base pairs is quite close to reality.
The periodic model avoids issues concerning the ends of the DNA strand
which are labeled as 5′ and 3′ depending on whether the strand ends at the
phosphate group or the sugar group respectively. The models were initially
constructed by first using the Amber program, and then imposing periodicity in
the axial direction (z-axis), before fully relaxing the models using VASP with
high accuracy. The x- and y-dimension of the simulation cell is set at 20 Å so
that there are no interactions between the dsDNA molecules from the adjacent
cells. To compensate for the effect of the solvents that neutralize the $[PO_4]^{-1}$
group, 20 Na ions were added to the two models and then the final equilibrium
structures were determined by minimizing the total energy as z varies. The
model shown in Fig. 10.8 has ten base pairs, a periodicity of 3.921 nm, a twist
angle of 36°, and 660 (650) atoms for the AT (GC) case (not shown).

The OLCAO method was used to investigate the electronic structure and
optical properties of the ten base pair AT and GC models, hereafter referred to
as the AT-10 and CG-10 models. Figure 10.9 shows the calculated total DOS of
the two models using a full basis set. Both are semiconductors with a HOMO–
LUMO gap of 3.30 eV and 2.90 eV respectively. The most important states are
those immediately below the HOMO and above the LUMO. These states come
mainly from the AT and GC pairs. Figure 10.10 shows the PDOS of the AT-10
model broken down into different functional groups. The PDOS of Na are all
in the unoccupied bands indicating that they are ionized by compensating for
the $(PO_4)^{-1}$ groups. Also calculated are the effective charges of every atom
in the model and the bond order values for every pair of atoms including the
H-bonds.

Fig. 10.9.

Calculated total DOS of at-10 and cg-10
models of DNA.

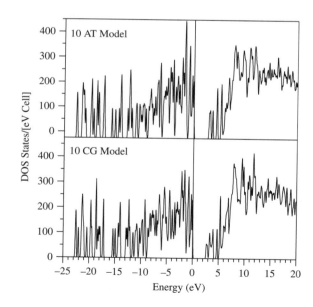

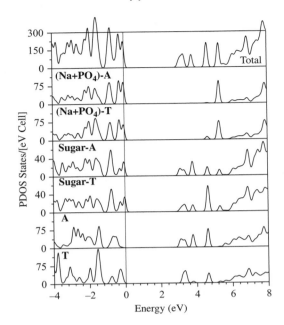

Fig. 10.10.
Calculated PDOS of different functional groups of the at-10 model.

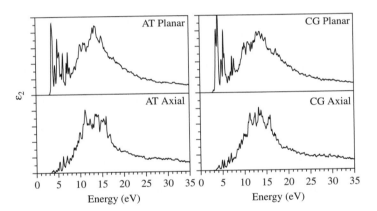

Fig. 10.11.
Calculated $\varepsilon_2(h\omega)$ in the axial and planar directions of the at-10 and cg-10 models of DNA.

The optical absorption curves for the AT-10 and GC-10 models were also calculated using the OLCAO method. Figure. 10.11 shows the imaginary part of the frequency-dependent dielectric functions resolved into the axial and planer components for the two models. The astonishing fact is that the anisotropy in the optical spectra, which show absorption in the first 5 eV range, is almost entirely absent from the axial direction. This anisotropy will have important implications in the long range van der Waals force calculations that will be further discussed in Chapter 12.

The elasticity of the dsDNA was also investigated by performing theoretical tensile experiments on supercomputers. The dsDNA model was sequentially elongated by changing the z-axis dimension at steps of 0.676 Å each using

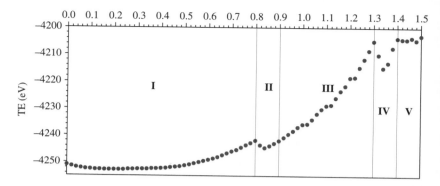

Fig. 10.12.
Variations of the total energy of the at-10 model of DNA as a function of stretching.

VASP up to a very large stretching and the resulting electronic structure and bonding at each step was calculated using the OLCAO method. Figure 10.12 shows the total energy vs. the z-axis length for the AT-10 model. These results show very interesting behavior of the changes occurring under the tensile strain that have a lot to do with the change in the geometry and the related H-bonding interaction. At high strain, a ring-opening mechanism occurs which may possibly explain some experiments associated with DNA over stretching in dsDNA.

The above results are from the beginning stages of DNA research using OLCAO and most of the results are still in preparation for publication. What we have shown is that the OLCAO method can be effectively applied to study DNA and other similar biomolecules. However, further consideration of the effects of solvents is needed. This will require the addition of water molecules to the model and an investigation of the changes in the electronic structure and optical absorptions. Detailed study of DNA models with specific base pairs and their modifications can lead to important insights that explain the character of DNA and its response to damage at the atomic level. Another possible avenue of study concerns the binding of proteins with DNA or the possible use of ssDNA wrapped around carbon nanotubes as a means to separate the CNTs with different chiralities (Tu et al., 2009). The use of DNA strands in nanotechnology and the self-assembly of biological systems in solutions is one emerging area of great promise. All these topics will require *ab initio* calculations of unprecedented complexity.

10.3 Collagen models

In this section, we discuss the application of the OLCAO method to another important biomolecule, collagen (Kuhn, 1987). Collagen molecules are the primary structural proteins found in all living animals. Together with mineral bioceramics such as hydroxyapatite (HAP) and tri-calcium phosphate (TCP) (see Chapter 7), these macromolecules constitute the major components of bones and teeth. The structure and properties of the triple helix chains in type-I collagen is an area of intense current interest because the interactions between collagen, mineral bioceramics, and their surrounding environment are important to human health but are still poorly understood. Collagen has a complicated triple helix structure that consists of peptide chains of many different amino acids. As in DNA, H-bonding plays a key role in the structure, properties, and functionality.

Collagen is a complex multifunctional protein of various forms based on different amino acid (residue) sequences (Ottani et al., 2001). The most abundant collagen molecule is type-I and it has a long rod-like triple helix structure consisting of three intertwined polypeptide chains of approximately 3000 Å in length and 15 Å in diameter. Most type-I collagen is in the heterotrimer $[\alpha 1(I)]_2[\alpha 2(I)]$ form where $\alpha 1(I)$ and $\alpha 2(I)$ are single chain molecules, although the homotrimer form $[\alpha 1(I)]_3$ can also be found *in vivo*. The primary structures of the α-chains in both types consist of amino acid sequences of the general triplet form $(Gly-X-Y)_n$. Although the residues X and Y can be any of the 20 natural amino acids, the presence of glycine (Gly) in every triplet in the chain is an absolute prerequisite for the correct formation of the triple helical structure (Ramshaw et al., 1998). Each α-chain is coiled into a helical secondary structure and the three separate chains are super coiled around a common axis into a triple helical tertiary structure. The collagen molecules are aligned to form microfibrils which are then bundled together to form collagen fibers (quaternary structure) that are the key to biomineralization.

Two models for type-I collagen molecules have been proposed based on x-ray crystallographic data on collagen-like model peptides (Wess et al., 1998) which are referred to as the 7/2 and 10/3 helices. They differ in the number of residues per turn around the central axis. In both models, the three α-chains are staggered by exactly one residue along the main helical axis and an angular displacement around the main axis. Figure 10.13 shows the

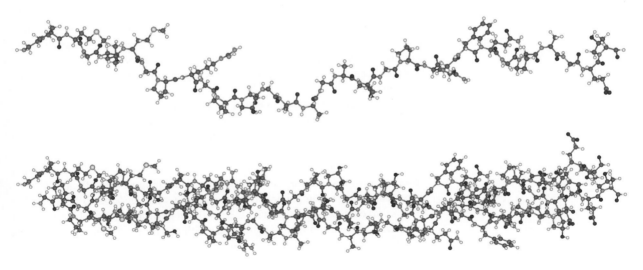

Fig. 10.13.

Top: single strand of $\alpha 2(I)$ chain in the 7–2 heterotrimer model of type I collagen. Bottom: model with all three strands in the triple helix.

structure of a single strand of the $\alpha2(I)$ chain and the three-chain helix structure in the 7–2 heterotrimer model. The amino acid sequence of $\alpha2(I)$ is: (Gly-Pro-Met) -(Gly-Leu-Met) -(Gly-Pro-Arg)-(Gly-Pro-Hyp)-(Gly-Ala-Ala)-(Gly-Ala-Hyp) -(Gly-Pro-Gln) -(Gly-Phe-Gln)-(Gly-Pro-Ala)-(Gly-Glu-Hyp). The two $\alpha1(I)$ chains have an amino acid sequence of: (Gly-Pro-Met)-(Gly-Pro-Ser) -(Gly-Pro-Arg)-(Gly-Leu-Hyp)-(Gly-Pro-Hyp)-(Gly-Ala-Hyp)-(Gly-Pro-Gln)-(Gly-Phe-Gln)-(Gly-Pro-Hyp)-(Gly-Glu-Hyp). This 7–2 hetero-trimer model contains a total of 90 residues, 1129 atoms, and 3240 valence electrons.

The properties of the collagen molecule are intimately related to its structure. The first triplet position must be occupied by a Gly residue with a single H-atom side chain whereas residues on the X and Y positions usually have side chains of different lengths with some of them being aromatic. Although the Gly-Pro-Hyp triplet is considered to be the most energetically stable, other residues with more substantial side chains can be accommodated within the triple helical structure with side chains directed away from the central axis. These charged side chains are easily accessible to the surrounding solvent to form additional water mediated intra-molecular bridges via H-bonding. The side chains are also accessible to other molecules or biominerals in the environment; thus the identity of the residues on the X and Y positions govern the interaction properties of the collagen molecule. Regions along the collagen triple helix with a high degree of substitution of Pro and Hyp for charged and hydrophobic residues often function as binding sites for other molecules or proteins. Finally, the interaction of the collagen molecule with crystalline bioceramics is also largely determined by the domains of specific residues on the X or Y position, and its attraction to a particular surface site. Specific substitutions or abnormal sequences in collagen can be related to certain diseases such as osteogenesis imperfecta which is caused by a single substitution of Gly resulting in the local deformation of the triple helix. The electronic influence of such substituted structures vs. the "normal" structure will be of great interest in pinning down how the defect affects the health of a living organism. Information on the electronic structure and bonding in collagen molecules can pave the way for understanding some of these intriguing phenomena.

The first step in conducting theoretical research on collagen molecules is to have plausible structure models at the atomistic level. This usually starts with the protein data base to construct an initial model that is then properly refined using molecular mechanics (MM) or similar techniques to incorporate specific features of the amino acid sequence that are consistent with experimental observations (Vesentini et al., 2005). Such models must be sufficiently large to be realistic yet simultaneously they must be small enough to be amenable to *ab initio* calculation. Even with the appropriate structure models of collagen molecules, there are practical difficulties with electronic structure calculations based on DFT. The most serious one is the presence of local charged units. Figure 10.14 shows sketches of two residues, Arg and Glu, in $\alpha1(I)$. Arg has a positively charged site NH^+ and Glu has a negatively charged CO^- site, while the entire molecule is electrically neutral. Such locally charged units are very common in polymers with radical centers. If they are separated by a distance on the order of 20 Å, they can cause difficulties in the convergence for the self-consistent field potential in the OLCAO calculation. Innovative and practical ways to address this problem need to be devised. The difficulty of local charged units can be avoided by adding a counter ion (Na) next to the CO^- unit in Glu and simultaneously removing an H in the N^+H_2 unit in Arg so that the number of electrons remains the same and the charge neutrality is maintained.

The OLCAO method has been applied to study the electronic structure of the 7–2 heterotrimer models. The calculations were done for both the individual chains as well as the entire triple helix. Na ions and H atoms were added or removed in the model to mitigate the effect of local charge discussed above. Some preliminary results from these calculations are presented here. Figure 10.15 shows the calculated total DOS for the single stranded $\alpha2(I)$ chain in the 7/2 heterotrimer model of Fig. 10.14. The HOMO–LUMO gap of this peptide is about 2.0 eV. The TDOS is further resolved into the PDOS of the 10 triplets in the $\alpha2(I)$ chain. It can also be seen that the HOMO–LUMO gaps for each of the triplets are different which may be related to the relative strengths in the peptide chain. The rich spectra contain information about the electronic structure and bonding in the molecule for individual triplets. For example, the Gly-Pro-Met triplet has a large presence at the top of the VB which determines the HOMO. The Gly-Pro-Ala and the Gly-Glu-Hyp triplets have very similar PDOS with large number of states immediately below the HOMO. It is possible to resolve the PDOS further down into individual residues or even atoms to obtain atomic details. Figure 10.16 shows the calculated charge transfer of each individual atom in the same strand from the Mulliken effective charge Q^*, and is expressed in a shaded representation (the original color version is more illustrative). It can be seen that there are differences in Q^* for atoms in different residues. Such information cannot be obtained from classical simulations but it is vital for explaining different geometric conformations as the result of intermolecular and intramolecular interactions.

The preliminary data presented above assures that the calculation of the electronic structure of complex collagen molecules by *ab initio* methods is possible and that analysis of the PDOS and effective charges can reveal intimate details. The nature of H-bonding and long range charge interactions is an extremely important issue in the fundamental study of collagen as well as for any other protein of biological interest. H-bonding is the origin of an attractive intermolecular force that is generally much weaker than ionic or covalent bonds and is responsible for maintaining the triple helix structure in collagen. So far, much of the discussion concerning H-bonding in collagen is based purely on geometric considerations (inter-atomic bond distances and bond angles) without any real input provided by *ab initio* quantum calculations. As has been demonstrated in the cases of the pure water model and the DNA model of last section, it is possible to obtain *ab initio* information for large systems with the OLCAO method. In the case of collagen, it is desirable to model the bridges between each of the polypeptide chains, to explore how the electronic structure evolves and to estimate the changes in the binding

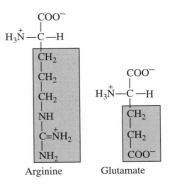

Fig. 10.14.
Sketch of residues ARG (left) and GLU (right) showing the positively charged N^+ site and the negatively charged O^- site respectively.

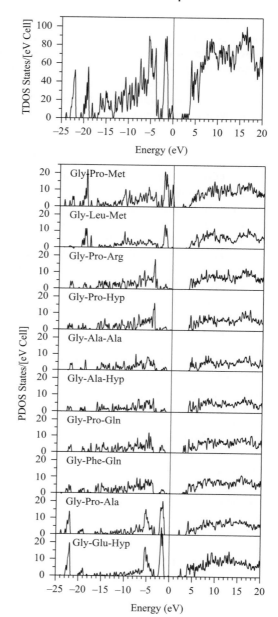

Fig. 10.15.

Calculated total DOS (top) and PDOS (bottom) for each triplet in the $\alpha2(I)$ chain. Note the difference in the HOMO–LUMO gaps for different triplets.

energy in different model structures. Additional water molecules can be added to the simulation cell to mimic the effect of solvents. It is also possible to calculate the optical absorption spectra of the collagen with and without solvents. Such calculations are within reach and they will help push collagen, and protein research in general, to a much higher level of accuracy and sophistication.

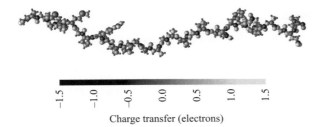

Charge transfer (electrons)

Fig. 10.16.
Ball and stick model of a single strand $\alpha2(I)$ with the size of the atoms represented by their covalent radii. The shade indicates the amount of charge transfer for each atom.

10.4 Other biomolecular systems

The examples described in the above three biomolecular systems amply demonstrate the effectiveness of the OLCAO method in investigating their electronic structure and spectroscopic properties. Of particular usefulness is a simple and direct way to obtain the charge transfer, intramolecular, and intermolecular bonding including hydrogen bonding. Such information is not easy to obtain using molecular dynamics, which is the main simulation method for complex biomolecular systems. It will be especially powerful if graphical presentations of the charge density and potential surfaces from the *ab initio* calculations can also be simultaneously presented. The present calculation has some limitations. The systems that can be treated are still limited to only several thousand atoms, far fewer than in real proteins and biomembranes. The use of the local density approximation in density functional theory also limits its accuracy. The structural models with specific atomic positions for the biomolecular system of interest are not always easy to construct. Connections to the experimental investigations cannot always be established, and interpretations that are vital for understanding biological functions and their impact on either biomedical applications or fundamental biological processes are not always easy to make. Still, over time, these limitations will likely be overcome as we gain more experience and develop new computational techniques.

It may be appropriate to speculate at this point about similar calculations on other bio-related systems. A particularly attractive system is the viral nano-particle or coated proteins such as the plant viruses BMV (Brome mosaic virus), CCMV (Cowpea Chlorotic Mottle Virus), CMV (Cytomegalovirus), or TMV (Tobacco mosaic virus) found either in dry form or in fluids. The fundamental long range interactions between them can be related to their self-assembly abilities. This is an area where quantum mechanical treatment is completely lacking. Another area is the interaction between different proteins or between proteins and bioceramics as in the case of collagen with hydroxyapatite. Also promising is the potential application of XAS spectral calculations on biomolecular systems. This subject has seen an increasing use of experimental research as an effective way to probe the geometric structure of biomolecules.

References

Banerjee, R. (ed.) 1999. *Chemistry and Biochemistry of B12* (New York: J. Wiley & Sons).

Bloomfeild, V. A., Crothers, D. M., & Tinoco, I. J. (2000), *Nucleic Acids, Structures, Properties, and Functions* (Sausalito: University Science Books).

Bonnett, R., Cannon, J. R., Johnson, A. W., Sutherland, I., Todd, A. R. & Smith, E. L. (1955), *Nature*, 176, 328–30.

Case, D. A., Darden, T. A., Cheatham, T. E., et al. (2010), *Amber 11* (San Francisco: University of California).

Hill, J. A., Pratt, J. M., & Williams, R. J. P. (1964), *Journal of the Chemical Society (Resumed)*, 5149–53.

Hodgkin, D. C., Pickworth, J., Robertson, J. H., Trueblood, K. N., Prosen, R. J., & White, J. G. (1955), *Nature*, 176, 325–8.

Kool, E. T. (2001), *Annual Review of Biophysics and Biomolecular Structure*, 30, 1–22.

Kratky, C., Faerber, G., Gruber, K., et al. (1995), *Journal of the American Chemical Society*, 117, 4654–70.

Kuhn, K. (1987), *Structure and Functions of Collagen Types* (London: Academic Press).

Ottani, V., Raspanti, M., & Ruggeri, A. (2001), *Micron*, 32, 251–60.

Ouyang, L., Randaccio, L., Rulis, P., Kurmaev, E. Z., Moewes, A., & Ching, W. Y. (2003), *Journal of Molecular Structure: THEOCHEM*, 622, 221–7.

Ramshaw, J. a. M., Shah, N., K. & Brodsky, B. (1998), *Journal of Structural Biology*, 122, 86–91.

Randaccio, L., Furlan, M., Geremia, S., Šlouf, M., Srnova, I., & Toffoli, D. (2000), *Inorganic Chemistry*, 39, 3403–13.

Tu, X., Manohar, S., Jagota, A., & Zheng, M. (2009), *Nature*, 460, 250–3.

Vesentini, S., Fitié, C. F. C., Montevecchi, F. M., & Redaelli, A. (2005), *Biomechanics and Modeling in Mechanobiology*, 3, 224–34.

Watson, J. D. & Crick, F. H. C. (1953), *Nature*, 171, 737–8.

Wess, T. J., Hammersley, A. P., Wess, L., & Miller, A. (1998), *Journal of Structural Biology*, 122, 92–100.

Application to Core Level Spectroscopy

11

Within the last quarter of a century, we have witnessed tremendous developments associated with experimental probes for the unoccupied states of solids and molecules using either photons or electrons. The theory and practice of x-ray absorption near edge structure (XANES) and electron energy loss near edge structure (ELNES) spectroscopy are amply documented (Egerton, 1996, Stöhr, 1992). The advances have been due to the availability of high intensity x-ray sources offered by many synchrotron radiation centers (SRC) worldwide, and the availability of high resolution transmission electron microscopy (HRTEM) and scanning transmission electron microscopy (STEM) in conjunction with electron energy loss spectroscopy (EELS). These are effective tools for the characterization of many different types of materials. To understand the underlying physics and to properly interpret the experimental data, many computational methods have been developed within the electronic structure theory of solids. The wide availability of computational resources has resulted in more realistic calculations for systems of greater complexity with increased accuracy. There are now several well established methods for XANES/ELNES calculation using different approaches and techniques (Blaha et al., 1990, Hatada et al., 2007, Moreno et al., 2007, Natoli et al., 1980, Rohlfing and Louie, 1998, Schwarz and Blaha, 2003, Shirley, 1998). In this chapter, we focus on the supercell OLCAO method which has been applied to a large number of pure and defect structure-containing materials. We present recent results and also discuss the prospect of extending such calculations to even more complex systems so that they may be used as a reliable predictive tool.

11.1 Basic principles of the supercell OLCAO method

The general theory for XANES/ELNES calculation in solids is based on quantum scattering theory (De Groot and Kotani, 2008, Egerton, 1996). The experimentally measured quantity is the inelastic partial differential scattering cross-section of the incoming particle (a photon or an electron):

$$\frac{d^2\sigma}{d\Omega dE} = \frac{1}{(\pi e a_0)q^2} IM\left\{\frac{-1}{\varepsilon(\vec{q}, \hbar\omega)}\right\} \qquad (11.1)$$

where $\varepsilon(\vec{q}, \hbar\omega)$ is the microscopic complex dielectric function that depends on both the wave vector and the frequency of the particle. For small momentum transfer and at energies far above the plasma frequency, the imaginary part (IM) of the inverse of $\varepsilon(\vec{q}, \hbar\omega)$ can be approximated as $\varepsilon_2(0, \hbar\omega)$ with no $\vec{q}$ dependence. The transition probability I per unit time for the inner shell core excitation within the dipole approximation can be reduced to the following simple expression according to the Fermi golden rule (Dirac, 1927).

$$I \propto \sum_n |\langle g \,|\vec{r}|\, f\rangle|^2 \, \delta(E_f - E_g - \hbar\omega) \qquad (11.2)$$

Here g and f stand for the initial and the final states with energies E_g and E_f respectively. The summation is over all n final states. The initial state g is the atomic-like core state of the target atom in the solid or molecule and f are the final states which span all conduction band (CB) states or unoccupied molecular orbitals. In the early days of XANES/ELNES calculation, the matrix elements in Eq. (11.2) were approximated by the orbital resolved local or partial density of states (PDOS) of the CB on the grounds that the atomic core level is highly localized and orthogonal to the final states. The dipole selection rule restricts the transition from a 1s core state ($\ell = 0$) to only the p states ($\ell = 1$) in the CB and the transition from a 2p core state ($\ell = 1$) can only be to s or d type states ($\ell = 0$ or 2). This approximation is a poor one because the modulation of peak intensities by the momentum matrix elements is not taken into account. Further the CB states of unoccupied molecular orbital states are generally delocalized and there is no precise way of decomposing them into different orbital components.

Another important issue with XANES/ELNES calculations for insulators is the core-hole effect. When an electron is excited from the inner core shell, it leaves behind a positively charged hole in the core (core-hole) which interacts with the excited electron in the CB. This is analogous to the excitonic effect in semiconductors. In metals, the core-hole effect is less important due to the effective screening by the CB electrons which reduces the interaction. Two different approaches have typically been used to account for the core-hole effect. The first is the Z + 1 approach which is based on the assumption that the presence of a core-hole can be mimicked by replacing the atom of atomic number (Z) with another atom of atomic number (Z + 1) (Robertson, 1983). The second approach is to use the Slater transition scheme where a half-electron is removed from the core orbital and put at the lowest unoccupied state (Tanaka and Adachi, 1996). A single diagonalization of the secular equation gives the transition energy ΔE, and the final spectrum is obtained by combining ΔE and the oscillator strength from the dipole matrix elements calculated between the initial and final states. Both approaches have their deficiencies. The Z + 1 approach cannot distinguish between transitions from different core levels in terms of its effect on the core-hole. The Slater transition scheme does not accurately reflect the actual interaction between the core-hole and the excited electron.

The OLCAO method was extended to the supercell OLCAO method more than ten years ago specifically for the purpose of performing XANES/ELNES spectral calculations (Mo and Ching, 2000). In this method, the core orbitals of the target atom whose absorption edges are to be calculated are retained while the core orbitals of all other atoms in the supercell are eliminated by orthogonalization. The dipole matrix elements of transition between the initial and final states are explicitly included, and the selection rules are automatically imposed by the symmetry of the wave functions. In the supercell OLCAO method, a different approach is used to account for the core-hole effect. The initial and final states are calculated separately. The initial state is the ground state of the supercell with the core states of the targeted atom retained throughout the orthogonalization procedure. The final states are obtained by placing a core electron in the lowest CB state and then solving the resulting Kohn–Sham equation self-consistently. The self-consistent solution of the final states accounts for all the multiple scattering effects and the interaction between the excited electron and the core-hole. The placement of the excited core electron into the CB is important because it maintains the charge neutrality condition. The core-hole effect cannot be accurately accounted for without using a supercell. If a small supercell is used, spurious interactions between the core-holes in the adjacent periodic cells may exist. This concept is illustrated schematically in Fig. 11.1. The size of the supercell is determined according to the shortest distance of separation between core-holes in adjacent cells. Thus, a cubic cell usually has a much smaller number of atoms in the supercell than an anisotropic elongated crystalline cell for the same distance of separation. In general, a distance of separation of about 9–10 Å between target atoms should be sufficient depending on the system to be studied. For highly anisotropic crystals, or for models involving microstructures or interfaces, supercells with up to several hundred atoms may be required. Fortunately, for a supercell of sufficiently large size, the corresponding BZ is very small. Therefore, one typically only needs to use a single k point for the SCF parts of the computation.

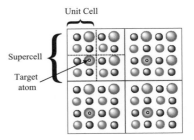

Fig. 11.1.
Supercell scheme for XANES/ELNES calculations.

The final calculation of the XANES/ELNES spectrum in the supercell OLCAO method entails the evaluation of Eq. (11.2) for the transition intensity between the initial core state and the final core-hole states which are calculated separately. The dipole matrix in Eq. (11.2) is equivalent to the momentum matrix element in the optical transition calculation. The explicit inclusion of the dipole transition matrix in the calculation of the XANES/ELNES spectrum is important because it provides more accurate amplitudes of transition in the spectral features. A Gaussian broadening with a FWHM of 1.0 eV is usually applied to the calculated spectra to account for the life-time broadening effect. The broadening of the spectrum can depend on the excitation energy and also on the instrumentation. This is a subject for future development. The specific procedures of the XANES/ELNES calculation using the supercell OLCAO method are summarized in the flow chart shown in Fig. 11.2.

There are several special advantages of the supercell OLCAO method for XANES/ELNES calculation which has led to its wide application. (1) The fundamental theory in the OLCAO method for electronic structure calculation is solidly based on DFT. Although many-body interactions are not considered, the separate calculation for the ground state and the final states is one step beyond the single particle approximation because the electron–hole interaction

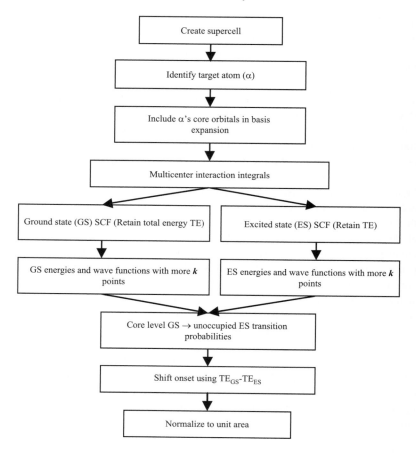

Fig. 11.2.
Flowchart for XANES/ELNES
calculation using the supercell OLCAO
method.

is explicitly accounted for in the final state calculation. (2) The OLCAO method is an all-electron method with the core states explicitly included. The physical presence of the core-hole in the method and its explicit interaction with the electron in the CB distinguishes itself from other approaches where the core is described by an effective potential. The full secular equation is diagonalized to obtain all electron states up to 40–50 eV above the absorption edge onset. Of course, the full diagonalization of a large secular equation for large complex systems can also be a computational burden. On balance, the supercell OLCAO method is still advantageous, as can be seen by its robust application to many complex systems. (3) The dipole matrix elements of transition between the initial and the final states are obtained from the *ab initio* wave functions and explicitly included in the spectral calculation. They automatically impose the selection rules for the transition and can also be resolved into Cartesian components to investigate the anisotropy of the XANES/ELNES spectra. The PDOS, which can be easily obtained in the OLCAO method, is only used to help in the interpretation of the final spectrum. (4) The transition energy ΔE for each absorption edge is obtained from the difference in the total energies of the ground state (an N electron system) and the final state (an

N-1 electron system with an extra electron in the CB). The ability to relate ΔE to experimental spectra is very important. The experimentally measured spectrum of a particular element in a crystal is the weighted sum of the spectra from all atoms of that element at non-equivalent sites. Each site will tend to have a slightly different local environment. When the spectrum of each atom is computed it can be seen that each spectrum is different with different edge onsets. Then, when the spectra are combined, a small difference in ΔE will result in a substantially different total spectrum from any of the individual spectra. This will be illustrated by specific examples later in the chapter. (5) The supercell OLCAO method is very versatile and can be applied to almost all elements in the periodic table and for any edges deep or shallow. It can be applied to almost any material system whether it is a metal, an insulator, an inorganic crystal, a biomolecule, having an open structure or a compact one, whether it contains numerous light atoms such as H or Li, or heavy atoms such as Au or rare earth elements. The method avoids the use of atomic radii in any parts of the calculation which could be problematic if the system under study is a complex one with different local bonding for the same element. (6) The use of local atomic orbitals and the efficient evaluation of multi-center integrals in the analytic forms described in Chapter 3 make the supercell OLCAO method highly efficient such that it can be applied to large systems.

11.2 Select examples

11.2.1 Simple crystals

In this subsection, we discuss the application of the supercell OLCAO method to simpler crystals. Since the early work by Mo and Ching on MgO, $MgAl_2O_4$, and α-Al_2O_3 (Mo and Ching, 2000), the method has been applied to many other crystals. We select a few of these for discussion focusing on some particular points to demonstrate the effectiveness of the method.

Figure 11.3 shows the effect of the supercell size on the O–K edge in MgO. It clearly shows the need for a sufficiently large supercell in order to avoid

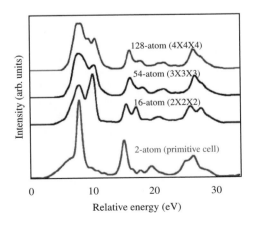

Fig. 11.3.
Calculated O–K edges in MgO of different supercell size.

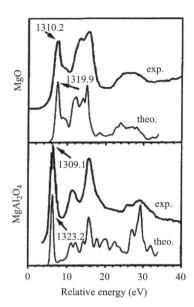

Fig. 11.4.
Calculated and measured Mg–K edge in
MgO and MgAl$_2$O$_4$.

the interactions between the core-holes in adjacent cells. Only at the size of $4 \times 4 \times 4$ with 128 atoms will the calculated spectrum be in good agreement with the measured data. Similar tests on many other crystals established a rule of thumb that the suitable supercell size should at least provide a distance of separation of about 9–10 Å between the core holes. The actual criterion could vary depending on the shape of the crystal and the type of atoms involved.

Figure 11.4 shows a comparison of the calculated Mg–K edges in MgO and spinel MgAl$_2$O$_4$. These two Mg–K spectra are quite different but both are in excellent agreement with the experimental data. The other edges, O–K, Mg–L$_{2,3}$, Al–K, and Al–L$_{2,3}$ in the three crystals all have similar good agreements. Another significant finding in this work is that the wave functions of the excited electron in the CB in the presence of the core-hole are only moderately localized and are significantly different from the CB wave functions obtained from the ground state calculations.

Other simple crystal examples are α-quartz (α-SiO$_2$) and stishovite SiO$_2$ (Mo and Ching, 2001). These two important crystals have very different crystal structures and different XANES/ELNES spectra especially for the O–K edge. The supercell OLCAO method was able to reproduce all the spectral details for the O–K, Si–K, and Si–L$_{2,3}$ edges. This is shown in Fig. 11.5. The usual interpretation by way of the orbital-resolved PDOS of the CB in the ground state calculation is unsatisfactory. This work later led to the prediction of very different spectra for the O–K, Si–K, and Si–L$_{2,3}$ edges in the new SiO$_2$ phase called the iota-phase (i-SiO$_2$) (Ching et al., 2005).

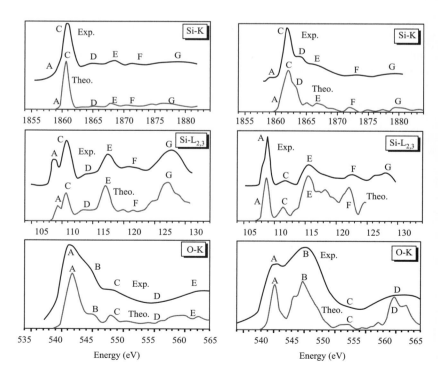

Fig. 11.5.
Calculated and measured O–K, Si–K, Si–L$_3$, and O–K edges in SiO$_2$. Left: α-quartz; right: stishovite.

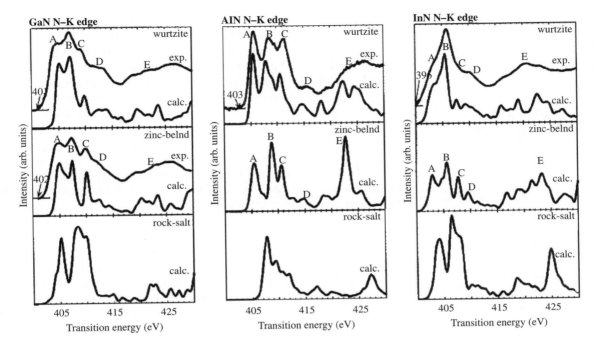

Fig. 11.6.
Calculated and measured N–K edges in GaN, AlN, and InN in wurtzite, zinc blend, and cubic structures.

Other examples of XANES/ELNES OLCAO calculations for simple crystals are for various phases of TiO_2, AlN, GaN, InN, and ZnO (Mizoguchi et al., 2004). Figure 11.6 shows the N–K edges in GaN, AlN, and InN in wurtzite, zinc-blend, and rock salt structures. The agreement with available experimental data is very satisfactory. Figure 11.7 shows the Zn–K, Zn–$L_{2,3}$, and O–K edges in the wurtzite phase of ZnO with the spectral data decomposed according to directions perpendicular and parallel to the crystalline c-axis. The agreement with the measured data from a polarized x-ray source in the two different directions is again very satisfactory and thus demonstrates the capability of the method to investigate the orientation dependence of the spectra. These and other examples for many different crystals clearly demonstrate the accuracy and the efficiency of the supercell OLCAO method.

An interesting example in the XANES/ELNES spectral calculation is the case of the three phases of $AlPO_4$ which are formed at different pressures (Pellicer-Porres et al., 2007). The most well-known phase is α-$AlPO_4$ (berlinite) with a trigonal structure that is isostructural with α-quartz so that both Al and P are tetrahedrally coordinated with O. At a pressure of 13 GPa, it transforms into an orthorhombic phase (o-$AlPO_4$) in which the Al ions become octahedrally coordinated but the P ions remain tetrahedrally coordinated. At an even higher pressure of 97.5 GPa, $AlPO_4$ transforms into a monoclinic phase (m-$AlPO_4$) in which both Al and P are octahedrally coordinated. This was the

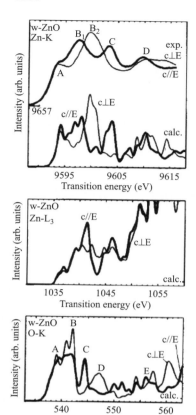

Fig. 11.7.
Calculated and measured Zn–K, Zn–L_3 and O–K edges in wurtzite ZnO crystal resolved into two perpendicular components.

first time the presence of six-fold coordinated P was reported. This is a rare example of structurally different crystals with the *same* formula unit, the *same* number of different types of ions, but distinctively different local units among them. According to the then prevailing notion of the so-called "finger printing" technique, the XANES/ELNES spectra of an ion can be predicted by the local nearest neighbor (NN) coordination. However, the supercell OLCAO calculation shows that this is not always true (Ching and Rulis, 2008b). Figure 11.8 is a side-by-side comparison of the calculated Al–K, Al–$L_{2,3}$, P–K, P–$L_{2,3}$, and O–K edges of Al, P, and O with the same local NN coordination from the three phases of $AlPO_4$. They are quite different and none of these comparisons supports the "finger printing" interpretation technique. The only experimental data we can locate is the P–K edge in α-$AlPO_4$ (Franke and Hormes, 1995) which is in good agreement with the calculations.

11.2.2   Complex crystals

The main advantage of the supercell OLCAO method is its applicability to complex crystals with large unit cells and many non-equivalent sites. Here we selectively discuss several such applications as illustrative examples. The successful synthesis of Si_3N_4 in the cubic spinel structure (γ-Si_3N_4) in 1999 (Zerr et al., 1999) has led to many theoretical and experimental studies of spinel nitrides. Three binary spinel nitrides γ-Si_3N_4, γ-Ge_3N_4, and γ-Sn_3N_4, have been successfully synthesized (Leinenweber et al., 1999, Serghiou et al., 1999, Shemkunas et al., 2002). Many of the early calculations using the supercell OLCAO method were on these compounds. The unique feature of spinel nitride is that the same cation occupies both the tetrahedral A site and the octahedral B site of the spinel lattice. We will first demonstrate the spectral differences between γ-Si_3N_4 and those of other compounds with slightly simpler structures. Figure 11.9 compares the calculated Si–$L_{2,3}$ edges with the measured spectra in γ-Si_3N_4 and hexagonal β-Si_3N_4 crystals. Both are in good agreement. The Si edges for γ-Si_3N_4 are the weighted sum of the spectra from the tetrahedral and the octahedral sites. Figure 11.10 shows a comparison of the N–K edge in the two crystals. This time the spectrum in the β-Si_3N_4 is the weighted sum of the two crystallographically non-equivalent N sites. Again, the agreement between the measured and the calculated spectra is very satisfactory. Figure 11.11 shows the calculated Si-K edges in five different crystals containing Si. The spectrum from γ-Si_3N_4 is distinctly different from other Si-containing crystals. These results clearly demonstrate the accuracy of the supercell OLCAO method in producing such spectra.

The only ternary compound that has been studied in considerable detail is the spinel nitride with Si and Ge as cations. There were controversies as to the preferred site for Si/Ge in the spinel lattice. XANES/ELNES spectra would be a useful technique to resolve some of these controversies. To this end, the XANES/ELNES spectra in the four spinel nitrides γ-Si_3N_4, γ-Si_3N_4, γ-$SiGe_2N_4$, and γ-$GeSi_2N_4$ were calculated using the OLCAO method (Ching and Rulis, 2009). Also calculated are the N–K edges in the four crystals which are shown in Fig. 11.12. The figure shows that the calculated N–K edges in

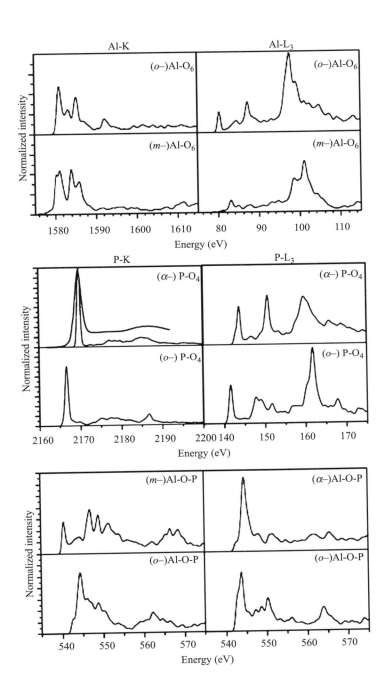

Fig. 11.8.
Comparison of the Al–K, Al–L$_3$, P–K, P–L$_3$, and O–K edges in α-AlPO$_4$, o-AlPO$_4$, and m-AlPO$_4$ crystals with different local bonding configurations.

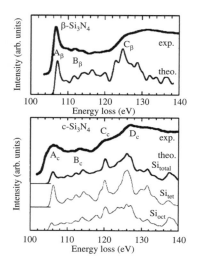

Fig. 11.9.
Calculated and measured Si–K, Si–L₃ in
γ- and β-Si₃N₄.

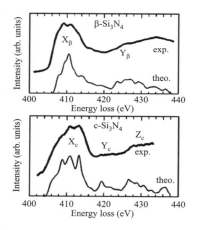

Fig. 11.10.
Calculated and measured N–K edges in
γ-Si₃N₄ and β-Si₃N₄.

the four spinel crystals have unambiguous and distinctively different features. Because there is only one unique site for N in the spinel nitrides, the use of the theoretical N–K edge spectra to interpret an experimentally measured one will likely be effective at identifying the correct phase in the ternary spinel nitrides. Within the first 25 eV from the edge onset, γ-SiGe₂N₄ has two prominent peaks A and B and a smaller peak C. On the other hand, γ-GeSi₂N₄ shows four well-resolved peaks A', A, B, and C within the same energy range with C being the most prominent. Also, the N–K edge in γ-GeSi₂N₄ has a steeper onset slope than that in γ-SiGe₂N₄.

Another example of XANES/ELNES calculation in complex crystals is the Y-Si-O-N system. The O–K, N–K, Si–K, Si–L₂,₃, Y–K, and Y–L₂,₃ edges in six binary (α-SiO₂, stishovite SiO₂, β-Si₃N₄, α-Si₃N₄, γ-Si₃N₄, Y_2O_3), three ternary (Si₂N₂O, Y₂Si₂O₇, Y₂SiO₅), and three quaternary (Y₂Si₃N₄O₃, Y₄Si₂O₇N₂, Y₃Si₅N₉O) crystals were calculated using the supercell OLCAO method and carefully analyzed (Ching and Rulis, 2008a). No meaningful correlation can be established between spectral features of the calculated spectra and the atomic environment based on the NN coordination which implies that a simple use of the "finger printing" technique based only on nearest neighbor coordination number is not valid in complex crystals.

More recently, the new beryllium phosphorus nitride (BeP₂N₄) has been synthesized at high temperature and high pressure (Pucher et al., 2010). BeP₂N₄ adopts the phenakite-type of structure isostructural with Be₂SiO₄ and β-Si₃N₄. It is classified as a true double nitride and not as a beryllium nitridophosphate because both Be and P retain coordination number 4. *Ab initio* calculations predicted the possible existence of a spinel-type phase of BeP₂N₄ at pressures around 24 GPa. The phenakite-type structure has a rhombohedral unit cell with 42 atoms in the primitive cell. There are two crystallographically non-equivalent P sites (P1 and P2) and both are tetrahedrally coordinated to four non-equivalent N atoms (N1, N2, N3, N4). The structure can be considered as consisting of corner-sharing BeN₄ and PN₄ tetrahedra. The predicted spinel-type structure has a face-centered cubic structure with a lattice parameters of 7.4654 Å. The positions for Be, P, and N are each uniquely defined with Be and P occupying the tetrahedral and octahedral sites of the spinel lattice, respectively. Figure 11.13 shows the calculated Be–K, P–K, P–L₂,₃, and N–K edges of BeP₂N₄ in the phenakite- (left panel) and spinel-type (right panel) phases using the supercell OLCAO method. For Be–K, the phenakite-type phase has a peak at 118 eV near the edge onset while the spinel-type phase has a very sharp peak above the edge onset at 119.3 eV. Such a large difference in the Be–K edge could be the most distinguishable feature for phase identification. The comparison of the N–K edges shown in Fig. 11.13 is between the single spectrum for the spinel-type phase and the averaged spectrum over four sites for the phenakite-type phase. There are considerable differences between the spectra from the four N sites (Ching et al., 2011). However, only the averaged spectrum will be observed experimentally. The averaged spectrum shows large differences from the single N–K edge of the spinel-type phase. Experimental measurements of XANES/ELNES spectra for the phenakite-type phase should be able to verify the theoretical predictions and the predicted spectra for the spinel-type phase can help to identify this yet to be synthesized high-pressure phase of BeP_2N_4.

11.2.3   Y–K edge in different local environments

From a computational stand point, defects and microstructures in crystals such as grain boundaries modeled by large super cells are not different from an extremely complicated crystal.

It is instructive to see the XANES/ELNES spectra of the same element in different crystals or in different local environments. Here we use the Y–K edge as an example. Figure 11.14 shows the calculated Y–K edge of: (a) a segregated Y ion in the grain boundary core of a $\Sigma 31$ GB in α-Al_2O_3; (2) the Y_2O_3 crystal in bixbyite structure; (3) the Yttrium alumina garnet (YAG) crystal in the garnet structure; and (4) Y in the $YBa_2Cu_3O_7$ (YBCO) high T_c superconductor. The $\Sigma 31$ GB model in that study had 700 atoms and contained two oppositely oriented GBs. The electronic structure and bonding in these crystals and GBs has been discussed in the previous chapters. The $\Sigma 31$ GB model is sufficiently large to serve as a supercell for the supercell OLCAO calculation. The Y ion is at the center of the 7-member rings of Al columns as observed from scanning transmission electron microscopy (STEM) experiments on bicrystals of α-Al_2O_3 doped with Y (Buban et al., 2006). The supercells used in the three

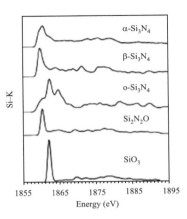

Fig. 11.11.
Comparison of calculated Si–K edges in five different crystals.

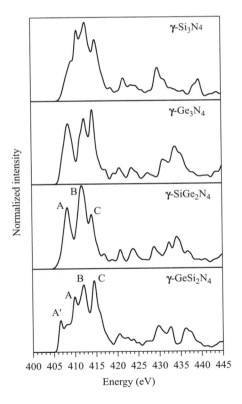

Fig. 11.12.
Comparison of calculated N–K edges in the four spinel nitrides.

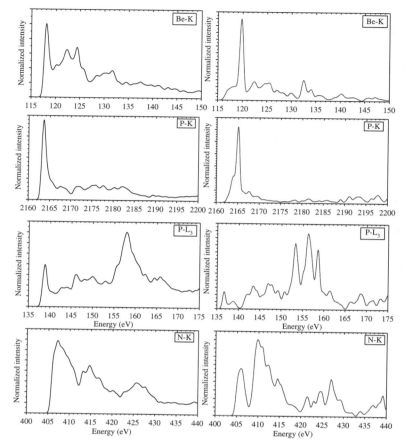

Fig. 11.13.
Calculated Be–K, P–K, P–L$_3$, N–K
edges in BeP$_2$N$_4$ in phenakite structure
(left) and spinel structure (right).

crystals of Y$_2$O$_3$, YAG, and YBCO have respectively 80, 160, and 117 atoms.
As can be seen, the Y–K edges in these four cases are completely different
because of the very different local environments of Y. In the Σ31 GB, Y is
bonded to six different O ions with bond lengths ranging from 2.14 Å to 3.00
Å. In Y$_2$O$_3$ there are two different Y sites, both are six-fold coordinated with
slightly different bond lengths. In YAG, Y is eight-fold coordinated with two
bond lengths of 2.30 Å and 2.43 Å. In YBCO, Y is also eight-fold coordinated
with Y-O bond lengths of 2.39 Å and 2.41 Å, similar to that in YAG. The large
differences in these spectra, especially in the case of the segregated Y in the
Σ31 GB, underscores the sensitive dependence of the spectrum of an ion to its
local environment, not just the number and type of NN atoms and their bond
lengths. Effects are also due to the presence of other ions beyond the NN atoms
such as in the YBCO superconductor.

11.2.4   Boron and boron-rich compounds

The electronic structures of boron and boron-rich compounds have been discussed in Section 5.6. Here we discuss the XANES and ELNES spectra of

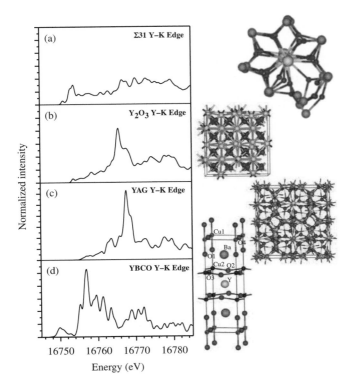

Fig. 11.14.
Comparison of the calculated Y–K edges in four systems. (a) Y at the core of the $\Sigma 31$ GB n alumina; (b) Y_2O_3; (c) $Y_3Al_5O_{12}$ (YAG); and (d) $YBa_2Cu_3O_7$ (YBCO). The structures of these four systems are sketched at the right.

these crystals using the supercell OLCAO method. The most fascinating part is their unusually complex spectra with multiple peaks even in the case of the elemental α-B_{12} crystal. This is attributed to the icosahedral B_{12} unit found in B and boron-rich compounds which, in α-B_{12}, has two non-equivalent sites (polar and equatorial) with 2-center 2-electron covalent bonding and unusual 3-center 2-electron bonding. Figure 11.15 shows the calculated B–K edges at the two different sites in α-B_{12}. It can be seen that the spectra at each site have multiple peaks and that their edge onsets are slightly different. When added together to form the total B–K edge in α-B_{12}, the spectra become even more complicated. The amazing part is that this calculated total B–K edge is in very good agreement with the measured data (Garvie et al., 1997) which is shown in Fig. 11.16. Almost all the peak structures are reproduced. Similar calculations have also been obtained in other B-rich compounds such as B_4C and $B_{12}O_2$ crystals. In spite of their complex spectral features, the calculated total B–K edge spectra are generally in good agreement with the measured ones with almost all the spectral features faithfully reproduced from the weighted summation of individual spectra. So, for the newly discovered high pressure phase (γ-B_{28}) where no measured data are available, the calculated spectrum serves as a prediction to be verified by experiments (Rulis et al., 2009b). There are many challenging opportunities for studying the XANES/ELNES spectra in other elemental B systems such as β-boron, tetragonal boron, or the many other

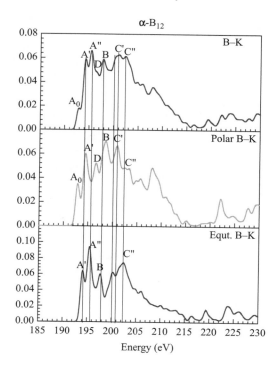

Fig. 11.15.
Calculated B–K edge in α-B_{12} at the polar and equatorial sites.

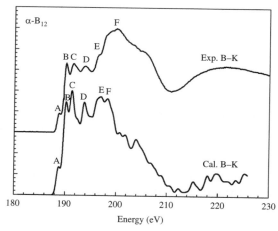

Fig. 11.16.
Comparison of the calculated and measured B–K edge in α-B_{12}.

boron-rich compounds. Theoretically, calculated spectra will serve as one of the characterization tools that can complement the experimental measurement whose data are likely to be difficult to interpret.

11.2.5 Substitutional defects in crystals

A notable example of the supercell OLCAO calculation for XANES/ELNES spectra is the application to identify ultra-dilute dopants in MgO (Tanaka

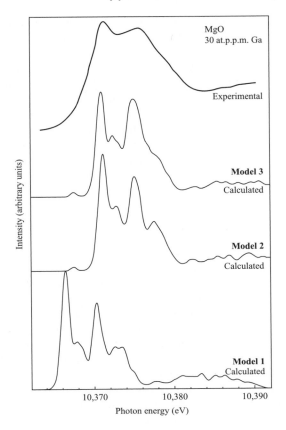

Fig. 11.17.
Calculated Ga–K edge in three defect models in MgO and compared with the measured spectrum.

et al., 2003). Substitutional Ga in MgO has a very low concentration of less than 1000 ppm. Its identification is beyond the reach of many other experimental probes and the local geometry of the impurity atom (Ga) in the MgO lattice is unknown. By assuming plausible models of the substituted Ga in MgO and calculating the Ga–K edge for each of these models, the one that matches well with the experiment will provide convincing evidence for the nature of its local geometry. Figure 11.17 shows the comparison of the Ga–K edge as measured by XANES and those calculated for three different models. Model 1 has a simple substitution of Mg by Ga. Models 2 and 3 contain two substituted Ga with one Mg vacancy at the nearest neighbor (NN) and next NN site respectively. Model 3 appears to have a better match than model 2, and model 1 can be ruled out. This example demonstrates the use of the calculated XANES spectra in conjunction with measurement to establish the local geometry of ultra-dilute dopants in oxides.

Another example impurity in crystals is the substitution of Ca by Zn in crystalline hydroxyapatite (HAP), which is a model for the major component of hard tissues in bones. There are considerable substitutional impurities in HAP, either natural or intentional, which affect their physical properties and physiological functions. Identification of the preferable sites of substitution

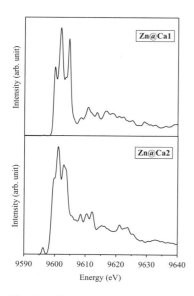

Fig. 11.18.
Calculated Zn–K edges for Zn substitution at Ca1 (top) and Ca2 (bottom) site in hydroxyapatite crystal.

and their local geometry is a subject of great importance. There are two Ca sites in the HAP crystal, the Ca1 and Ca2 sites, as has already been discussed in Section 7.5. It is not clear which site the Zn impurity would prefer to occupy in HAP. XANES/ELNES spectroscopy in conjunction with the calculated spectrum, such as the Zn–K edge, is ideally suited to answer this question. Figure 11.18 shows the calculated Zn–K edge for substitution at Ca1 and Ca2 in a $2 \times 2 \times 2$ supercell of 352 atoms (Matsunaga, 2008). By comparing this with the measured spectrum (Matsunaga et al., 2010), it can be easily concluded that the substitution site Ca2 is the right one in spite of the low energy resolution of the experimental data. Similar investigations have been carried out with other common substitutional impurities in HAP using the OLCAO method.

11.2.6 Biomolecular systems

In Chapter 10 about biomolecular systems, we described the electronic structure and bonding in five vitamin B_{12} cobalamins (Cbls) and double stranded DNA. Measurement of the XANES spectra of biomolecular systems is one of the fastest developing areas in the study of complex biomolecular materials. Here we describe efforts to use the OLCAO method to perform such calculations. In biomolecular systems, the focus will be on a particular atom of interest. In the case of cobalamins, the central focus is the octahedral Co (III) ion at the center of the corrin ring to which the cofactor R (R = CN, NH_3, Ado, or OH) is attached. Figure 11.19 shows the calculated Co–K edge in CNCbl in comparison with the measured data before irradiation (Champloy et al., 2000). The motivation of this experimental study was to investigate if the bond elongation of Co is due to the x-ray induced reduction of the cofactor's Co center. Measurement of Co K-edges before and after irradiation may shed some light on this issue. In spite of the low experimental energy resolution, there is a very decent agreement between the calculated and the measured spectra. Not only is the double peak structure and its slopes faithfully reproduced, but also in agreement is the presence of two pre-peaks between 7670 eV and 7680 eV as well as some other minor structures in the spectrum. This example gives us confidence to use the supercell OLCAO method to calculate the XANES spectra in other complex biomaterials and biomolecules.

Fig. 11.19.
Calculated (lower) and measured (upper) Co–K edge in CNCbl.

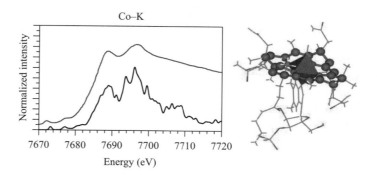

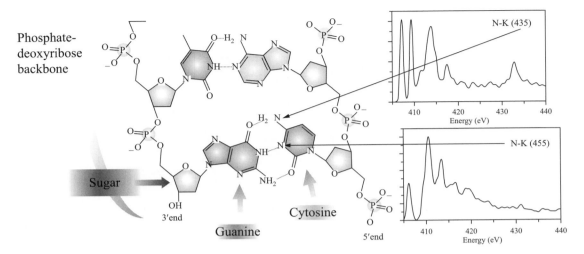

Fig. 11.20.
Calculated N–K edges at two different sites in a b-DNA model.

A preliminary calculation of the N–K and O–K edges in the B-DNA model described in the previous chapter has also been carried out. The result is illustrated in Fig. 11.20 for two XANES spectra at two different N-sites in the cytosine base of the CG-10 model. The first N atom (labeled 435) is the one involved in H bonding with a H from the guanine base, and the second N atom (labeled 455) is attached to the 6-atom ring in cytosine but not involved in any H-bonding. They have very different N–K edges in all three energy ranges: (1) region with energy less than 410 eV at the pre-edge or near edge onset; (2) energy range from 410 eV to 425 eV; and (3) energy range above 425 eV. Although these results are very preliminary and may require many more calculations on other sites for a more in-depth analysis, they clearly demonstrate that the supercell OLCAO method can compute spectral edges in biomolecular systems that are very sensitive to different types of inter-atomic bonding.

11.2.7   Application to grain boundaries and surfaces

ELNES spectroscopy using HRTEM has been used to study grain boundaries and other microstructures in ceramics for quite some time. Naturally, it is attractive to build grain boundary models in the form of a supercell and then to calculate the ELNES spectra of the atoms at the grain boundaries. One of the earliest such calculations using the OLCAO method was on a {122} $\Sigma = 9$ GB model in β-SiC (Rulis et al., 2004). This was discussed in subsection 9.4.2 and Fig. 9.16 shows the labeling of the atoms in that model. The special feature of this model is the presence of the so-called Si-Si and C-C "wrong bonds" in SiC due to the creation of the 5-member and 7-member rings at the GB which have

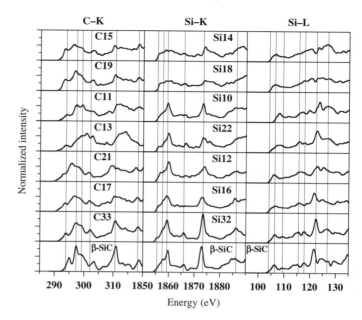

Fig. 11.21.
Calculated C–K, Si–K, Si–L$_3$ edges in the non-polar SiC GB model for selected atoms. The corresponding spectra from the bulk crystalline β-SiC are shown (see cited reference for details).

been experimentally observed (Tanaka and Kohyama, 2002). These wrong Si-Si (C-C) bonds have large deviations in bond length from the bulk Si-C bonds in β-SiC and this leads to different electronic structures and XANES/ELNES spectra. Figure 11.21 shows the calculated C–K, Si–K, and Si–L$_{2,3}$ edges for selected wrong bond atoms (Si14, Si18, C15, C19), GB region atoms (Si10, Si22, Si12, S16, C11, C13, C21, C17), and bulk crystal region atoms (Si32, C33). When compared with the corresponding spectra of perfect β-SiC crystal, it is clear that the wrong bonded atoms have spectra that are very different from those of the bulk atoms. Furthermore, for atoms in the GB region that do not have the wrong bonds, their spectra also deviate slightly from those in the bulk due to large bond angle variations. This clearly demonstrates that the XANES/ELNES spectra of atoms depend critically on their local bonding environments. Similar calculations on the other GB model with polar interfaces give similar conclusions.

ELNES calculations for surface models using the supercell OLCAO model have also been attempted. Specific examples are the (001) surface models of FAP and HAP crystals (Rulis et al., 2007). The calculated O–K and Ca–K edges of the surface atoms differ significantly from the same type of atoms in the bulk region. This is illustrated in Fig. 11.22 for Ca–K edges on different layers in the model for the (001) surface of FAP. The Ca atoms on the surface show a much wider distribution of spectral weights than Ca atoms at the subsurface level or in the bulk region. These differences are in addition to the difference for the Ca1 and C2 sites in the crystal. The variations are much smaller for the O–K edges at different layers (Rulis et al., 2007).

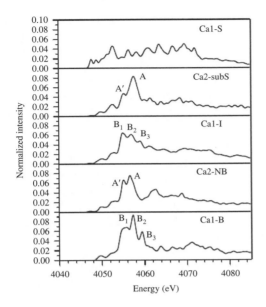

Fig. 11.22.
Calculated Ca–K edges for Ca at different locations in a fluorapatite (FAP) surface model.

11.2.8   Application to intergranular glassy films

In subsection 8.3.2, we discussed the electronic structure of a prismatic model for an intergranular glassy film (IGF) in β-Si_3N_4. The ELNES spectra for all atoms within the IGF region have been calculated using the supercell OLCAO method. With 907 atoms in the model, this is the largest calculation applied to complex structures. The extreme sensitivity of the calculated ELNES spectra to the local atomic environment in the targeted region of the material makes it a powerful tool for characterization. This is particularly useful when the exact environment has some degree of uncertainty. However, finding correlations between specific local atomic environments and the resultant computed spectra is still a daunting task. Experimentally, this is still an extremely difficult region in which to obtain reliable spectra with reasonable resolution.

The O–K, Si–K, Si–$L_{2,3}$, and N–K edges for all the atoms within the IGF region are calculated and the local atomic environments of these atoms are properly documented (Rulis and Ching, 2011). Here we present only the averaged spectra as shown in Fig. 11.23. Also shown are the same Si and N edges for the atoms in the bulk region of the model which are similar to those of the perfect β-Si_3N_4 crystal. For the O–K edge, the comparison is with that of SiO_2.

The averaged IGF spectra have substantially fewer sharp features because the averaging process tends to eliminate fine details. For the N–K edge spectra, the overall shape of the bulk-like spectra can still be identified in the averaged IGF spectra, but the resemblance is not very strong in spite of the fact that each of the individual N–K edge spectra is similar to that of the bulk but with differences in fine structure. The Si–K and Si–L edges also have striking differences between the bulk-like and IGF region. In the IGF region the Si

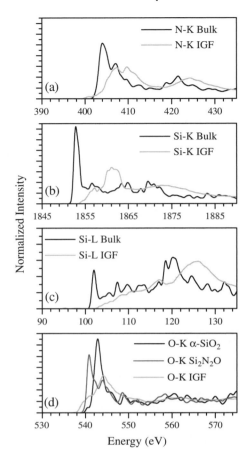

Fig. 11.23.
Site averaged N–K, Si–K, Si–L$_3$, and O–K edges in the prismatic IGF model in β-Si$_3$N$_4$ (see cited reference for details).

atoms have bonding configurations of the type: 4O, 2N2O, 3N, 4O, and 1N2O. Although many individual spectra retain an appearance similar to that of the bulk-like spectra, including some subtle features, the slight shifts in the energy onset values cause the first narrow peak of each spectrum to misalign with the others resulting in the elimination of the leading peak when they are combined. This could explain the experimentally observed decrease in intensity of the leading peak of the Si–L edge in such IGF systems (Gu et al., 1995). For the O–K edge, most of the O atoms in the IGF region assume the simplest bridging structure which is also the most flexible. Again, the overall shape of the spectrum from each atom tends to be similar, but the fine structures tend to be different. It fact, no clear trend in either the peak structure, number of peaks, or energy onset exists when the spectra are ordered by Si-O-Si bond angle or Si-O bond lengths.

The general process of interpreting and comparing computed ELNES spectra obtained from models of complex structured systems is quite difficult and the method of finger printing is not likely to be useful for interpreting measured spectra in regions with large structural variation. A key difficulty associated with the measurement of ELNES spectra in IGFs is the problem of spatial resolution which poses a serious technical challenge since the IGF is rather small in size and inhomogeneous in nature. Subtracting the bulk spectra is a useful tool for solving part of the problem, but is of less utility when studying variation within the IGF itself. A theoretical calculation on individual atoms in the IGF with specific local bonding enables us to understand the reasons that lead to such experimental difficulties.

11.2.9   Statistical description of O–K edges in bulk water

In Section 8.4, we discussed the electronic structure and hydrogen bonding network in a model of bulk water with 340 water molecules. Here, we focus on the calculation of the O–K edges for *all* 340 oxygen atoms in this model using the supercell OLCAO method and comparing them with the measured XANES spectra of water. This data provides us with a sufficiently large sample of O–K edge spectra to investigate meaningful correlations of spectral features with the hydrogen bond (HB) structure in water. Figure 11.24 (a) compares the calculated results with experimental x-ray Raman scattering (XRS) (Wernet et al., 2004) and x-ray absorption spectra (XAS) (Myneni et al., 2002, Rulis and Ching, 2011). After aligning the main peak, the agreement is very good, showing a pre-edge peak at ∼535 eV, a shoulder-like structure near 537 eV, a main peak at 538 eV, and a post-edge broad peak between 540 and 541 eV.

The 340 O–K edge spectra are divided into four groups according to the HB numbers and shown in Fig. 11.24 (b). It is clear that the pre-peak originates mostly from the H_2O molecules with 2 HBs and not from the over-bonded group with 5 HBs. This is consistent with other findings on the relation between the pre-peak intensity and broken HBs. On the other hand, the spectral features in the post-edge region are more affected by the fully-bonded molecules. It should be pointed out that the theoretical curve is the superposition of many individual curves with different edge onsets which tends to smooth out the sharp features. This is illustrated in Fig. 11.25 where we display the 11 O–K edge spectra for the group with 2 HBs. All spectra have a strong pre-peak but a different edge onset. When added together the result is a less prominent pre-peak. The difference in their edge onsets is related to their different local geometries: the intramolecular covalent bonding (H-O-H bond angle and O-H bond length) and the intermolecular HBs. These results indicate that accurate calculations on many XANES spectra which are sensitive to the local structural environments provide a meaningful way for statistical analysis, from which the elucidation of subtle points of the structural properties of complex materials becomes possible.

11.3   Spectral imaging

11.3.1   Introduction

The previous sections provided substantial evidence for the effectiveness of the supercell OLCAO method for XANES/ELNES calculation. At the heart of the

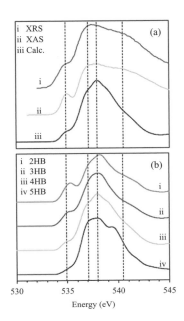

Fig. 11.24.
Top: comparison of calculated and measured O–K edge in water. Bottom: calculated O–K edges grouped according to number of H-bonds (see cited reference for details).

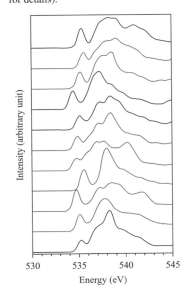

Fig. 11.25.
Variations of the calculated O–K edges within the group of water molecules with only 2 H-bonds.

method is its efficiency at computing the spectrum for each atom in a given model. This leads to an important application that uses the calculated data to construct three-dimensional images, or spectral images (SIs), for visualization and better understanding of the calculated results. In the past few decades, data sets have become substantially more complex and have thus demanded better visualization techniques. A now commonplace type of data set is the function field which can be understood in analogy to scalar and vector fields. (Anderson et al., 2007) A function field of calculated ELNES spectra can serve as a data set for theoretical SI. One problem shared by both experimental and theoretical SI is how such a multi-dimensional data set should be visualized and interpreted. There are several existing methods to extract images based on plotting the intensity from each spectrum integrated over a chosen energy range, or by using multiple linear least-squares fitting or other finger printing type techniques. However, their effectiveness is quite limited, especially for complex multi-component systems. Clearly there is an urgent need for visualization techniques that can be more effective for this type of data set. A powerful approach that we have adopted is to use spectral differences to map the function field into a scalar field. This approach allows for the extraction of specific spectral features and enables correlation with atomic scale structures much more easily than other techniques.

11.3.2   Procedures for SI

We briefly outline the procedures for theoretical SI based on *ab initio* data in conjunction with the technique of function field visualization. A single function from a set of functions is chosen as a target and then each function from the other points of a three-dimensional uniform grid of points are compared to the target using a weighted Euclidian difference. The resultant scalar value at each point represents the dissimilarity between the function at that point and the target point. Because the difference is weighted, it is possible to emphasize particular parts of the function. In using the *ab initio* XANES/ELNES data, details of electronic bonding can be extracted by a judicious choice of the function field range. The stepwise procedures for the creation of the function field are succinctly summarized as follows:

(1) Atomistic model construction: To apply the SI technique to real microstructures in a material, atomistic models of microstructures in the form of large periodic models must be constructed first. This step is tantamount to theoretical sample preparation and must be done with great care.

(2) Spectral calculation and data collection: The ELNES spectra for each atom-edge and for every atom α in the model are calculated and collected. All the spectra $S_{\alpha', n}$, which have an energy index (E_n), are then normalized to unit area within that energy range. The function field depends on the choice of the ELNES spectra (K–, L–, M– . . . edges) and the specific energy range. Because different edges reflect different types of conduction band states, and the spectrum in different energy ranges reflect the characteristic features for

specific interactions, the electronic structure information is contained in the SI for a particular spectral feature.

(3) Creation of the function field: A three-dimensional set of regularly spaced real space mesh points $(\vec{r}_i)$ is defined. The spectrum P at each mesh point $(P(\vec{r}_i, E_n))$ is computed as a weighted sum of the calculated spectra $(S_{\alpha',n})$ of the nearby atoms positioned at $\vec{\tau}_\alpha$ according to:

$$P(\vec{r}_i, E_n) = \sum_{\alpha'=1}^{N_\alpha} \left[w_{i,\alpha',n} * S_{\alpha',n} \right] \tag{11.3}$$

where the weighting factor $w(i, \alpha', n)$ is defined according to a Gaussian function $\exp(-\sigma(\vec{r}_i - \vec{\tau}_\alpha)^2)$ with the restriction that the separation $(\vec{r}_i - \vec{\tau}_\alpha)$ is less than a predetermined value, say 4Å. N_α is the total number of atoms within the region making a contribution to $P(\vec{r}_i, E_n)$ and σ is chosen so that the full-width-at-half-maximum (FWHM) of the Gaussian function is about 2 Å. The sum of all the weighting factors adds to one. These parameters are chosen to be consistent with the spreading of the electron beam and the spot size. However, they neglect the difference in relative signal strength from positions over the atomic columns and positions between them in order to enhance the ELNES spectral response.

(4) Conversion of function field to scalar field: A mesh point $\vec{r}_0$ from the three-dimensional model structure is chosen which serves as the reference spectrum. In the example given below, the target spectrum is chosen such that $\vec{r}_0$ is a point in the bulk crystal region far from the defect area. We then evaluate the Euclidean difference $(D_i(P_0, P))$ at each mesh point according to:

$$D_i(P_0, P_i) = \sqrt{\sum_{n'=1}^{n} \left(P_{i,n'} - P_{0,n'} \right)^2} \tag{11.4}$$

The scalar field $D_i(P_0, P_i)$ is then averaged over one of the dimensions and plotted with a color code indicating the value of D_i which results in a two-dimensional spectral image. The Euclidean difference can be computed over the entire energy range $(1..n)$ or over a subset of that range to focus on specific visually identified features of the spectra, or other specific selections.

11.3.3 Application to a Si defect model

The theoretical SI technique was applied to a planar {113} extended defect model in crystalline Si (Rulis et al., 2009a). The 180-atom supercell model for the defect is shown in Fig. 11.26. This model contains 8-, 7-, 6-, and 5-membered rings but has no dangling bonds. Such passive or inactive defects exist in crystalline samples used in many microelectronic devices and their detection is often a great challenge. Also shown in Fig. 11.26 is the calculated Si–K edge in the bulk region with six different energy windows. Figure 11.27

shows the SI obtained for the function fields that correspond to the Si–K edge in the six energy windows with a bulk region spectrum as the target for comparison. (Note: The figure in the original reference (Rulis et al., 2009a) is colored for better visualization.) The images show the correlation between structural variation and spectral feature variation. The energy windows with varying ranges identify features in the edge spectra that depend on specific electron states in the conduction band. For the passive defect in the present model the variations in the electron states, and hence the edge's spectra, are small. However, the SI still shows in vivid detail how the differences can be delineated using the *ab initio* data, which when cast into different energy panels, give different spatial distributions. The largest contrasts in Fig. 11.27 are panels (a), (c), and (e) where the SI shows different patterns of intensity variation in the same region of the defective area. Further, we can say that strong variations in the leading edge are primarily localized to specific atomic sites while the variations at higher energies are more evenly distributed. The implication is that the leading Si–K edge is an acute measure of specific local structural variations whereas the spectrum at higher energies responds to a wide variety of structural variations with relatively equal variations in intensity. The nature of the deviations can be further probed by selecting the target spectrum from a region with a large variation. The resulting SI would then show how the variations themselves tend to vary. It is therefore possible to analyze the ELNES spectra of complete structural units without the need to find impossible correlations between individual bond lengths, bond angles, and nearest and second nearest neighbor species of each atom in the model. Similar SI has also been obtained by using the Si–$L_{2,3}$ edge as the function field. It has also been applied to study the effect of B doping at different locations in the same model by using the B–K edges as the functional field (Rulis et al., 2009a).

The passive defect and the B-doped systems considered here represent a substantial challenge to microscopy technology because of the inherent difficulty in detecting the subtle differences in the material's electronic structure. The above example shows the advantage of using *ab initio* data and function field visualization to obtain spectral images. The theoretical technique can, in principle, have almost unlimited resolution and can access almost all elements in the periodic table including elements with low Z values (low scattering power) and any core edges (K, L, M, etc.). These spectra are not mixed even if two edges from different atoms are close in energy. Spectral data can be calculated in three dimensions and viewed along any orientation for systems or combinations of systems that are otherwise inaccessible (such as B substitution separately on each Si site (Rulis et al., 2009a)). Once the data are obtained, there is a variety of ways to exploit them in order to maximize the details of the SI for a particular goal including: peak position, absorption edge, peak spread, peak strength, and the Euclidean difference used here. This technique can be used to separate the dynamical and intrinsic parts of the experimental SI. The SI data can be interpreted using simultaneously calculated electronic structure and bonding information. The present example demonstrates that differences in the spectra between the same elements with slightly different environments can be fully mapped out. In particular, the combination of *ab*

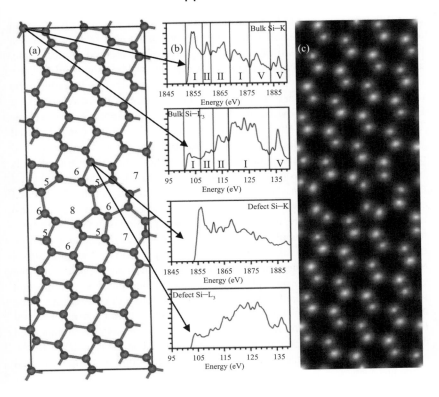

Fig. 11.26.
Illustration of the principal of spectral imaging technique applied to a planar defect model of Si (see cited reference for details).

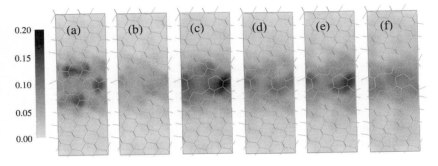

Fig. 11.27.
Calculated spectral images for the model in Fig. 11.26 with different function fields (see cited reference for details).

initio SI and experimental high resolution STEM images provides a powerful tool for performing advanced structural, chemical, and electronic analysis of complex materials.

11.4 Further development of the supercell OLCAO method

In spite of the many advantages of the supercell OLCAO method, there is still ample room for improvement. There are also cases where the agreement between experiment and calculation is not always satisfactory, and it is not clear if there are some fundamental limitations in the method or if it is related to the experimental side such as the nature of the samples that were used. We point out a few directions for future development of the method. For example, at present the $L_{2,3}$ edges of the transition metals cannot be calculated because the spin-orbit interaction is not included. A proper treatment may require the inclusion of multiplet effects in some way. For magnetic oxides, the underlying LDA theory needs to be extended to account for the intra-atomic correlations. The generalization with the inclusion of spin-orbital splitting could be used for x-ray magnetic circular dichroism calculations which can now be accessed by modern STEM. In certain material systems where the dipole transition is forbidden or weak, it would be desirable to extend the calculation beyond the dipole approximation. There are also questions about putting the excited electron from the core level to the bottom of the conduction band in the present implementation. Is this the best description of what happens in the experimental process? Perhaps it would be more realistic to spread the excited electron throughout the entire CB region. This is particularly important for applications to biological systems since there will be only excited energy levels, not the conduction bands. The proper placement of the excited electron is important for the accuracy of the final spectrum. Another issue worth considering is the proper development of the energy dependent broadening procedure to account for the life time broadening which could improve the agreement with measured data. The theoretical spectral imaging technique based on the calculated *ab initio* data that can complement the experimental microscopy imaging discussed in the last section is only at its beginning. However, the greatest opportunity for the supercell OLCAO method is with biomaterials and biomolecules where the structures are far more complicated and the need for theoretical input is urgent. The localized orbital description for the electronic structure in such systems is very natural. Obviously, such extensions will require much effort and many resources with an expectation of encountering many technical difficulties. Nevertheless, these are not insurmountable obstacles.

In conclusion, we are optimistic that the supercell OLCAO method is one of the most competitive methods for studying core level spectroscopy in materials, especially those with complex structures. This success is mainly due to the method's ability to efficiently calculate many spectra using large supercells. In complex crystals and non-crystalline materials, there would be many non-equivalent atomic sites for a given element. In these cases, the weighted average of the spectra from different sites should be used to compare with the measured ones. Depending on the actual variations of their local environment, these spectra can differ widely and so the final weighted sum will as well. The often celebrated finger printing technique which is based on the local nearest neighbor coordination of an ion is of very limited use in such cases. A common argument for using the finger printing technique is that it can reflect the oxidation state of an ion. However, the assignment of oxidation state or charged state with an integer value is something of an oversimplification from the stand point of *ab initio* calculations. The last point is that the main purpose of XANES/ELNES calculation is for the characterization of structurally complicated systems such as those involving defects, interfaces, or grain boundaries where experimental investigations face many challenges and the interpretations are difficult. However, spectral calculation of structurally complicated multi-component systems can be overwhelming and intractable. Thus, the best strategy would be to carefully model the anticipated structure through large-scale modeling, followed by the calculation of the XANES/ELNES spectra of atoms of interest, and then comparing with experiments to assess the validity of the model constructed.

References

Anderson, J., Gosink, L., Duchaineau, M., & Joy, K. (2007), Feature Identification and Extraction in Function Fields. *In:*Fellner, D. & Moller, T., (eds.) *Eurographics/IEEE VGTC Symposium on Visualization* (Norrkoping, Sweden: The Eurographics Association), 195.

Blaha, P., Schwarz, K., Sorantin, P., & Trickey, S. B. (1990), *Computer Physics Communications*, 59, 399–415.

Buban, J. P., Matsunaga, K., Chen, J., et al. (2006), *Science*, 311, 212–15.

Champloy, F., Gruber, K., Jogl, G., & Kratky, C. (2000), *Journal of Synchrotron Radiation*, 7, 267–73.

Ching, W. Y., Ouyang, L., Rulis, P., & Tanaka, I. (2005), *Phys. Status Solidi B*, 242, R94–6.

Ching, W. Y. & Rulis, P. (2008a), *Phys. Rev. B*, 77, 035125/1–035125/17.

Ching, W. Y. & Rulis, P. (2008b), *Phys. Rev. B*, 77, 125116/1–125116/7.

Ching, W. Y. & Rulis, P. (2009), *J. Phys: Condens. Matter*, 21, 104202/1–104202/16.

Ching, W. Y., Aryal, S., Rulis, P., & Schnick, W. (2011), *Physical Review B*, 83, 155109.

De Groot, F. & Kotani, A. (2008), *Core Level Spectroscopy of Solids*(Boca Raton: CRC Press).

Dirac, P. A. M. (1927), *Proceedings of the Royal Society of London. Series A*, 114, 243–65.

Egerton, R. F. (1996), *Electron Energy-Loss Spectroscopy in the Electron Microscope* (New York: Plenum Press).

Franke, R. & Hormes, J. (1995), *Physica B: Condensed Matter*, 216, 85–95.

Garvie, L. a. J., Hubert, H., Petuskey, W. T., Mcmillan, P. F., & Buseck, P. R. (1997), *Journal of Solid State Chemistry*, 133, 365–71.

Gu, H., Ceh, M., Stemmer, S., Müllejans, H., & Rühle, M. (1995), *Ultramicroscopy*, 59, 215–27.

Hatada, K., Hayakawa, K., Benfatto, M., & Natoli, C. R. (2007), *Physical Review B*, 76, 060102.

Leinenweber, K., O'keeffe, M., Somayazulu, M., Hubert, H., Mcmillan, P. F., & Wolf, G. H. (1999), *Chemistry—A European Journal*, 5, 3076–78.

Matsunaga, K. (2008), *Physical Review B*, 77, 104106.

Matsunaga, K., Murata, H., Mizoguchi, T., & Nakahira, A. (2010), *Acta Biomaterialia*, 6, 2289–93.

Mizoguchi, T., Tanaka, I., Yoshioka, S., Kunisu, M., Yamamoto, T., & Ching, W. Y. (2004), *Phys. Rev. B*, 70, 045103/1–045103/10.

Mo, S.-D. & Ching, W. Y. (2000), *Phys. Rev. B*, 62, 7901–7.

Mo, S.-D. & Ching, W. Y. (2001), *Appl. Phys. Lett.*, 78, 3809–11.

Moreno, M. S., Jorissen, K., & Rehr, J. J. (2007), *Micron*, 38, 1–11.

Myneni, S., Luo, Y., Näslund, L. Å., et al. (2002), *J. Phys: Condens. Matter*, 14, L213.

Natoli, C. R., Misemer, D. K., Doniach, S., & Kutzler, F. W. (1980), *Physical Review A*, 22, 1104.

Pellicer-Porres, J., Saitta, A. M., Polian, A., Itie, J. P., & Hanfland, M. (2007), *Nat Mater*, 6, 698–702.

Pucher, F. J., Römer, S. R., Karau, F. W., & Schnick, W. (2010), *Chemistry—A European Journal*, 16, 7208–14.

Robertson, J. (1983), *Physical Review B*, 28, 3378.

Rohlfing, M. & Louie, S. G. (1998), *Phys. Rev. Lett.*, 80, 3320.

Rulis, P., Yao, H., Ouyang, L., & Ching, W. Y. (2007), *Phys. Rev. B*, 76, 245410/1–245410/15.

Rulis, P., Lupini, A. R., Pennycook, S. J., & Ching, W. Y. (2009a), *Ultramicroscopy*, 109, 1472–78.

Rulis, P., Wang, L., & Ching, W. Y. (2009b), *physica status solidi (RRL)—Rapid Research Letters*, 3, 133–35.

Rulis, P. & Ching, W. (2011), *Journal of Materials Science*, 46, 4191–8.

Schwarz, K. & Blaha, P. (2003), *Computational Materials Science*, 28, 259–73.

Serghiou, G., Miehe, G., Tschauner, O., Zerr, A., & Boehler, R. (1999), *The Journal of Chemical Physics*, 111, 4659–62.

Shemkunas, M. P., Wolf, G. H., Leinenweber, K., & Petuskey, W. T. (2002), *Journal of the American Ceramic Society*, 85, 101–4.

Shirley, E. L. (1998), *Phys. Rev. Lett.*, 80, 794.

Stöhr, J. (1992), *Nexafs Spectroscopy* (Berlin; New York: Springer-Verlag).

Tanaka, I. & Adachi, H. (1996), *Physical Review B*, 54, 4604.

Tanaka, I., Mizoguchi, T., Matsui, M., et al. (2003), *Nat Mater*, 2, 541–45.

Tanaka, K. & Kohyama, M. (2002), *Philosophical Magazine A*, 82, 215–29.

Wernet, P., Nordlund, D., Bergmann, U., et al. (2004), *Science*, 304, 995–99.

Zerr, A., Miehe, G., Serghiou, G., et al. (1999), *Nature*, 400, 340–42.

Enhancement and Extension of the OLCAO Method

12

The OLCAO method has undergone many evolutionary steps from the time that the name "OLCAO" was given to the present. Each step improved some aspect of the method such that it became more versatile, more efficient, or more easy to use. Accordingly, as we look toward to the future development of the OLCAO method, this chapter will be divided into sections based on the same three categories of versatility, efficiency, and ease of use. While these sections are not completely orthogonal, the degree of overlap is relatively small. Many of the topics listed in this chapter are under current and active development. This list of topics should not be considered exhaustive but rather representative of the types of enhancements and extensions that are being made or can be made to the method.

12.1 Versatility

The ability of a program to accurately compute a wide array of useful results for a wide array of possible inputs defines our concept of versatility. The versatility of the OLCAO method can thus be enhanced by defining an improved and more flexible basis set, or by implementing algorithms for the computation of Van der Waals forces, or by providing a framework for adding more complex exchange correlation functionals, or by introducing practical schemes to address some of the drawbacks in the local density approximation of density functional theory, and so on. In this section we will visit a collection of different issues that are associated with enhancing the versatility of the OLCAO method.

12.1.1 The OLCAO basis set

The OLCAO basis set of contracted Gaussian orbitals has been time tested and evolutionarily improved to work very well for a wide range of configurations of the atoms in the first four rows of the periodic table of the elements. Some of the elements in the remaining rows of the periodic table have also been considered and tested equally well, but not all. In particular, those elements with f-electrons in the valence shell and those with large relativistic effects associated with the core electrons remain at the forefront of future development. For the well tested first four rows however, many excellent results have been obtained for an exceptionally diverse set of systems. These results have all relied upon a slowly evolving basis set that did not require dramatic alteration each time a new type of material system was investigated. This characteristic is emblematic of the basic principles of the OLCAO method which are to be robust, fast, and reasonably accurate. Evolutionary improvements can continue to be made whereby the number of terms and the minimum and maximum values of the Gaussian exponential coefficients are improved through some suitable scheme, however, more dramatic improvements to the basis and the way that it is used are also under active consideration because of the transformative role that it could play in extending the capabilities of the OLCAO method.

The form of the basis functions were described in detail in Section 3.1 and were presented mathematically through Eqs (3.1), (3.2), and (3.3). Without repeating all of the details, the essential concept is that the basis set is fixed for each element, and the set of exponents are all the same for the Gaussians used for each orbital of a given element. The key differences for each orbital are that the coefficients for the Gaussians are different and that the spherical harmonic functions used to make the orbitals are different. In the non-relativistic approach a single orbital is defined in terms of the n, l, and m quantum numbers typically associated with the solution to the Schrodinger equation for an electron in a spherical potential. The basic concepts for the construction of the basis functions were described in Section 3.1 using two approaches. We now consider these approaches with respect to the construction of more complex basis sets.

The first approach is to solve a single atom eigenvalue problem within the same density functional formalism and with a basis of single Gaussian functions. This produces a set of coefficients for each atomic orbital, but some care should be taken because the electron wave function for an isolated atom is not likely to be the best representation of the electron wave function in a solid. In particular, the isolated atom's wave function will be much more extended. A possible correction would be to remove or modify the first (broadest) term in the Gaussian expansion of each orbital so as to reduce the range of the wave function. The only true complication is that orthonormality should be preserved. Because this procedure for constructing the basis uses the isolated atom model it may have further limitations for constructing a basis for the solid state wave function. This is because the same element may take on substantially different electronic configurations for different systems. For example, transition metal oxides with partially filled d-bands can be insulators while relatively pure transition metal materials are excellent conductors. If the basis functions for these two cases were distinguished in some way (e.g., more or less long range) the solid state wave function may become more accurate. Dynamically requesting or recognizing the optimal form for the basis function is a challenging but potentially very useful advance.

The DFT based basis construction scheme can be extended in a rather straightforward way to include the scalar relativistic formalism. This has been implemented in the OLCAO method but has not yet been widely used or deeply tested. The scalar relativistic correction (mass velocity and Darwin terms) will tend to modify the radial distribution of the existing Schrodinger style orbitals to make them more compact. The practical effect will be a more accurate calculation of the system total energy and the energy levels of occupied orbitals in heavy atoms. Once the existing implementation is fully tested an extension to the creation of atomic orbital basis functions to include the fully relativistic representation could be the next step. This would be a powerful enhancement and probably of much greater practical use because of the explicit inclusion of the spin-orbit splitting effect. An example exploitation of this effect is x-ray circular magnetic dichroism. This calculation is analogous to the x-ray absorption near edge structure calculations that were discussed in Chapter 11 except that what is observed is the difference between spectra obtained from spin-orbit split core level electron excitations. Computing a basis set with such properties has been initiated by Tatewaki (Tatewaki and Mochizuki, 2003), but would need modification for use with the OLCAO program to ensure that all the contracted Gaussians have the same exponential coefficients.

The second approach for constructing the atomic orbital basis set is to fit the Gaussian series to a numerically computed set of radial functions. In this case the same issues of controlling the extent of the radial functions and including scalar or fully relativistic corrections apply. The advantage of this approach is that it can make use of other highly accurate programs (GRASP2K, ADPACK, etc.) (Jönsson et al., 2007) for computing the Schrodinger, scalar relativistic, and fully relativistic wave functions of various atoms. The disadvantage is that the fitting procedure will need to be closely monitored because the fitting accuracy will depend on the number of points used and other parameterized constraints. Additionally, the issue of orthonormality will need to be addressed.

In some cases a third option for basis construction is available but it has yet to be integrated into the existing scheme. The concept is that individual supplemental Gaussians could be added, such as only the xyz component of an f-electron orbital. This would only modestly increase the computational burden, but would provide substantial extra freedom compared to adding an entire new shell. Another example for enhanced orbital hybridizations would be to add some d-electron orbital components to say O or other second row elements. Again, this would need to be integrated into the existing scheme and it would only become most useful if it were automated to some degree because of the complexity of dealing with systems composed of hundreds of atoms.

12.1.2 The OLCAO potential and charge density representation

The representation of the electronic potential and the charge density in the OLCAO method was detailed in Chapter 3, but efforts continue to refine and improve it. One particular key advantage that should be retained in any improvements is that the potential and charge density both share the same set of Gaussian functions for their definition. The only difference between them is the set of leading coefficients for the Gaussian functions. One specific deficiency of the existing representation is that the potential and charge are formed from a linear combination of spherical Gaussian functions and thus do not represent the lobed distributions associated with electrons in p-type or higher ℓ quantum number orbitals as well as it could. This will have an impact on the electronic

SCF convergence rate and also on the ability to accurately calculate forces. Fortunately, the existing scheme that uses a superposition of spherical Gaussians is actually quite effective (see discussion in Chapter 3) and so this deficiency has not had a severe impact, as demonstrated throughout the previous chapters by the quality of the agreement between computed results and experiments. Still, a clear improvement would be to also include p-type Gaussians in the definition of the potential and charge density function. While this would surely increase the accuracy of the calculation it is not without cost. The computation of the three-center interaction integrals would become more complex because of new possible combinations of angular types (e.g., three p-type Gaussians or a p-type and two d-type Gaussians instead of always having at least one s-type). Further, the spherical symmetry of the charge associated with a particular atomic site would be lost which could cause complications in the evaluation of the long range Coulomb interactions. In the end, it is the proper balance between the efficiency and improved accuracy that one has to consider for these additional implementations.

12.1.3   Relativistic OLCAO

As was alluded to in subsection 12.1.1, the inclusion of relativistic effects is critical for the proper calculation of the properties of materials with heavy elements. Two different approaches can be taken to include relativistic effects as explained in Section 4.5. The first is the scalar relativistic approach and the second is the fully relativistic approach. In either case the inclusion begins with an enhanced basis set that is derived under the appropriate scalar or fully relativistic conditions. For scalar relativistic calculations, the orbital basis functions are essentially the same as for the non-relativistic case except that the orbitals they describe tend to be more compact near the nucleus. Further, because only the Darwin term and the mass velocity term are included in this approach while the spin-orbit coupling term is not included, the two spin-orbital basis functions are the same. Hence, it is possible to perform scalar relativistic paramagnetic calculations, but this is not likely to be very useful because many of the interesting materials where relativistic corrections are important are also magnetic so that a paramagnetic approximation would be poor. A key step that has not yet been performed is a detailed analysis of the quality of the scalar relativistic basis set in comparison to the description of the wave functions obtained with highly accurate atomic structure programs such as Grasp2K (Jönsson et al., 2007). Within the OLCAO program itself the most important modifications would be to the construction of the Hamiltonian to include the new terms and the evaluation of the relativistic exchange-correlation functional (Macdonald and Vosko, 1979). Once these steps are taken, the calculation proceeds much as a traditional non-relativistic calculation would for properties calculations such as the band structure, density of states, bond order, charge transfer, and optical properties.

However, for a fully relativistic description the situation quickly becomes more complex because there must now be four different components to describe one electron orbital. That is, there is a mixture of spin-up and spin-down positive energy solutions with spin-up and spin-down negative energy solutions for a single orbital. The construction of a basis set for use in the OLCAO method for this level of the theory is not complete and, hence, this is an area of active development. Once the basis is developed, the inclusion of fully relativistic effects must also be carried through to the main program and the properties calculations. Some of the complicating factors are that the minor component and major component of a particular orbital have different angular characters to them (e.g., the minor component of the $2p_{1/2}$ orbital has an s-type angular character while the minor component of the $2p_{3/2}$ has a d-type angular character). Many of the difficulties such as this are merely technical in nature and with diligent programming and careful attention they can be overcome. Other issues are more problematic such as the evaluation of the kinetic energy contribution to the total energy. In this case, the term contains a fourth derivative which demands a high degree of accuracy in the wave function to evaluate accurately. This issue is also present for the scalar relativistic calculation and so alternate approaches for evaluating the kinetic energy term are under active development.

12.1.4 Exchange-correlation functionals

One of the more prominent components of density functional theory is, of course, the functional that is used for computing the exchange and correlation components of the total energy. The OLCAO method at present has only a limited number of exchange-correlation functionals implemented in the code. The local density approximation (LDA) has multiple only slightly different parameterizations including those from Wigner (Wigner, 1934), Hedin-Lundqvist (Hedin and Lundqvist, 1971), Gunnarsson-Lundqvist (Gunnarsson and Lundqvist, 1976), von Barth-Hedin (Barth and Hedin, 1972), and Ceperley-Alder (Ceperley and Alder, 1980). The last three of these functionals have been extended to the level of the LSDA, but beyond that the OLCAO method does not have an exceptionally well tested and commonly usable exchange correlation functional.

The next step beyond the LDA is of course the exchange-correlation functional of generalized gradient approximation (GGA), but this approach has not been fully integrated into the existing OLCAO suite. Considerable work has been done to develop the method for OLCAO, but as will be discussed later, the difficulties of development in an academic setting have prevented it from being thoroughly tested and then more broadly used. Work is currently ongoing to rectify this situation and includes an implementation of the generalized gradient approximation into the main code base. Along a similar line, the LSDA $\pm$ U (Anisimov et al., 1991) functional has also been implemented by past researchers, but it too has not yet been brought into the main line of the source code. Then, in conjunction with the development of a relativistic OLCAO method as proposed in Section 12.1.3 it is necessary to develop an implementation of a successful relativistic exchange-correlation functional. There have been many interesting and useful developments in the area of exchange-correlation functionals in recent years. One of the more successful functionals is the B3LYP (Becke, 1993) hybrid functional because of its enhanced applicability to biological systems. The OLCAO method has made good use of the LDA for many years and has produced the bulk of its results with it, but the need for greater accuracy in many situations is very clear so that considerable effort is being applied to develop and integrate more functionals into the main code base.

12.1.5   Magnetism and non-collinear spin polarization

Currently, the spin-polarized version of the OLCAO method uses a spin collinear approach for computing the non-spin degenerate properties of materials. In the basis set for expansion of the solid state wave function, the spin quantum number s is explicitly recognized, but the description of the spin-up and spin-down basis functions is identical. There are a number of ways that the treatment of magnetic systems can be improved in the current OLCAO implementation. The SCF convergence of magnetic systems is initiated with a small "kick" that will split the two otherwise identical spin states. Once the kick has been performed on the first iteration of the SCF cycle it is no longer needed, but the convergence tends to be rather slow and methods should be implemented to improve the SCF convergence of magnetic systems. One approach is a dynamic mixing factor that will adjust automatically depending on the stability of the convergence. It may also be possible to take better control of the initial kick such that the charge distribution of different atoms is kicked differently so that an informed user can select the most appropriate starting conditions for the SCF process. It is also likely that the more accurate charge and potential functions discussed in Section 12.1.2 would significantly improve the convergence rate for the magnetic systems.

However, it is clear that a more accurate non-collinear spin-polarized approach would also be helpful for a more detailed and accurate analysis of spin glasses, some magnetic crystals, and magnetic systems with spin frustration. To do this it will be necessary to have each of the current spin states doubled so as to represent a linear combination of the up and down spin states.

12.1.6   Configuration interaction

One of the most vexing and persistent challenges in electronic structure theory is the determination of the solid state wave function that correctly describes the unoccupied or virtual states beyond the essentially one-electron approach of the DFT, or the inclusion of the many-body electron–electron interaction. The roots of the problem are the probabilistic character of the wave function itself and the anti-symmetry condition that is demanded for systems with multiple electrons. Crudely, the electrons correlate their motions so as to avoid each other and, under excitation, the electrons can be in any of a large number of different possible configurations with different probabilities. The Hartree–Fock method for computing ground state energies represents the zeroth order approximation $|\Psi_0\rangle$ to the exact total solid state wave function $|\Phi_0\rangle$. To account for dynamical and static correlation effects, the conceptually simplest approach is to expand the exact total solid state wave function as a series of variations on the zeroth order approximation. The variations are different configurations of the electrons such that in some cases a single electron has been promoted into the unoccupied states and in others two electrons have been promoted, then three, and four. Following the notation of Szabo and Ostlund (Szabo and Ostlund, 1996) we identify the exact solid state wave function with the following expansion:

$$|\Phi_0\rangle = c_0 |\Psi_0\rangle + \sum_{ar} c_a^r |\Psi_a^r\rangle + \sum_{\substack{a < b \\ r < s}} c_{ab}^{rs} |\Psi_{ab}^{rs}\rangle +$$

$$\sum_{\substack{a < b < c \\ r < s < t}} c_{abc}^{rst} |\Psi_{abc}^{rst}\rangle + \sum_{\substack{a < b < c < d \\ r < s < t < u}} c_{abcd}^{rstu} |\Psi_{abcd}^{rstu}\rangle + \dots \quad (12.1)$$

Here, the terms of the expansion are referenced to the Hartree–Fock ground state but the first summation includes all possible configurations where one electron has been promoted from state a into state r. The same applies to the next summation except that every possible combination of two electrons being promoted to unoccupied states is considered without double counting. The remaining terms are considered in a similar way. This expansion is a mathematically complete set that then forms the exact solid state wave function. The only problem is that there are too many terms to compute in a reasonable period of time for even small molecular systems. A number of truncations to the series expansions must therefore be employed. For the reference Hartree–Fock ground state a minimal basis is frequently used that typically does not include many of the unoccupied states. Then, the Slater determinate based expansion of the solid state wave function is typically truncated to include only the singly and doubly excited configurations.

Considerable effort has been put into determining ways to reduce the number of terms required and yet still obtain an accurate result. Some special care must be taken in any approach because with the large number of terms involved it may be possible to neglect important ones. One method, detailed by Ogasawara et al. (Ogasawara et al., 2000), is of particular interest because it used a hybrid of DFT and CI. The OLCAO method is well suited to this approach and it has some advantages over the DFT method employed in this study because of its increased speed and accuracy. There are certain key modifications to the OLCAO program that would need to be made. (1) The introduction of two-electron integrals, which are four-center Gaussian integrals. This is a complex task because as the angular types (s, p, d, f) of the different Gaussian based orbitals become more complex the formulae become more cumbersome. Deriving the formulae by hand is impossible, but recursive techniques have been developed (Obara and Saika, 1986) and could prove useful. (2) The form of the electronic potential will need to be adjusted to allow for some portion of the correlation energy to be expressed through the CI formalism and the rest of it to be expressed by the density functional. (3) Then the solid state wave function must be expanded as a linear combination of Slater determinates.

(4) There are a number of other steps and modifications that will have to follow a similar line as is detailed by Ogasawara et al. (Ogasawara et al., 2000) before an accurate DFT + CI combination can be implemented within OLCAO.

It is hoped that the development of a configuration interaction method within OLCAO will enable computations for much larger systems than have been treated via CI in the past and that the results will be more accurate. The OLCAO method has some inherent speed and because analytical functions are used it has reasonably good accuracy. Such a potent addition to the OLCAO method would be of great value.

12.1.7   Hamaker constants and long-range van der Waals–London interaction

One of the areas where the OLCAO method has great utility is in the calculation of optical spectra of specific material systems for the purpose of estimating the long-range van der Waals (vdW)–London interactions between macroscopic objects. This is an area of special interest in nanoscale science (French et al., 2010), and a practical method that can effectively link microscale calculations based on quantum mechanics to macroscale objects is highly desirable. Classical simulations generally fail for non-bonded interactions involving long-range electrostatic and dispersive forces, and different types of complicated interactions which collectively determine the diffusive motion of biomolecules or colloidal particle suspensions. The mathematical formulism for vdW–Ld forces for simple geometric objects (block, sphere, cylinder, etc) has been worked out (Parsegian, 2005) using the continuum theory of Lifshitz (Lifshitz, 1956). The theory has its roots in electrodynamics and quantum field theory where the interaction originates from microscopically induced dipoles related to the electronic structure of the materials (Abrikosov et al., 1963, Casimir, 1948, Landau et al., 1984)

A practical approach for evaluating vdW–Ld forces in complex biomolecular or colloidal systems based on the supercell-OLCAO method is emerging. In this approach, the vdW–Ld forces are calculated via the evaluation of Hamaker coefficients (Hamaker, 1937) between simple geometric objects with an intervening medium, such as the case of two infinite slabs with water or a vacuum between them. In the early days, the required information on the oscillating dipole interaction was approximated by the refractive index of the materials and the medium (Tabor and Winterton, 1969), or a few carefully selected frequencies of oscillations (Ninham and Parsegian, 1970). More recently, progress has been made in ceramic materials by using the full optical spectrum obtained from experimental vacuum ultraviolet (VUV) spectroscopy or the valence band electron energy loss spectroscopy (VBEELS). (French, 2000) The London dispersion functions are then fed into standard formulas that have been derived for specific geometries to obtain the Hamaker coefficients which are the scaling factors for the vdW–Ld force. The full spectrum approach has significantly improved accuracy and has enabled realistic estimation of the dispersive forces for real materials. It was soon realized that one may be able to use theoretically calculated optical spectra instead of experimentally measured ones to evaluate the Hamaker constants. The feasibility of this approach has been successfully applied to both the metallic and semi-conducting single wall carbon nanotubes (SWCNT) and multi-walled carbon nanotubes (MWCNT) of different chiralities (Rajter et al., 2008, Rajter et al., 2007a, Rajter et al., 2007b).

The approach used is a step-by-step strategy. First, we need to construct the appropriate structural models of biomolecular systems. This will be followed by full quantum mechanical calculations of the electronic structure at the level of density functional theory. The calculated energies and wave functions will be used to obtain the *ab initio* optical properties. These data will then be used to evaluate the Hamaker coefficients for specific geometric objects in order to estimate the macroscopic van der Waals forces. Each step has their respective challenges and obstacles that need to be overcome. These steps are succinctly described below.

For a proper description of the oscillating dipole in the Lifshitz formulism, a complex frequency is defined as $\hat{\omega} = \omega + i\xi$. The calculated $\varepsilon(\hbar\omega) = \varepsilon_1(\hbar\omega) + i\varepsilon_2(\hbar\omega)$ (a function of real frequency ω) is transformed into the so-called vdW–Ld dispersion spectrum (a function of imaginary frequency ξ) by KK conversion.

$$\varepsilon(i\xi) = 1 + \frac{2}{\pi} \int_0^\infty \frac{\omega\varepsilon_2(\omega)}{\omega^2 + \xi^2} d\omega \tag{12.2}$$

The magnitude of $\varepsilon(i\xi)$ describes the material's response to the fluctuating dipoles up to a given frequency. The Lifshitz formulation for a particular geometrical configuration is used to calculate the Hamaker coefficient A. A is then multiplied by the proper geometric scale factor g to estimate the interaction energy or force $G = A \cdot [g/\ell^n]$. Here ℓ^n is a scaling factor, ℓ is the distance of separation between two objects, and n is an integer. Essentially, the Hamaker coefficient A is an interaction strength that takes into account the material properties of the two objects within a given geometrical configuration and it is completely independent of g and ℓ.

The calculation of the Hamaker coefficient A in the non-retarded limit can be fairly involved. For a simpler case of two planar isotropic blocks (denoted as left L and right R) separated by a medium m, it can be shown that

$$G = A^{NR} \cdot [g/(12\pi\ell^2)]; \quad A^{NR} = \frac{3k_bT}{2} \sum_{n=0}^{\infty}{}' \Delta_{Lm} \bullet \Delta_{Rm} \tag{12.3}$$

where the Δ_{Lm} and Δ_{Rm} terms are given by:

$$\Delta_{Lm}(i\xi_n) = \frac{\varepsilon_L(i\xi_n) - \varepsilon_m(i\xi_n)}{\varepsilon_L(i\xi_n) + \varepsilon_m(i\xi_n)}; \quad \Delta_{Rm}(i\xi_n) = \frac{\varepsilon_R(i\xi_n) - \varepsilon_m(i\xi_n)}{\varepsilon_R(i\xi_n) + \varepsilon_m(i\xi_n)} \tag{12.4}$$

The summation in (12.3) is over the so-called discrete Matsubara frequencies $\xi_n = (2\pi k_b T/\hbar)n$ with n ranging from 0 to ∞. The prime on the summation indicates that for the first term (n = 0) there is a multiplicative factor of 0.5.

The above outlined approach for evaluating vdW–Ld forces is not without difficulties. First, there is a need for realistic atomic-scale modeling of large biological molecules that can be used for *ab initio* electronic and optical properties calculations. Second, how can we approximate the shapes of biological entities in order to make use of the analytical formulas derived for simple geometrical objects? At the initial level of study, we can approximate a double helix of DNA as a cylindrical object (with further mixing rules for the dielectric functions), a biomembrane can be considered as a pair of plane-parallel blocks, and a large protein as a spherical object. In this way, the analytic formulas that have been derived for geometric objects can be applied. Third, the technical difficulties in the electronic structure and optical properties calculations (presence of gap states, need for counter ions, role of H_2O molecules, issues related to H-bonding, effect of many-electron interaction in the excitation spectrum etc.) cannot be underestimated. The limitations of various approximations involved must be carefully assessed. Last but not least, there should be bench marks for the validation of the calculations. These difficulties, given enough time and resources, are not insurmountable obstacles.

12.2 Efficiency

To make a program scientifically interesting it is only necessary that it contain algorithms and methodologies for calculating the result of a thought experiment. However, a functioning approach, by itself, is not typically sufficient to make the program useful in the practical sense. The remaining necessary ingredient is for the approach to have a wise balance between efficiency and accuracy (i.e., inclusion of more detailed physical theories and the rigorousness of the methods used to compute them). While the previous section described ways that the OLCAO method could be extended to improve its accuracy, this section will describe a number of ways that it can be modified or extended so as to improve its efficiency. Before jumping in to the discussion, we must first recognize that efficiency comes in multiple forms and that these forms may sometimes be in competition with each other. The starkest example of this issue is efficient programmability versus efficient execution. As we proceed through this section we should recognize that efficiency achievements in one area may come at the expense of efficiency in other areas. A balanced approach to efficiency must be at the forefront of the minds of the user and the developer at all times so as to avoid large scale flaws in the structure of the program suite. While some implementation and programming details and philosophies are included in the next sections the intent is to present the users of the OLCAO suite with a slightly deeper understanding of development directions so that they can make the most efficient use of the program for the architecture that they are running the program on.

12.2.1 The memory hierarchy

A key attribute for the efficient execution of any program is how well it makes use of the memory hierarchy present in modern computer systems. Reading, writing, and operating on information is the primary task of the central processing unit's (CPU's) core(s) and the rate at which it is able to perform that task depends in large part on the rate at which the necessary information can be presented to it. In the simplified case of serial operation there is a general inverse relationship between the amount of information stored by some physical mechanism and the speed with which it can be accessed. There are only a few registers per CPU core, the local cache for the CPU core (itself often composed of multiple levels) is relatively small compared to the main memory, and the main memory is now dwarfed by the storage capacity of hard disks or other similar technologies. Requests for access to each of those information stores must traverse a communication bus with different response times, the registers being the shortest and the hard disk being longer by orders of magnitude. With the development of multiple core CPUs, the use of multiple CPUs in a given network node, and the construction of supercomputers with many thousands of nodes, the situation becomes worse and more complex because the limited resources for communication to and from the stored information must now be shared and coordinated.

An extension to the OLCAO program that is under active development is the instrumentation of techniques to measure and monitor the performance of the program with respect to the frequency that the program must access information at the different levels of the memory hierarchy. By understanding the performance constraints of the program it will be possible to improve the program. Improvements can be made through more careful flow control and organization of the data structures in the program. This will enhance use of temporal and spatial data locality which is very important for proper use of the CPU cache. Methods for improving the program through parallel programming will be discussed in a later section, but the information gleaned from this type of performance monitoring will be very valuable when making decisions about how to most effectively parallelize the program suite. This will be a critical tool for users to gauge the performance of the program on their machine and to assist them in adjusting program execution parameters to enhance that performance.

12.2.2   Modularization

Modularization in this context represents the concepts of data encapsulation and internal interfaces that are often seen as core components of object oriented program (OOP) design, but it leaves out additional OOP concepts such as inheritance and polymorphism. In its essence we see modularization as the process of writing a program such that the data and the subroutines that operate on that data are co-located in one place in the program code. Further, the other sections of the program code are shielded from direct interaction with the data and can only access it through a well-defined internal interface. Modularization is often seen as a key element for teams to have efficient program development.

The OLCAO program suite was originally written within a highly procedural structure in the days before object oriented concepts were widely practiced. However, in recent years a series of changes have been made to the code base to improve its readability and organization. The code has been rewritten in a progressively more encapsulated style so that data structures and the code sections that operate on them are stored together in Fortran 90 modules. The encapsulation is not total and it essentially overlies a procedural philosophy but the ground rules for the sharing of data between program units has been established and the code is now more compartmentalized than at any point in its development history. Throughout these modifications it has retained its high level of performance. New developers now have a reasonably well organized set of modules through which they can interface to add new features, but they must still learn a host of internal details to make efficient use of them. Hence, room for improvement remains.

At present an informal rule exists stating that data stored within one module shall not be *modified* by another module or program subroutine, but modules and subroutines can have read access to the data of other modules. In this way information is free to be exchanged, but modifications to data are localized to aid in debugging. An excellent modification to the program would be to formalize this rule to prevent the accidental modification of data. Presently, the Fortran 90/95 specification does not permit variables defined within modules to be accessed in an explicit read-only state by other modules or subroutines but still be modifiable internally to the module. Direct access to variables and sub-routines in a module is an either all (public) or nothing (private) proposition. As a result, the only way to strictly enforce the principle of public-read and private-read/write is with rigid documentation requirements and peer review of code. On balance, this may not be a bad thing because it will enhance communication between developers. This is a very challenging modification because it requires coordinated agreement and vigilance from mistake-prone humans as opposed to computer enforced rules of syntax. An alternative modification, which may be more challenging, would be to perform a detailed study of the program data structures to determine if a better modular structure exists so that such read-only data sharing is not even necessary. This is a more risky thing to do because there is currently no proof that this will actually enhance the performance (and it may even reduce it), and it may not improve the program organization enough beyond its current condition to be regarded as an enhancement of readability and so on.

12.2.3   Parallelization

The development of parallel programming methods and parallel computational architectures has had a profound impact on the capabilities of many applications. Unfortunately, the OLCAO method has not yet been developed to take advantage of the promise of parallel computing, but it is definitely ripe for it. Many different parts of the OLCAO program suite possess a high level of independence and can thus be executed in parallel. Better yet, the efficiency gain for many of the parallelizable code segments should be relatively large because it is expected that there will be a low level of inter-process communication required. This combination of independent tasks and low communication needs also points to the potential for a high level of scalability across many processing units.

To understand the approach to parallelization that must occur with the OLCAO program it is important to first understand the architecture of the computers that it is being parallelized for. The current state-of-the-art in massively parallelized supercomputers relies on a hierarchy of computing and communication units. At the top is the so-called node level. The nodes are connected together through a complex interconnection network that allows any node to communicate with any other node and with dedicated hard disk based data storage nodes. The speed of communication tends to be the fastest between neighboring nodes and it decreases as the message must traverse more links in the network to reach its final destination. In other words, the interconnection network is not a direct all-to-all topology so care must be taken to limit communication across nodes with a bias against more distant nodes. Within each node are a number of CPUs (e.g., 8 or 16) that share a single block of main memory (e.g., 8 or 32 GB worth). Within each CPU are a number of CPU cores (e.g., 2 or 8) that may or may not share certain resources such as cache and the main memory bus. Efficient use of this hierarchy by both developers and users requires that the tasks be broken down into chunks that are as small as possible so that many cores can be applied to each task, but not so small that the inter-process communication required to coordinate the solution becomes a burden. The parallelization of the OLCAO suite will use the traditional tools of MPI and OpenMP for parallelization across nodes and within nodes respectively.

As is detailed in Appendix C, OLCAO calculations are divided amongst a linear sequence of discrete executable programs. Each of these programs has components that require their own unique approach to parallelization. The first program, OLCAOsetup.exe, is responsible for (1) computing two- and three-center Gaussian integrals; (2) creating the data structures for fast computation of the charge density on a non-uniform atom centered spherical mesh; and (3) creating data structures for the Ewald summation procedure needed for computation of the long range Coulomb interaction. Typically $1 > 3 > 2$ in terms of computational requirements, but for certain cases $3 > 1 > 2$. The parallelization of 2 and 3 is relatively straightforward across the number of unique atom centered potential sites in the system so that it scales well as the system size increases. The parallelization of 1 also scales with system size, but it is a more complex calculation that deserves special treatment. The three-center interaction integrals are the dominant component of this part of the calculation. Each term in the potential function (Eq 3.10) must be integrated against all pairs of orbitals from all atoms. If the division of labor is done simply by assigning a single processing unit to compute the integrals for one term of the potential function then a memory deficiency will quickly arise. Each core on the node will have a substantial memory requirement which, if multiplied by the number of cores that share the main memory, will exhaust the available memory. However, if the responsibility for computing the interaction integrals for a given term in the potential are instead assigned on a per-node basis instead of a per-core basis the memory problem is averted. The price to pay is that all the cores on the node must be coordinated in their calculation of that set of three-center integrals which can require a higher degree of inter-process communication. Fortunately, inter-process communication within a node is cheap compared to the cost of inter-node communications. Hence, the three-center integral calculations can be divided on a per-node basis with all the cores on each node working together. The user must be keenly aware of the number of cores on a node and the memory available to each node to be able to understand any monitoring of the program performance.

The second program, OLCAOmain.exe, performs the iterative steps of the SCF process. Here, solving the eigenvalue problem for a large complex system is by far the most computationally intensive portion. Unfortunately the complexity of this calculation scales as $O(N^3)$ as long as the matrix remains dense and worse, it does not scale all that well at large processor counts because of the demands of inter-process communication. Fortunately, the OLCAO method uses a very compact basis set and so even if the scaling is poor the total amount of work to be done will not be excessive compared to other similarly purposed methods. The venerable ScaLAPACK library for parallelizing the solving of eigenvalue equations will be the primary tool. However, the possibility also exists that as the system size increases a threshold may be passed where the matrix becomes sufficiently sparse that other diagonalization methods that are more efficient and scalable can be used.

Once the OLCAOsetup.exe and OLCAOmain.exe calculations are done, the bulk of the computational work is also typically complete. Although it is possible to obtain electronic structure data directly from these programs it is often more favorable to use only the converged SCF potential and then to recompute certain details with a greater number of k points or a different basis set. This recomputation will also require parallelization, but the issues associated with parallelizing it are similar to the issues for parallelizing OLCAOsetup.exe and OLCAOmain.exe so that once thoee are solved the remaining parts are relatively simple variations. For example, the OLCAOband.exe program is essentially just a slight variation on the component of OLCAOmain.exe that is responsible for solving the secular equation. Hence, the same parallelization techniques used in OLCAOmain.exe can be reused for OLCAOband.exe. Another example is the OLCAOintg.exe program which is responsible for computing the multi-center integrals. This algorithm is a somewhat more complex variation on the algorithm used by OLCAOsetup.exe, but it relies on the same principles and so it will follow a similar path when being parallelized. The only substantially new pieces of code to be parallelized in the post SCF part are the programs for properties calculation such as OLCAOoptc.exe or OLCAO-dos.exe. In these cases there exist a large number of different parallelization routes that can be exploited and which tend to have only limited inter-process communication. For the DOS calculation the evaluation of the DOS and PDOS for each state in the solid state wave function can be done independently of the others and hence leads to a high degree of parallelism. Other viable approaches include parallelization across k points and parallelization across each component of each state. This last approach will require more inter-process communication, but because the memory requirements can be reduced it will likely be possible to keep the divided labor on a single compute node for many circumstances, thus avoiding the bottleneck of needing to communicate between processes on different nodes. The OLCAOoptc.exe program has another potential route to parallelization. In this case the parallelization can be across the number of initial states for electron transition. As this number can be quite large for big systems we anticipate that this division of labor would be quite effective. A final program where parallelization will be necessary is in the OLCAOwave.exe program which is used to compute values of the wave function or charge density on a three-dimensional uniform grid for the purpose of visualization. Here, the computation is done entirely in real space and it can be very easily divided into independent parts that can be stitched together for output. Parallelization can occur along both the number of grid points and along the atoms/orbitals that contribute to the wave function values at those grid points without much inter-process communication until all the values are computed and the results must be written to disk for the user.

Separate from the traditional approach of using MPI and OpenMP there do exist a few other ways that the OLCAO method can be trivially parallelized. In particular one that has already been used to great effect is with the spectral imaging calculations mentioned in Chapter 11. Here, a large number of independent XANES/ELNES calculations are performed to form a function field data set. Because these spectra are independent of each other they can be computed in parallel. On supercomputer machines where there are often constraints in terms of a maximum run time and a minimum number of CPU cores this approach can still be performed, but with a little extra care by dividing the computation into components and by bundling the jobs into appropriately sized groups. While this approach is not as technical as other more fundamental techniques, it did serve its purpose well. We mention this as a reminder that substantial efficiency gains do not always require highly detailed approaches for every type of problem.

The task of parallelizing the OLCAO program suite is a large and complex job that will be manpower intensive, but the gains to be made in terms of the size and type of system that will then be addressable far outweigh those costs. Hence, the project is currently underway and we anticipate that such a version will be available for use in relatively near term future releases of the program.

12.3 Ease of use

An issue that is frequently underappreciated by developers of scientific programs is that of ease of use. This follows naturally from the academic development environment of many scientific programs where the programmer may be the only one who knows the details of what they are programming and how to use the program after they are done. Worse, the programmer may even be the only one who ever uses a particular functionality. Fortunately, in recent years the ease of use issue is being more widely recognized because of its exceptional hidden power. That is, when a program is easy to use, it will attract users and other developers. These people will help improve the program by complaining about aspects that do not live up to their expectations and by contributing code for needed features or fixes for bugs. A key motivation for the writing of this book is to help prepare the OLCAO method for a much wider distribution to interested scientists. The OLCAO program suite has undergone substantial transformations in recent years that have made it much more user friendly. However, much work will remain to be done even after the release.

If we assume that a user understands the capabilities of the OLCAO program suite with regards to the materials problem that they are interested in, we consider the concept of "ease of use" to be divided into two major categories. The first is the ease with which users can interface with the program to create input, run and monitor jobs, and analyze computed results. The second is the ease with which third party programs can interface with the program.

12.3.1 User interface and control

While substantial advances have been made in recent years, spending development time on the user interface and control systems could still be very fruitful. The current user interface is entirely text based and command line driven. Two somewhat complex scripts, called "makeinput" and "olcao," and which are detailed in Appendix C, are used to create the input files and execute specific computational tasks. These scripts are powerful simplifying agents for dealing with the OLCAO program, but they are in need of some polishing to make use of them by non-experts more intuitive, robust, and less error prone. As mentioned before, this is not typically a high priority for academic research programs, but it is an important issue for the overall success of the program because it will broaden the base of users. This in turn will attract more expert developers to help exploit the inherent efficiencies and capabilities of the method that have not yet been exploited. Specific areas of improvement include more error checking, input sanity checking, more detailed error reporting, better tutorials, user guides, and user references, more detailed internal documentation and consistent use of terminology within the code. While many of these things already exist in the user interface scripts and programs, they need to be improved into a solid framework so that future development does not require an extensive implementation effort and a consistent programming style can be maintained.

Unfortunately, an often quoted law must always be kept in mind, "Investment in reliability will increase until it exceeds the probable cost of errors, or until someone insists on getting some useful work done." While ease of use is an important component of the program it is not, in itself, the goal of the program. Perhaps one of the most effective ways to improve the ease of use of the program is to sit down with the new user and instruct them on the capabilities and limitations of the program in terms of both functionality and the user interface. Text tutorials are a workable substitute for distributing this information to a large number of people, but they are also often insufficient because of a lack of immediate feedback. One way to improve the user experience is through the use of video tutorials and on-line workshops. With such an approach it is possible to provide a more nuanced explanation about how to use the program compared to a flat text file. The creation of an on-line forum and workshop with video tutorials for instructing new users about how to use OLCAO is under active development.

12.3.2   Interaction with third party software

The issues in subsection 12.3.1 that are about building a framework for the ways that the user directly interacts with OLCAO can be extended to include the ways that developers of third party applications will directly interact with OLCAO. Specifically, the Fortran 90 source code of the OLCAO program is in need of a polished framework so that when it is desired to have other programs integrate or interact with the OLCAO program they will have a clear and consistent interface through which to do so. As has been well demonstrated in the commercial market for consumer applications, a powerful and easy to use application program interface (API) will make an application much more attractive to developers.

There are also many other ways that the OLCAO program can be enhanced for interaction with third party software. In the following paragraphs we describe three examples that are under consideration or active development.

The first is with respect to the input file. The skeleton input file is described in some detail in Appendix C, but it is not one of the common standard file formats for describing solid state or molecular systems. The ability to convert data to the format of the OLCAO skeleton file from other popular formats (e.g., protein data bank (PDB) or crystallographic information file (CIF)) and vice versa is crucial for its integration into the wider community of programs that compute materials properties from the atomic structure. Programs such as Open Babel (Guha et al., 2006, Hutchison et al.) exist to facilitate this goal of program data interchange, but the OLCAO program's input file will need to be either adapted to match an existing format, or it will need to be added to the collection of data files that are treatable by programs such as Babel.

The second enhancement for OLCAO would be to improve the format of the computed output such that other programs can more easily access and use the data. This can be done both through better documentation and also through a more uniform description of the computed results. Specifically, it would be much more convenient if all of the output were stored in well organized data structures instead of within a flat text file. Presently, much of the intermediate data computed in the OLCAO program is stored using the hierarchical data format version five (HDF5), but none of the computed results take advantage of the organizational capabilities of this file format. It would be extremely efficient if the computed results were also stored in the HDF5 format as can be demonstrated with the following example: when results such as the partial density of states are resolved into orbital and spin components and then written to a flat text file, the organization of the data in memory must be reconstructed by the program that reads in the data. For large or complex data sets (i.e., systems with a large number of different types of atoms such as is found in biological systems with several thousand atoms) this reconstruction process can be a barrier to accessing the data. If the data structure could be retained in the stored file, then another program that wishes to access that data or a specific subset of it will have a much easier time finding it.

12.3.3   Data visualization

Another interaction with third party software that is extremely important for enabling productive use of the program is the interaction with visualization and data plotting software. At present, the main data analysis programs used in conjunction with OLCAO are Gnuplot, Origin, and OpenDX. There exists a script for making quick and simple plots of basic data (e.g. DOS, PDOS, optical properties, XANES/ELNES spectra, bond order, charge transfer, etc.) in Gnuplot, and there are also a few scripts for plotting that data in more complex and 'publication ready' formats within Origin. For visualizations of three-dimensional data sets such as charge density, the wave function itself, or the potential function, we have been using a program written within OpenDX. However, in all of these cases, there is some room for improvement. As may be expected, none of the scripts is exceptionally robust and they could each use some polish. For example, the Gnuplot scripts and the Origin scripts need to be updated to work effectively on multiple different versions of the software. Better error reporting and fault tolerance needs to be implemented. The programs for preparing the data in OpenDX format and the programs for plotting the data within OpenDX are powerful but it would appear that OpenDX is not installed or actively supported at many national level supercomputer centers. Thus, its utility is somewhat restricted and so an important step would be to modify the programs and scripts so as to produce data in a format that is amenable to other more widely supported visualization programs such as VisIt.

Development, enhancement, and extension of any program such as this one takes place on multiple fronts and requires many different skill sets beyond the traditionally anticipated ones of physics, chemistry, and materials science. Serious proficiency in computer programming and computer science is increasingly becoming a core asset for students or researchers that want to develop scientific simulations. An understanding of how people will actually use the program and even mundane sounding issues such as how to efficiently distribute, compile, and install the application on a wide range of production level supercomputers with different architectures are becoming increasingly important. While this book strives to document the successful application of the OLCAO method to many classes of systems, this chapter has tried to point out the promising future of the method in equally many directions. What is needed for this method and, likely also for many other scientific applications, is a cadre of dedicated researches and developers to use this program and to work to enhance it so that it can fulfill its role as powerful tool in the advancement of human knowledge and understanding. Gaining such a collection of users and developers will probably be the most substantial advancement, enhancement, or extension that can be applied to this method.

References

Abrikosov, A. A., Gorkov, L. P., & Dzyaloshinskii, I. E. (1963), *Methods of Quantum Field Theory in Statistical Physics* (Englewood Cliffs: Prentice-Hall).

Anisimov, V. I., Zaanen, J., & Andersen, O. K. (1991), *Physical Review B*, 44, 943.

Barth, U. V. & Hedin, L. (1972), *Journal of Physics C: Solid State Physics*, 5, 1629.

Becke, A. D. (1993), *The Journal of Chemical Physics*, 98, 1372–77.

Casimir, H. B. G. (1948), *Proc. Koninklike Nederlandse Akademie Van Wetenschappen*, 51, 793.

Ceperley, D. M. & Alder, B. J. (1980), *Physical Review Letters*, 45, 566.

French, R. H. (2000), *Journal of the American Ceramic Society*, 83, 2117–46.

French, R. H., Parsegian, V. A., Podgornik, R., et al. (2010), *Reviews of Modern Physics*, 82, 1887.

Guha, R., Howard, M. T., Hutchison, G. R., et al. (2006), *Journal of Chemical Information and Modeling*, 46, 991–98.

Gunnarsson, O. & Lundqvist, B. I. (1976), *Physical Review B*, 13, 4274.

Hamaker, H. C. (1937), *Physica*, 4, 1058–72.

Hedin, L. & Lundqvist, B. I. (1971), *Journal of Physics C: Solid State Physics*, 4, 2064.

Hutchison, G. R., Morley, C., James, C., Swan, C., De Winter, H. & Vandermeersch, T. *The Open Babel Package*[Online]. Available: http://openbabel.sourceforge.net/.

Jönsson, P., He, X., Froese Fischer, C., & Grant, I. P. (2007), *Computer Physics Communications*, 177, 597–622.

Landau, L. D., Lifshitz, E. M., & Pitaevskii, L. P. (1984), *Electrodynamics of Continuous Media* (Oxford: Elsevier Butterworth-Heinemann).

Lifshitz, E. M. (1956), *Sov. Phys. JETP*, 2, 73.

Macdonald, A. H. & Vosko, S. H. (1979), *Journal of Physics C: Solid State Physics*, 12, 2977.

Ninham, B. W. & Parsegian, V. A. (1970), 52, 4578–87.

Obara, S. & Saika, A. (1986), *The Journal of Chemical Physics*, 84, 3963–74.

Ogasawara, K., Ishii, T., Tanaka, I., & Adachi, H. (2000), *Physical Review B*, 61, 143.

Parsegian, V. A. (2005), *Van Der Waals Forces: A Handbook for Biologists, Chemists, Engineers, and Physicists* (New York: Cambridge University Press).

Rajter, R., French, R. H., Podgornik, R., Ching, W. Y., & Parsegian, V. A. (2008), *J. Appl. Phys.*, 104, 053513/1–053513/13.

Rajter, R. F., French, R. H., Ching, W. Y., Carter, W. C., & Chiang, Y. M. (2007a), *J. Appl. Phys.*, 101, 054303/1–054303/5.

Rajter, R. F., Podgornik, R., Parsegian, V. A., French, R. H., & Ching, W. Y. (2007b), *Phys. Rev. B*, 76, 045417/1–045417/16.

Szabo, A. & Ostlund, N. S. (1996), *Modern Quantum Chemistry: Introduction Ot Advanced Electronic Structure Theory* (New York: Dover Publications Inc.).

Tabor, D. & Winterton, R. H. S. (1969), *Proceedings of the Royal Society of London. Series A, Mathematical and Physical Sciences*, 312, 435–50.

Tatewaki, H. & Mochizuki, Y. (2003), *Theoretical Chemistry Accounts: Theory, Computation, and Modeling (Theoretica Chimica Acta)*, 109, 40–42.

The Open Babel Package, version 2.3.1 http://openbabel.org [accessed Jan 2012].

Wigner, E. (1934), *Physical Review*, 46, 1002.

Appendix A
Database for Atomic Basis Functions

The choice of basis set for wave function expansion in a quantum mechanical method is of paramount importance and is usually a defining characteristic of the method. OLCAO is no different. In OLCAO the wave function is expanded in terms of atomic orbitals which are themselves expanded as Gaussian functions with appropriate spherical harmonic multipliers for different angular momentum orbitals. Further, the set of Gaussian functions for a particular element are the same for *all* orbitals of that element while the coefficients of the Gaussian functions differ for orbitals with different n and l quantum numbers. The orbitals of a given element are divided into two groups called core and valence where the valence orbitals are orthogonalized against the core orbitals as described in Section 3.5 so that the core orbitals are not part of the final secular equation. The particular boundary for dividing the core and valence orbitals can be flexible to accommodate various special cases. A final characteristic of the OLCAO basis set is that unoccupied orbitals representing possible excited states are included for each element. This appendix details the characteristics of the OLCAO basis set for each element in the default basis set database provided with the OLCAO distribution.

One of the complexities of using a basis set of atomic orbitals is that there is no single parameter that can be modified to systematically improve the accuracy of the wave function expansion as exists with plane wave basis set based methods. Instead, the options available to atomic orbital based methods include tweaking the atomic orbital definitions and adding more orbitals. By increasing the number of atomic orbitals associated with a particular atomic type the wave function expansion will gain both more variational freedom and access to higher energy states. Although orbitals cannot be added indefinitely because of the effects of over completeness this is not a serious deficiency because the accuracy, variational freedom, and number of states achieved with the default basis set is already quite sufficient for the vast majority of cases. For OLCAO, the atomic orbital definitions for the most commonly used elements have been extensively tested over many years and have proven to be robust and transferrable to a wide variety of systems. The default orbital definitions provided in the data base rarely need any modification for standard usage. However, the number of orbitals that each atom contributes to the system's basis set is important for different types of calculations (e.g. Mulliken effective charge and bond order calculations give more consistent results with a more localized basis that does not include any fully unoccupied orbitals).

To make using different sized basis sets for different types of calculations as simple as possible the OLCAO package has three types of basis set defined for each atom. The minimal basis set includes only up to the valence shell of the highest occupied orbital of each atom. The full basis set extends the minimal basis set by adding one additional shell of orbitals. The extended basis set adds one more additional shell of orbitals to the full basis set and thus contains the largest number of orbitals. By default the appropriate basis set will be used for the requested calculation. The details of this arrangement are provided in Appendix C.

In Table A.1 the atomic orbital basis set definition for each atom in the default data base is provided. The orbitals for each element are categorized into either the core or valence section and the valence orbitals are further divided into minimal, full, and extended basis categories. The notation used in the table for the different basis sets is that each successive basis is an extension of the previous one. The minimal basis is an extension of the core, the full basis is an extension of the minimal basis and the extended basis is an extension of the full basis. Descriptions of the contracted Gaussian orbitals are provided by listing the number (N) of Gaussian functions for each angular momentum orbital type and the maximum values of the exponential coefficients (α). The minimum exponential α for all elements is 0.12 and so this value is not shown in the table. All the intermediate α values between the minimum and maximum terms follow the increasing progression of a geometric series with constant ratio: $(\alpha_{max}/\alpha_{min})^{1/(N-1)}$. When "d" or "f" angular momentum orbitals are part of the basis they may use fewer terms than the "s" or "p" type orbitals which always use the same number of terms. The notation employed in Table A.1 for the maximum α follows what is commonly found in computational work so that $5e4 = 5.0 \times 10^4$.

In general we can note a few typical rules of thumb that the basis sets follow. The atomic basis functions represent the characteristics of different elements and as such they tend to follow many of the same patterns present in the periodic table of the elements. With each additional row the number of Gaussian terms increases by approximately 1 and the maximum value of the Gaussian α coefficient increases by approximately 1/2 an order of magnitude (with row one to row two being an exception along with an increase midway through row four). The minimum Gaussian α coefficient is a constant value of 0.12 for all elements. Unsurprisingly, the number of core orbitals increases as the atomic Z number increases because more orbitals are below the threshold for being considered as a core orbital. That threshold is in the range of −30 to −32 eV. The number of Gaussian functions used to expand the "d" and "f" type orbitals increases with atomic Z number as well, but the pattern is less tied to the rows of the periodic table. Instead, the number of Gaussian functions for "d" and "f" type orbitals increases when another orbital of that type is added to the valence minimal basis, which could occur mid-row. All the parameters have been chosen through an empirical process and while physical reasoning was employed there is no physically principled motivation for the exact choice of particular parameters.

Table A.1. Data base of atomic orbital basis set definitions for each element.

Z	Name	Core	Valence: MB/FB/EB	N (s p)	N (d,f)	Max α
1	H	-	1s/2s2p/3s3p	16	-,-	1e4
2	He	-	1s/2s2p/3s3p	16	-,-	1e4
3	Li	1s	2s2p/3s3p/4s4p	20	-,-	5e4
4	Be	1s	2s2p/3s3p/4s4p	20	-,-	5e4
5	B	1s	2s2p/3s3p/4s4p	20	-,-	5e4
6	C	1s	2s2p/3s3p/4s4p	20	-,-	5e4
7	N	1s	2s2p/3s3p/4s4p	20	-,-	5e4
8	O	1s	2s2p/3s3p/4s4p	20	-,-	5e4
9	F	1s	2s2p/3s3p/4s4p	20	-,-	5e4
10	Ne	1s	2s2p/3s3p/4s4p	20	-,-	5e4
11	Na	1s-2s 2p	3s3p/4s4p3d/5s5p4d	21	12,-	1e5
12	Mg	1s-2s 2p	3s3p/4s4p3d/5s5p4d	20	12,-	1e5
13	Al	1s-2s 2p	3s3p/4s4p3d/5s5p4d	20	12,-	1e5
14	Si	1s-2s 2p	3s3p/4s4p3d/5s5p4d	21	12,-	2e5
15	P	1s-2s 2p	3s3p/4s4p3d/5s5p4d	21	12,-	1e5
16	S	1s-2s 2p	3s3p/4s4p3d/5s5p4d	21	12,-	1e5
17	Cl	1s-2s 2p	3s3p/4s4p3d/5s5p4d	21	12,-	1e5
18	Ar	1s-2s 2p	3s3p/4s4p3d/5s5p4d	21	12,-	1e5
19	K	1s-3s 2p	4s3p3d/5s4p4d/6s5p5d	22	16,-	5e5
20	Ca	1s-3s 2p	4s3p3d/5s4p4d/6s5p5d	22	16,-	5e5
21	Sc	1s-3s 2p	4s3p3d/5s4p4d/6s5p5d	22	16,-	5e5
22	Ti	1s-3s 2p-3p	4s4p3d/5s5p4d/6s6p5d	22	16,-	5e5
23	V	1s-3s 2p-3p	4s4p3d/5s5p4d/6s6p5d	22	16,-	5e5
24	Cr	1s-3s 2p-3p	4s4p3d/5s5p4d/6s6p5d	22	16,-	5e5
25	Mn	1s-3s 2p-3p	4s4p3d/5s5p4d/6s6p5d	22	16,-	5e5
26	Fe	1s-3s 2p-3p	4s4p3d/5s5p4d/6s6p5d	22	16,-	5e5
27	Co	1s-3s 2p-3p	4s4p3d/5s5p4d/6s6p5d	22	16,-	5e5
28	Ni	1s-3s 2p-3p	4s4p3d/5s5p4d/6s6p5d	22	16,-	5e5
29	Cu	1s-3s 2p-3p	4s4p3d/5s5p4d/6s6p5d	22	16,-	5e5
30	Zn	1s-3s 2p-3p	4s4p3d/5s5p4d/6s6p5d	23	17,-	1e6
31	Ga	1s-3s 2p-3p	4s4p3d/5s5p4d/6s6p5d	23	17,-	1e6
32	Ge	1s-3s 2p-3p 3d	4s4p/5s5p4d/6s6p5d	23	17,-	1e6
33	As	1s-3s 2p-3p 3d	4s4p/5s5p4d/6s6p5d	23	17,-	1e6
34	Se	1s-3s 2p-3p 3d	4s4p/5s5p4d/6s6p5d	23	17,-	1e6
35	Br	1s-3s 2p-3p 3d	4s4p/5s5p4d/6s6p5d	23	17,-	1e6
36	Kr	1s-3s 2p-3p 3d	4s4p/5s5p4d/6s6p5d	23	17,-	1e6
37	Rb	1s-4s 2p-3p 3d	5s4p4d/6s5p5d/7s6p6d	24	18,10	5e6
38	Sr	1s-4s 2p-3p 3d	5s4p4d/6s5p5d/7s6p6d	24	18,10	5e6
39	Y	1s-4s 2p-3p 3d	5s4p4d/6s5p5d/7s6p6d	24	18,10	5e6
40	Zr	1s-4s 2p-3p 3d	5s4p4d/6s5p5d/7s6p6d	24	18,10	5e6
41	Nb	1s-4s 2p-3p 3d	5s4p4d/6s5p5d/7s6p6d	24	18,10	5e6
42	Mo	1s-4s 2p-4p 3d	5s5p4d/6s6p5d/7s7p6d	24	18,10	5e6
43	Tc	1s-4s 2p-4p 3d	5s5p4d/6s6p5d/7s7p6d	24	18,10	5e6
44	Ru	1s-4s 2p-4p 3d	5s5p4d/6s6p5d/7s7p6d	24	18,10	5e6
45	Rh	1s-4s 2p-4p 3d	5s5p4d/6s6p5d/7s7p6d	24	18,10	5e6
46	Pd	1s-4s 2p-4p 3d	5s5p4d/6s6p5d/7s7p6d	24	18,10	5e6
47	Ag	1s-4s 2p-4p 3d	5s5p4d/6s6p5d/7s7p6d	24	18,10	5e6
48	Cd	1s-4s 2p-4p 3d	5s5p4d/6s6p5d/7s7p6d	24	18,10	5e6
49	In	1s-4s 2p-4p 3d	5s5p4d/6s6p5d/7s7p6d	24	18,10	5e6
50	Sn	1s-4s 2p-4p 3d	5s5p4d/6s6p5d/7s7p6d	24	18,10	5e6

51	Sb	1s-4s 2p-4p 3d-4d	5s5p/6s6p5d/7s7p6d	24	18,10	5e6
52	Te	1s-4s 2p-4p 3d-4d	5s5p/6s6p5d/7s7p6d	24	18,10	5e6
53	I	1s-4s 2p-4p 3d-4d	5s5p/6s6p5d/7s7p6d	24	18,10	5e6
54	Xe	1s-4s 2p-4p 3d-4d	5s5p/6s6p5d/7s7p6d	24	18,10	5e6
55	Cs	1s-4s 2p-4p 3d-4d	5s6s5p5d/7s6p6d/8s7p7d	26	20,10	1e7
56	Ba	1s-4s 2p-4p 3d-4d	5s6s5p5d/7s6p6d/8s7p7d	26	20,10	1e7
57	La	1s-5s 2p-4p 3d-4d	6s5p5d/7s6p6d/8s7p7d	26	20,10	1e7
58	Ce	1s-5s 2p-4p 3d-4d	6s5p5d4f/7s6p6d/8s7p7d	26	20,10	1e7
59	Pr	1s-5s 2p-4p 3d-4d	6s5p5d4f/7s6p6d/8s7p7d	26	20,10	1e7
60	Nd	1s-5s 2p-4p 3d-4d	6s5p5d4f/7s6p6d/8s7p7d	26	20,10	1e7
61	Pm	1s-5s 2p-4p 3d-4d	6s5p5d4f/7s6p6d/8s7p7d	26	20,10	1e7
62	Sm	1s-5s 2p-4p 3d-4d	6s5p5d4f/7s6p6d/8s7p7d	26	20,10	1e7
63	Eu	1s-5s 2p-4p 3d-4d	6s5p5d4f/7s6p6d/8s7p7d	26	20,10	1e7
64	Gd	1s-5s 2p-4p 3d-4d	6s5p5d4f/7s6p6d/8s7p7d	26	20,10	1e7
65	Tb	1s-5s 2p-4p 3d-4d	6s5p5d4f/7s6p6d/8s7p7d	26	20,10	1e7
66	Dy	1s-5s 2p-4p 3d-4d	6s5p5d4f/7s6p6d/8s7p7d	26	20,10	1e7
67	Ho	1s-5s 2p-4p 3d-4d	6s5p5d4f/7s6p6d/8s7p7d	26	20,10	1e7
68	Er	1s-5s 2p-4p 3d-4d	6s5p5d4f/7s6p6d/8s7p7d	26	20,10	1e7
69	Tm	1s-5s 2p-4p 3d-4d	6s5p5d4f/7s6p6d/8s7p7d	26	20,10	1e7
70	Yb	1s-5s 2p-4p 3d-4d	6s5p5d4f/7s6p6d/8s7p7d	26	20,10	1e7
71	Lu	1s-5s 2p-4p 3d-4d	6s5p5d4f/7s6p6d/8s7p7d	26	20,10	5e7
72	Hf	1s-5s 2p-5p 3d-4d 4f	6s6p5d/7s7p6d/8s8p7d	28	22,12	5e7
73	Ta	1s-5s 2p-5p 3d-4d 4f	6s6p5d/7s7p6d/8s8p7d	28	22,12	5e7
74	W	1s-5s 2p-5p 3d-4d 4f	6s6p5d/7s7p6d/8s8p7d	28	22,12	5e7
75	Re	1s-5s 2p-5p 3d-4d 4f	6s6p5d/7s7p6d/8s8p7d	28	22,12	5e7
76	Os	1s-5s 2p-5p 3d-4d 4f	6s6p5d/7s7p6d/8s8p7d	28	22,12	5e7
77	Ir	1s-5s 2p-5p 3d-4d 4f	6s6p5d/7s7p6d/8s8p7d	28	22,12	5e7
78	Pt	1s-5s 2p-5p 3d-4d 4f	6s6p5d/7s7p6d/8s8p7d	28	22,12	5e7
79	Au	1s-5s 2p-5p 3d-4d 4f	6s6p5d/7s7p6d/8s8p7d	28	22,12	5e7
80	Hg	1s-5s 2p-5p 3d-4d 4f	6s6p5d/7s7p6d/8s8p7d	28	22,12	5e7
81	Tl	1s-5s 2p-5p 3d-4d 4f	6s6p5d/7s7p6d/8s8p7d	28	22,12	5e7
82	Pb	1s-5s 2p-5p 3d-4d 4f	6s6p5d/7s7p6d/8s8p7d	28	22,12	5e7
83	Bi	1s-5s 2p-5p 3d-4d 4f	6s6p5d/7s7p6d/8s8p7d	28	22,12	5e7
84	Po	1s-5s 2p-5p 3d-4d 4f	6s6p5d/7s7p6d/8s8p7d	28	22,12	5e7
85	At	1s-5s 2p-5p 3d-4d 4f	6s6p5d/7s7p6d/8s8p7d	28	22,12	5e7
86	Rn	1s-5s 2p-5p 3d-4d 4f	6s6p5d/7s7p6d/8s8p7d	28	22,12	5e7
87	Fr	1s-5s 2p-5p 3d-4d 4f	7s6p5d/8s7p6d/9s8p7d	30	24,14	5e7
88	Ra	1s-5s 2p-5p 3d-4d 4f	7s6p5d/8s7p6d/9s8p7d	30	24,14	5e7
89	Ac	1s-6s 2p-5p 3d-5d 4f	7s6p5d/8s7p6d/9s8p7d	30	24,14	5e7
90	Th	1s-6s 2p-6p 3d-5d 4f	7s7p6d5f/8s8p7d6f/9s8p8d7f	30	24,14	5e7
91	Pa	1s-6s 2p-6p 3d-5d 4f	7s7p6d5f/8s8p7d6f/9s8p8d7f	30	24,14	5e7
92	U	1s-6s 2p-6p 3d-5d 4f	7s7p6d5f/8s8p7d6f/9s8p8d7f	30	24,14	5e7
93	Np	1s-6s 2p-6p 3d-5d 4f	7s7p6d5f/8s8p7d6f/9s8p8d7f	30	24,14	5e7
94	Pu	1s-6s 2p-6p 3d-5d 4f	7s7p6d5f/8s8p7d6f/9s8p8d7f	30	24,14	5e7
95	Am	1s-6s 2p-6p 3d-5d 4f	7s7p6d5f/8s8p7d6f/9s8p8d7f	30	24,14	5e7
96	Cm	1s-6s 2p-6p 3d-5d 4f	7s7p6d5f/8s8p7d6f/9s8p8d7f	30	24,14	5e7
97	Bk	1s-6s 2p-6p 3d-5d 4f	7s7p6d5f/8s8p7d6f/9s8p8d7f	30	24,14	5e7
98	Cf	1s-6s 2p-6p 3d-5d 4f	7s7p6d5f/8s8p7d6f/9s8p8d7f	30	24,14	5e7
99	Es	1s-6s 2p-6p 3d-5d 4f	7s7p6d5f/8s8p7d6f/9s8p8d7f	30	24,14	5e7
100	Fm	1s-6s 2p-6p 3d-5d 4f	7s7p6d5f/8s8p7d6f/9s8p8d7f	30	24,14	5e7
101	Md	1s-6s 2p-6p 3d-5d 4f	7s7p6d5f/8s8p7d6f/9s8p8d7f	30	24,14	5e7
102	No	1s-6s 2p-6p 3d-5d 4f	7s7p6d5f/8s8p7d6f/9s8p8d7f	30	24,14	5e7
103	Lr	1s-6s 2p-6p 3d-5d 4f	7s7p6d5f/8s8p7d6f/9s8p8d7f	30	24,14	5e7

The development of this data base of basis function definitions has been a long and evolutionary process where new ideas are still under active development. It is extremely important to note that while the data base contains definitions up to $Z = 103$ this does not mean that all atoms from hydrogen to lawrencium have been subjected to the same rigorous level of testing. Further, many of the heavier elements lack the inclusion of important physical effects such as relativity in the default data base. As such, while they are included by default they should be considered more as place holders for future development than as established definitions that are ready for general use.

Appendix B
Database for Initial Atomic Potential Functions

As described in the opening paragraph of Appendix A, the choice of basis set is of paramount importance and is a defining characteristic of a particular quantum mechanical electronic structure method. Of no less importance and no less unique to a given method is the choice of how to define the system's electronic potential function. In OLCAO the potential function is an effective potential that can be described as a summation of atom-centered Gaussian functions that includes all electron–electron, electron–nucleus, and exchange-correlation terms within it. There are many advantages to using Gaussian functions for the potential. All the multi-center interaction integrals involving the potential and the basis functions can be computed with efficient analytic formulas. It is possible to compute all the necessary multi-center interaction integrals for each term in the potential function only once and then update the potential in the SCF iterations simply by changing the coefficient for each term independently. If the charge density is cast into the same set of Gaussian functions with different coefficients then the total number of multi-center integrals that need to be computed is greatly decreased. These characteristics help make the OLCAO method particularly suited to calculating the properties of large complex systems. This appendix details the characteristics of the default OLCAO potential function for each element in the data base provided with the OLCAO distribution.

Much as with the basis set described in Appendix A, a critical complexity exists for the potential function. There is no single parameter that can be modified to systematically improve the accuracy of the potential function. Simply adding more terms or increasing the range of the terms will not necessarily improve the accuracy of the computed results. There are three parameters that define the potential function: the minimum and maximum Gaussian exponential coefficients (α_{min}, α_{max}) and the number of terms N_{pot} in the element's potential function. In the early days of the OLCAO method there was no way to guide the choice of these parameters except experience, physical intuition, and the actual computed results that could be compared to experimental data. Somewhat recently this problem has been circumvented with a more direct approach for identifying the quality of the potential function. As described in Section 4.10 it is possible to perform geometry optimization of the cell parameters and internal coordinates of many simple crystals by linking OLCAO to GULP and using the conjugate gradient method on finite differences in total energy. This capability allows for the comparison of different potential definitions by performing an objective comparison of the lattice parameters and internal coordinates to known experimental values. In such a way the potential functions for many elements have been optimized to give the best agreement in the largest number of different crystal systems. In the few cases where different potential functions were necessary to achieve optimal results in different crystal systems the average between the two that maximizes the accuracy for both systems was chosen. In this way the potential function for each element is quite transferrable to different crystal systems with very little cost to the overall accuracy of the calculation. This also frees the user from worries about the robustness of the potential function description. The default potential function definitions provided in the database rarely need any modification for standard usage and in almost all cases will provide reliable, robust service. However, the ability to modify a particular potential definition on the command line while preparing input for OLCAO still exists so that users can change α_{min}, α_{max}, and N_{pot} for a particular element for testing purposes.

In Table B.1 the potential function definition for each atom in the default data base is provided. Descriptions of the Gaussian functions are provided by listing the number (N_{pot}) of Gaussian functions for each element and the minimum and maximum values of the exponential coefficients (α). Just as with the basis set Gaussians, all the intermediate α values between the minimum and maximum terms follow the increasing progression of a geometric series with constant ratio: $(\alpha_{max}/\alpha_{min})^{1/(N-1)}$. The notation employed in Table B.1 for the maximum α follows what is commonly found in computational work so that $5e4 = 5.0 \times 10^4$.

The potential functions follow some notable trends that derive from the form of the periodic table of the elements. Atoms in the same column of the periodic table tend to have similar chemical behaviors and hence atoms of the same column will tend to have similar potential functions. The α_{min} value for atoms in column one (alkali atoms with one electron in the outer shell) will be smaller indicating a more broad and far reaching potential while those on the other end of the periodic table (atoms with more localized nearly closed shell structures) will have an α_{min} value that is substantially larger (perhaps three times). The α_{max} value tends to increase by an order of magnitude approximately once per row. The number of terms tends to increase with the atomic Z number but it is not clearly tied to any characteristics of the periodic table. These slight variations are the result of extensive testing and go to show that each atom has its own unique characteristics and one should not be too quick to lump superficially similar atoms into one group.

The development of this data base of potential function definitions has been a long and evolutionary process where new ideas are still under active development. There also exists a situation that is the same as was mentioned for the atomic basis set in Appendix A and the repetition of the observation is probably warranted. It is extremely important to note that while the data base contains definitions up to $Z = 103$ this does not mean that all atoms from hydrogen to

Table B.1. Data base of atomic potential definitions for each element.

Z	Name	N_{pot}	Min α	Max α
1	H	6	0.2	1e3
2	He	10	0.3	1e3
3	Li	16	0.1	1e6
4	Be	16	0.2	1e6
5	B	16	0.2	1e6
6	C	16	0.2	1e6
7	N	16	0.25	1e6
8	O	16	0.3	1e6
9	F	16	0.3	1e6
10	Ne	16	0.3	1e6
11	Na	16	0.1	1e6
12	Mg	16	0.15	1e6
13	Al	16	0.15	1e6
14	Si	20	0.15	1e6
15	P	20	0.15	1e6
16	S	20	0.2	1e6
17	Cl	20	0.25	1e6
18	Ar	20	0.3	1e6
19	K	20	0.08	1e6
20	Ca	22	0.08	1e6
21	Sc	22	0.135	1e6
22	Ti	22	0.15	1e7
23	V	22	0.15	1e7
24	Cr	22	0.15	1e7
25	Mn	23	0.15	1e7
26	Fe	23	0.15	1e7
27	Co	22	0.15	1e7
28	Ni	27	0.15	1e7
29	Cu	27	0.15	1e7
30	Zn	26	0.12	1e7
31	Ga	25	0.15	1e7
32	Ge	24	0.1	1e7
33	As	21	0.2	1e7
34	Se	23	0.22	1e7
35	Br	24	0.25	1e7
36	Kr	26	0.3	1e7
37	Rb	26	0.1	1e7
38	Sr	24	0.1	1e7
39	Y	24	0.15	1e7
40	Zr	26	0.1	1e7
41	Nb	26	0.15	1e7
42	Mo	26	0.15	1e7
43	Tc	26	0.15	1e7
44	Ru	26	0.15	1e7
45	Rh	26	0.15	1e7
46	Pd	26	0.15	1e7
47	Ag	26	0.15	1e7
48	Cd	28	0.15	1e7
49	In	32	0.12	1e8
50	Sn	34	0.15	1e8

(*continued*)

Table B.1. Continued

Z	Name	N_{pot}	Min α	Max α
51	Sb	34	0.1	1e8
52	Te	34	0.1	1e8
53	I	34	0.25	1e8
54	Xe	34	0.3	1e8
55	Cs	32	0.1	1e8
56	Ba	31	0.1	1e8
57	La	32	0.15	1e8
58	Ce	32	0.15	1e8
59	Pr	32	0.15	1e8
60	Nd	32	0.15	1e8
61	Pm	32	0.15	1e8
62	Sm	32	0.15	1e8
63	Eu	32	0.15	1e8
64	Gd	32	0.15	1e8
65	Tb	32	0.15	1e8
66	Dy	32	0.15	1e8
67	Ho	32	0.15	1e8
68	Er	32	0.15	1e8
69	Tm	32	0.15	1e8
70	Yb	31	0.15	1e8
71	Lu	35	0.15	1e8
72	Hf	38	0.15	1e8
73	Ta	42	0.15	1e8
74	W	50	0.15	1e8
75	Re	42	0.15	1e8
76	Os	42	0.15	1e8
77	Ir	42	0.15	1e8
78	Pt	42	0.15	1e8
79	Au	40	0.15	1e8
80	Hg	42	0.15	1e8
81	Tl	42	0.15	1e8
82	Pb	42	0.15	1e8
83	Bi	40	0.15	1e9
84	Po	42	0.15	1e9
85	At	42	0.15	1e9
86	Rn	42	0.2	1e9
87	Fr	42	0.1	1e9
88	Ra	42	0.1	1e9
89	Ac	42	0.1	1e9
90	Th	42	0.15	1e9
91	Pa	42	0.15	1e9
92	U	42	0.15	1e9
93	Np	42	0.15	1e9
94	Pu	42	0.15	1e9
95	Am	42	0.15	1e9
96	Cm	42	0.15	1e9
97	Bk	42	0.15	1e9
98	Cf	42	0.15	1e9
99	Es	42	0.15	1e9
100	Fm	42	0.15	1e9
101	Md	42	0.15	1e9
102	No	42	0.15	1e9
103	Lr	42	0.15	1e9

lawrencium have been subjected to the same rigorous level of testing. Further, many of the heavier elements lack the inclusion of important physical effects such as relativity in the default data base. As such, while they are included they should be considered more as place holders for future development than as established definitions that are ready for general use.

Appendix C
Current Implementation of the OLCAO Suite

C.1 Introduction

This appendix will provide the reader with an introductory reference to the OLCAO computational package; however, it will not be possible to delve too deeply into the complex internals. Instead, we strive for a broad overview of the entire package with pointers into the source distribution for those who wish to explore it in deeper detail, modify it to suit a particular problem of interest, or invest in coupling it to other programs. The OLCAO package was developed over many years in a somewhat *ad hoc* way consistent with an academic environment, however, in recent years a concerted effort has been made to eliminate inconsistencies, improve program interfacing and integration, and consolidate functionality. This multi-year effort has produced a much more user friendly and powerful package. Considerable complexity remains due both to its heritage and its goal of applicability to complex systems with thousands of atoms and hence the development task is still ongoing. However, the OLCAO package has clearly reached a stage where we believe it can and should reach a wider audience.

The OLCAO suite comprises three parts that all rely on a command line interface. The parts are identified as input generation, program execution, and results analysis. The input generation component is built around the concept of a skeleton input file and an input generation script called "makeinput." The program execution component centers on a command line script, "olcao," that will run a sequence of executable Fortran 90 programs to compute a specifically requested result. The results analysis component itself comprises numerous scripts and programs for assisting in the collection, organization, and display of computed data in a wide variety of ways. The actual display of plotted data is typically carried out with free third party software such as gnuplot and OpenDX or proprietary but popular programs such as Origin. Many of the results analysis scripts will prepare data specifically for plotting using these programs. These aspects (preparation, execution, and analysis) of the OLCAO suite are further detailed in the next three sections.

C.2 Input generation

The input files for the programs of the OLCAO suite are somewhat large and complex in their entirety but they can be derived from a much simpler form (called a skeleton input file) with the aid of an input generation script called "makeinput" that will be discussed later. The skeleton input file is, as the name implies, minimalistic in form and content. It describes only the most essential attributes of a system to be computed. The name of the skeleton file is "olcao.skl" and its rigid line-by-line form, which was adapted from the GULP program by Julian Gale, is provided in Table C.2.1. The file is composed of keyword/value pairs and keyword section definitions that follow a simple but strict ordering. Although the skeleton file follows a rigid format it is flexible enough to accommodate a very wide array of atomic systems including complex crystals, models containing microstructures and defects, non-crystalline materials, models containing surfaces, and isolated molecular systems.

Table C.2.1. The olcao.skl skeleton input file format specification.

Keyword	Value	Comments
"title"	None	Start of title section.
None for arbitrary number of lines.	Title, comments, and description of the input file contents.	Use any number of lines until the "end" keyword.
"end"	None	Ends the title section.
"cell" or "cellxyz"	None	Start of the cell definition section.
If "cell" next 2 lines:	$a\ b\ c\quad \alpha\ \beta\ \gamma$ $(abc)_1\ (xyz)_1\ (abc)_2\ (xyz)_2$	Define the unit cell in magnitude/angle form with all values in angstroms and degrees. Also define relative orientation of a,b,c axes to x,y,z axes.
If "cellxyz" next 3 lines:	a_x, a_y, a_z b_x, b_y, b_z c_x, c_y, c_z	Define the unit cell in Cartesian vector format with all values in angstroms.
"frac" or "cart"	Integer number of atoms listed (N_a).	Number of atoms listed in the file and the form (fractional or Cartesian) of their coordinates.
None for next N_a lines.	Element abbreviation, optional species number, and atomic coordinates in chosen form.	Examples: Si 0.000 0.000 0.000 O7 0.5 0.5 0.5
"space"	Space group designation.	Number or Hermann–Mauguin name in ASCII form.
"supercell"	Integer numbers $C_a\ C_b\ C_c$.	Number of cells to replicate in each of the a, b, c directions.
"prim" or "full"	None	Based on cell defined by "cell" line above. "full" uses given cell. "prim" converts given cell to a primitive cell if possible. Does not convert primitive cells into full cells.

Table C.2.2. Example olcao.skl skeleton input file for fluorapatite.

title

Fluorapatite (FAP), from J. Y. Kim, R. R. Fenton, B. A. Hunter, and B. J. Kennedy, Aust. J. Chem. 53, 679 (2000).
This is a perfect crystal.
end
cell
  9.3475   9.3475   6.8646   90.0   90.0   120.0
  1 1 2 2
frac 7

Ca1	core	0.3333333333	0.6666666667	0.0011000000
Ca2	core	0.2390000000	−0.0114000000	0.2500000000
P	core	0.3964000000	0.3700000000	0.2500000000
O1	core	0.3259000000	0.4852000000	0.2500000000
O2	core	0.5909000000	0.4695000000	0.2500000000
O3	core	0.3380000000	0.2552000000	0.0707000000
F	core	0.0000000000	0.0000000000	0.2500000000

space 176
supercell 1 1 1
prim

An example skeleton input file is provided in Table C.2.2 for the fluorapatite crystal system. This skeleton input file was input by hand from information in the journal article noted in the title section, but there are numerous other ways to create skeleton files. Online databases such as the Protein Data Bank or the Crystallography Open Database can be used as sources of information and converted to the skeleton format through the automated scripts pdb2skl and cif2skl that are included in the OLCAO package. Once created there are also many methods for manipulating the contents of the file in powerful and automated ways such as though the modStruct Perl script which is also included as part of the OLCAO package. The creation of skeleton input files for more complex systems is beyond the scope of this work and falls into the category of system modeling.

Once the desired skeleton file has been created, it is converted via the makeinput script into a set of input files that are necessary and sufficient for running an OLCAO calculation. This process is designed to be a relatively simple black box type of task even for large complex systems. However, making effective use of the power in the makeinput script requires an understanding of its internal details and advanced options. Because this is an ever evolving process the most reliable and up-to-date source for such detailed information is the OLCAO manual included with the current distribution as well as the documentation internal to the script itself. In this appendix the core information will be presented to initiate an understanding and provide a foundation for those with further interest. The key input files, output files, command line parameters, and data bases that are associated with the makeinput script are illustrated in a schematic non-comprehensive way in Fig. C.2.1 with files

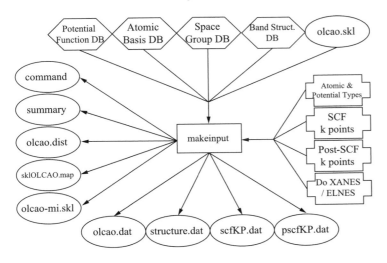

Fig. C.2.1.
Schematic of the makeinput script.

indicated by ellipses, data bases by hexagons, and command line parameters by cross boxes. Output files on the bottom are necessary for running an OLCAO calculation. Output files on the left are for record keeping and structural analysis. The makeinput script requires access to four data bases and the skeleton file as input shown at the top. Command line options on the right can be used to control the content of the output files in powerful ways. For example the command line parameter to prepare for a XANES/ELNES calculation can create multiple sets of input files in specific subdirectories where each member of the set belongs to a large set of specific atoms or atomic types. Only the central aspects that will be useful to a user approaching OLCAO as a black box are shown. The schematic does not include all command line options possible for makeinput nor does it show all the files produced by makeinput (particularly any intermediate files or those produced by advanced options). A detailed description of all options is beyond the scope of this discussion. However, they can be viewed in the on-line manual or by issuing the script command with the "-help" option. Table C.2.3 presents a short collection of example calls to the makeinput script for various common situations.

Certain output files of the makeinput script deserve some comment on their content and structure because they are each used in almost every OLCAO calculation. Familiarity with these files is a prerequisite for efficient and effective use of the OLCAO suite. These files are the k point files, the structure file, the potential file, and the primary input data file. There are two k point files (kp-scf.dat and kp-pscf.dat) that define (possibly) different k point meshes for the self-consistent field (SCF) and post-SCF (PSCF) parts of an OLCAO calculation according to command line options for makeinput. The meshes follow the scheme of Monkhorst and Pack. The details of the k point files are given in Table C.2.4.

The structure file, named structure.dat, defines the lattice parameters and lists data for each atomic and potential site. The site data include an index number, a number identifying which atomic or potential type the site is asso-

Table C.2.3. Example "makeinput" commands for a variety of typical ways that one would want to use to create OLCAO input.

Command	Effect
makeinput –cif	Produce input files with 1 general k point for both SCF and post-SCF parts. Also produce a crystallographic information file for visualizing the crystal structure in a third party program.
makeinput –kp 2 3 4	Use a k point mesh of $2 \times 3 \times 4$ for both SCF and post-SCF parts. (Used e.g. for a complex perfect crystal.)
makeinput –scfkp 1 1 1 – pscfkp 3 3 3 –reduce	Use a Γ k point for the SCF part and a $3 \times 3 \times 3$ mesh for the post-SCF part. Reduce the number of potential types. (Used e.g., for a large molecule or defect containing system.)
makeinput –scfkp 1 1 1 – pscfkp 3 3 3 –xanes	Use k points as above and prepare separate input for XANES computation of one atom of each non-equivalent elemental species.
makeinput –scfkp 1 1 1 – pscfkp 3 3 3 –xanes – atom 1..5 11	Use k points as above and prepare separate input for XANES computation of atoms 1, 2, 3, 4, 5, and 11.

Table C.2.4. Structure and content of the k point input files: kp-scf.dat and kp-pscf.dat. Each k point for the scf and post-scf parts of the OLCAO suite are listed in reciprocal space a, b, c coordinates (often called b_1, b_2, b_3).

Line Number	Content
1	Tag: "NUM_BLOCH_VECTORS"
2	Integer number of k points. (N_k)
3	Tag: "NUM_WEIGHT_KA_KB_KC"
	Index number
Each of the next N_k lines contains the following space separated content.	k point weighting factor, (Sum of all weights = 2)
	k point coordinate in reciprocal lattice a vector (b_1)
	k point coordinate in reciprocal lattice b vector (b_2)
	k point coordinate in reciprocal lattice c vector (b_3)

ciated with, the Cartesian coordinates of the site in atomic units, and the elemental name of the atom associated with the site. A detailed description of the structure file is given in Table C.2.5.

The potential file, named scfV.dat, contains the description of the electronic potential used by OLCAO which is a summation of atom-centered Gaussian functions. In the OLCAO method the charge density is cast into a function of atom-centered Gaussians that use the same Gaussians as the potential function but with different coefficients. The coefficients for the total charge density, valence charge density, and spin charge density difference are included for reference purposes in the scfV.dat file. Initially, before the SCF iterations, these three charge density values are all zero. After the SCF iterations the converged potential and charge density coefficients are written into a file of identical structure to scfV.dat but with a slightly different name (detailed in section C.3). The initial zero valued charge density coefficients are not problematic because the SCF cycle starts from an initial potential and derives the charge density at a later step in the SCF cycle and so does not rely on any initial charge density

Table C.2.5. Structure and content of the structure.dat input file. The Cartesian coordinates of each atom and potential site are listed along with associated meta-data.

Line Number	Content
1	Tag: "CELL_VECTORS"
2	Lattice vector a in x, y, z Cartesian components
3	Lattice vector b in x, y, z Cartesian components
4	Lattice vector c in x, y, z Cartesian components
5	Tag: "NUM_ATOM_SITES"
6	Integer number for number of atomic sites (N_a)
7	Tag: "NUM_TYPE_X_Y_Z_ELEM"
	Integer index number
	Atomic type number
Each of the next N_a lines contains the following space separated content.	Cartesian x coordinate
	Cartesian y coordinate
	Cartesian z coordinate
	Element abbreviation
8 + N_a	Tag: "NUM_POTENTIAL_SITES"
9 + N_a	Integer number for number of potential sites (N_p)
10 + N_a	Tag: "NUM_TYPE_X_Y_Z_ELEM"
	Integer index number
	Potential type number
Each of the next N_p lines contains the following space separated content.	Cartesian x coordinate
	Cartesian y coordinate
	Cartesian z coordinate
	Element abbreviation

Table C.2.6. Structure and content of the scfV.dat potential file. Each potential term of the potential function is given in an ordered list grouped by potential types. The total charge density, valence charge density, and spin-density difference data are included in this file initially as zeros, but after SCF convergence a new file is written with an identical structure where the charge density data is included.

Line number	Content
1	Tag: "NUM_TYPES"; Integer number for number of potential types.
2	Integer number for the number of terms in the first type (N_{terms}).
	Coefficient for the potential Gaussian of this term
	Exponential α for the Gaussian of this term
Each of the next N_{terms} lines contains the following space separated content.	Coefficient for the total charge density Gaussian
	Coefficient for the valence charge density Gaussian
	Coefficient for the spin charge density difference
Remaining Lines	Repeat content of lines 2 through N_{terms} of each type for each potential type.

value. A detailed description of the form and content of the potential file is given in Table C.2.6.

The primary input data file, called olcao.dat, can be divided into a title prefix and three main sections. The form of the title prefix is shown in Table C.2.7. The first section defines the basis functions of the system's atomic types. The

Table C.2.7. Form of the olcao.dat title prefix.

Line Number	Content
1	Tag: "TITLE"
Next N_t Lines	Name, source, and any other description for this system on any number of lines (N_t).
2 + N_t	Tag: "END_TITLE"

Table C.2.8. Structure and content of the first part of the olcao.dat input file.

Section line #	Content
1	Tag: "NUM_ATOM_TYPES".
2	Integer number of atomic types (N_{at}).
3	Tag: "ATOM_TYPE_ID_SEQUENTIAL_NUMBER".
4	Integer element ID number.
	Integer atomic species ID number within the associated element ID group.
	Integer atomic type ID number within the associated species ID group.
	Integer atomic type ID number ordered sequentially for all system types.
5	Tag: "ATOM_TYPE_LABEL".
6	Character string uniquely identifying this atomic type. It is a concatenation of the element name, species ID number, and the type ID number.
7	Tag: "NUM_ALPHAS_S_P_D_F".
8	Four integer numbers indicating the number of Gaussian functions that atomic orbitals of each angular momentum quantum number will use in their expansion. The list is ordered (s, p, d, f) and the number of Gaussians must be decreasing. The largest number of Gaussian functions is called N_α.
9	Tag: "ALPHAS"
Next $N_\alpha/4$ lines	Coefficients in the Gaussian exponents at a rate of 4 per line.
10 + $N_\alpha/4$	Tag: "NUM_CORE_RADIAL_FNS"
11 + $N_\alpha/4$	Three integer numbers indicating the number of atomic orbitals that are designated as core orbitals for each of the three basis cases: minimal, full, and extended.
12 + $N_\alpha/4$	Tag: "NL_RADIAL_FUNCTIONS"
13 + $N_\alpha/4$	Two integer numbers. The first integer indicates the number of components associated with this particular atomic orbital. In non-relativistic calculations, each atomic orbital has only one component. The second integer is a code identifying which basis sets this atomic orbital should be used for: 1 = MB, FB, EB; 2 = FB, EB; 3 = EB.
14 + $N_\alpha/4$	Five integer numbers. The first integer is the n quantum number, the second is the l quantum number, the third is 2 * the j quantum number, the fourth is the number of electron states in this atomic orbital, and the fifth and last is an index number identifying the current component.
Next $N_\alpha/4$ lines	Coefficients of the Gaussian functions at a rate of 4 per line.
	Repeat from tag NL_RADIAL_FUNCTIONS for remaining core radial basis functions.
	Repeat from tag NUM_CORE_RADIAL_FNS except with "VALE" instead of "CORE" for all the valence orbitals.
	Repeat from tag ATOM_TYPE_ID_SEQUENTIAL_NUMBER for all remaining atomic types.

Table C.2.9. Structure and content of the second part of the olcao.dat input file. This part deals with the expression of the potential function.

Section Line #	Content
1	Tag: "NUM_POTENTIAL_TYPES"
2	Integer number of potential types (N_{pot}).
3	Tag: "POTENTIAL_TYPE_ID__SEQUENTIAL_NUMBER"
4	Integer element ID number.
	Integer potential species ID number within the associated element ID group.
	Integer potential type ID number within the associated species ID group.
	Integer potential type ID number ordered sequentially for all system types.
5	Tag: "POTENTIAL_TYPE_LABEL"
6	Character string uniquely identifying this potential type. It is a concatenation of the element name, species ID number, and the type ID number.
7	Tag: "NUCLEAR_CHARGE__ALPHA"
8	Two real numbers. The first is the nuclear charge for this potential type and the second is the exponential coefficient in the Gaussian function used to model the nuclear charge distribution.
9	Tag: "COVALENT_RADIUS"
10	Real number representing a cutoff distance in atomic units for controlling different sampling rates of the spherical mesh for the exchange-correlation potential evaluation.
11	Tag: "NUM_ALPHAS"
12	Integer number of Gaussian functions for this potential type.
13	Tag: "ALPHAS"
14	Two real numbers representing the minimum and maximum valued coefficients in the exponential Gaussian terms.
	Repeat from tag POTENTIAL_TYPE_ID__SEQUENTIAL_NUMBER for all remaining potential types.

Table C.2.10. Structure and content of the third and final part of the olcao.dat input file. This part deals with control parameters for each individual program in the OLCAO suite.

Section Line #	Content
1	Tag: "NUM_ANGULAR_SAMPLE_VECTORS"
2	Integer indicating the number of rays that are used to define the radial mesh for numerical exchange-correlation evaluation.
3	Tag: "WTIN_WTOUT"
4	Two real numbers that can be used to apply different weights to separate regions of the atom centered spherical mesh that is used for numerically evaluating the charge density.
5	Tag: "RADIAL_SAMPLE-IN_OUT_SPACING"
6	Three real numbers that define different regions in the atom centered spherical mesh and also define the degree to which points in the mesh should be separated.
7	Tag: "SHARED_INPUT_DATA"
8	Tag: "BASIS_FUNCTION_AND_ELECTROSTATIC_CUTOFFS"
9	Two real numbers used to define the maximum range beyond which certain calculations are deemed to be negligible
10	Tag: "NUM_STATES_TO_USE"
11	Three integers, one each for the three basis set choices, that define the number of states to record in the solid state wave function.
12	Tag: "NUM_ELECTRONS"
13	Integer number of valence electrons in the system.
14	Tag: "THERMAL_SMEARING_SIGMA"
15	Real number in units of eV that defines the temperature to use for the purpose of thermal smearing of the electron distribution in states near the top of the valance band or Fermi level. Note: 1 eV = 11604.505 K.

Table C.2.11. Example olcao.dat input file for crystalline silicon. This olcao.dat input file was input automatically generated by the makeinput script from a simple olcao.skl file.

```
TITLE
Crystalline Silicon
END_TITLE
```

```
NUM_ATOM_TYPES
1
ATOM_TYPE_ID__SEQUENTIAL_NUMBER
 1   1   1   1
ATOM_TYPE_LABEL
si1_1
NUM_ALPHA_S_P_D_F
 21   21   12    0
ALPHAS
 0.12000000E + 00    0.24562563E + 00    0.50276626E + 00    0.10291023E + 01
 0.21064492E + 01    0.43116493E + 01    0.88254299E + 01    0.18064598E + 02
 0.36976070E + 02    0.75685587E + 02    0.15491933E + 03    0.31710133E + 03
 0.64906844E + 03    0.13285654E + 04    0.27194143E + 04    0.55663154E + 04
 0.11393581E + 05    0.23321296E + 05    0.47735901E + 05    0.97709673E + 05
 0.20000000E + 06
NUM_CORE_RADIAL_FNS
 3   3   3
NL_RADIAL_FUNCTIONS
 1   1
 1   0   0   2   1
  0.20224320E − 05   −0.11772981E − 04    0.31599443E − 04   −0.13701126E − 04
 −0.37771061E − 03   −0.10203811E − 01   −0.71772721E + 00   −0.51621824E + 01
 −0.12488962E + 02   −0.16727944E + 02   −0.16426865E + 02   −0.13599426E + 02
 −0.10441508E + 02   −0.74813896E + 01   −0.55293424E + 01   −0.35667066E + 01
 −0.30769484E + 01   −0.11888129E + 01   −0.23798723E + 01    0.53238664E + 00
 −0.20168466E + 01
 1   1
 2   0   0   2   1
 −0.22356584E − 04    0.11125771E − 03   −0.47126646E − 01   −0.66130559E + 00
 −0.21788738E + 01   −0.25458103E + 01   −0.39283300E + 00    0.31156516E + 01
  0.53931962E + 01    0.56617663E + 01    0.49543957E + 01    0.38194257E + 01
  0.28633595E + 01    0.20038244E + 01    0.14823704E + 01    0.94324913E + 00
  0.82064362E + 00    0.31227729E + 00    0.63283812E + 00   −0.14222559E + 00
  0.53480044E + 00
 1   1
 2   1   2   6   1
 0.23836036E − 04    0.16174360E − 03    0.36515487E − 01    0.47477314E + 00
 0.22563512E + 01    0.57962046E + 01    0.10072202E + 02    0.12917543E + 02
 0.12677324E + 02    0.11158363E + 02    0.85836697E + 01    0.65469523E + 01
 0.45883788E + 01    0.33947580E + 01    0.22318523E + 01    0.17309191E + 01
 0.10389629E + 01    0.88798721E + 00    0.51251153E + 00    0.33127889E + 00
 0.47849771E + 00
```

NUM_VALE_RADIAL_FNS
 2   5   8
NL_RADIAL_FUNCTIONS
 1   1
 3   0   0   2   1
  0.37025872E + 00   0.23322659E + 00   0.42086535E + 00   −0.78821213E + 00
−0.89529614E + 00   −0.11628782E + 01   0.20862018E − 01   0.85189985E + 00
  0.17063889E + 01   0.15883816E + 01   0.14966144E + 01   0.10581068E + 01
  0.86184102E + 00   0.54810327E + 00   0.44949458E + 00   0.25332138E + 00
  0.25003675E + 00   0.79387325E − 01   0.18916434E + 00   −0.44924818E − 01
  0.15518037E + 00
 1   2
 4   0   0   2   1
  0.17453941E + 01   −0.45732442E + 01   0.39246496E + 01   −0.31156687E + 01
  0.39555762E + 01   −0.16488301E + 01   0.20575545E + 01   −0.23254371E + 01
−0.22590176E − 01   −0.22885283E + 01 −0.33773040E + 00   −0.14874734E + 01
−0.14663289E + 00   −0.84598331E + 00 −0.19089389E − 01   −0.45891512E + 00
  0.73193668E − 02   −0.21301259E + 00 −0.44098773E − 01   −0.23137457E − 01
−0.10131509E + 00
 1   3
 5   0   0   2   1
0.15801987E + 01   −0.79548674E + 01   0.15825594E + 02   −0.14367039E + 02
0.85486234E + 01   −0.95904156E + 01   0.62507901E + 01   −0.37624179E + 01
0.61503130E + 01   −0.98817584E + 00   0.44093069E + 01   −0.55740528E + 00
0.26730678E + 01   −0.51135219E + 00   0.15617856E + 01   −0.43586390E + 00
0.92130726E + 00   −0.33977182E + 00   0.55124390E + 00   −0.22960266E + 00
0.26094514E + 00
 1   1
 3   1   2   6   1
  0.15887642E + 00   0.27780095E − 01   0.40461519E + 00   −0.26664027E + 00
−0.30851015E + 00   −0.18678203E + 01 −0.22243863E + 01   −0.35970246E + 01
−0.26866326E + 01   −0.31620356E + 01 −0.15684578E + 01   −0.21288180E + 01
−0.49373705E + 00   −0.14865191E + 01   0.20219177E + 00   −0.12244133E + 01
  0.62687306E + 00   −0.11368823E + 01   0.78705828E + 00   −0.83816143E + 00
  0.28541998E + 00
 1   2
 4   1   2   6   1
0.44420760E + 00   −0.13845441E + 01   0.11677417E + 01   −0.16812967E + 01
0.25108564E + 01   −0.46223764E + 00   0.49911823E + 01   0.80501825E + 00
0.60129596E + 01   −0.26787827E + 00   0.55213897E + 01   −0.20843629E + 01
0.52328582E + 01   −0.36616986E + 01   0.55115458E + 01   −0.50005205E + 01
0.61424202E + 01   −0.59814561E + 01   0.62383332E + 01   −0.50115067E + 01
0.28067503E + 01
 1   3
 5   1   2   6   1
0.37406391E + 00   −0.23951077E + 01   0.57616585E + 01   −0.52938942E + 01
0.47099047E + 01   −0.88710311E + 01   0.23280084E + 01   −0.12691675E + 02
0.28086294E + 01   −0.13230890E + 02   0.61766915E + 01   −0.13167499E + 02
0.98050895E + 01   −0.14151231E + 02   0.13199527E + 02   −0.16117304E + 02
0.16307341E + 02   −0.17974474E + 02   0.17160162E + 02   −0.14604605E + 02
0.74473568E + 01

(*continued*)

```
  1   2
  3   2   4   10   1
0.70755268E − 01    −0.70266486E − 01    0.25497220E + 00    −0.18933548E + 00
0.61915066E + 00    −0.46734690E + 00    0.13076032E + 01    −0.12696668E + 01
0.25083394E + 01    −0.29564711E + 01    0.39583342E + 01    −0.29215271E + 01
  1   3
  4   2   4   10   1
0.93668565E − 01    −0.43314686E + 00    0.35986475E + 00    −0.10873779E + 01
0.83694797E + 00    −0.24137964E + 01    0.21685098E + 01    −0.48988032E + 01
0.56141398E + 01    −0.91034396E + 01    0.10123547E + 02    −0.87905870E + 01
NUM_POTENTIAL_TYPES
  1
POTENTIAL_TYPE_ID__SEQUENTIAL_NUMBER
  1   1   1   1
POTENTIAL_TYPE_LABEL
si1_1
NUCLEAR_CHARGE__ALPHA
14.000000 20.000000
COVALENT_RADIUS
1.000000
NUM_ALPHAS
16
ALPHAS
1.500000e-01 1.000000e + 06
NUM_ANGULAR_SAMPLE_VECTORS
100
WTIN_WTOUT
0.5 0.5
RADIAL_SAMPLE-IN_OUT_SPACING
0.1 3.5 0.8
SHARED_INPUT_DATA
BASISFUNCTION_AND_ELECTROSTATIC_CUTOFFS
   1.00000000E-16 1.00000000E-16
NUM_STATES_TO_USE
8 20 20
NUM_ELECTRONS
8
THERMAL_SMEARING_SIGMA
0 0
MAIN_INPUT_DATA
LAST_ITERATION
50
CONVERGENCE_TEST
0.0001
CORRELATION_CODE
1
FEEDBACK_LEVEL
2
RELAXATION_FACTOR
0.2
EACH_ITER_FLAGS__TDOS
0
```

```
NUM_SPLIT_TYPES__DEFAULT_SPLIT
0 0.01
TYPE_ID__SPIN_SPLIT_FACTOR
PDOS_INPUT_DATA
   0.01 0.1      ! PDOS Delta Energy, PDOS sigma broadening
   -30 30        ! PDOS EMIN and EMAX
   0             ! Flag for all atom PDOS
BOND_INPUT_DATA
   3.5     ! MAXIMUM BOND LENGTH
   1       ! Flag for all atom BOND
SYBD_INPUT_DATA
   7,300,0            ! IF IFSYK=1, READ: # SYMMETRY K,# K GENERATED,IFCAR
   0.0 0.0 0.0        ! GAMMA
   0.5 0.0 0.5        ! X=(0.0,1.0,0.0)*2*PI/A
   0.5 0.25 0.75      ! W=(0.25,1.0,0.0)*2*PI/A
   0.5 0.5 0.5        ! L=(0.5,0.5,0.5)*2*PI/A
   0.0 0.0 0.0        ! GAMMA
   0.375 0.375 0.75   ! K=(0.75,0.75,0.0)*2*PI/A
   0.5 0.5 1.0        ! X'=(1.0,1.0,0.0)*2*PI/A
PACS_INPUT_DATA
   0             ! Excited atom number
   0.01 0.5      ! PACS delta Energy, PACS sigma factor
   5 50          ! Energy slack before onset, energy window
   0             ! Number of possible core orbitals to excite
OPTC_INPUT_DATA
   45       ! OPTC energy cutoff
   100      ! OPTC energy trans
   0.01     ! OPTC delta energy
   0.1      ! OPTC broadening
SIGE_INPUT_DATA
   5        ! SIGE energy cutoff
   0.3      ! SIGE energy trans
   0.001    ! SIGE delta energy
   0.1      ! SIGE broadening
WAVE_INPUT_DATA
   10 10 10 !          a,b,c # of mesh points
   -100000.0 100000.0  ! min,max range of energy
   0                   ! 0=Psi^2; 1=Rho
   1                   ! 1=3D+1D; 2=3D; 3=1D.
END_OF_DATA
```

second section helps define the potential functions of the system's potential types. The last section contains a set of program parameters that control the behavior of each possible step in a calculation (e.g., convergence criteria, Gaussian interaction cutoff values, output form for DOS and bond order calculations, etc.). These three sections are detailed in Tables C.2.8, C.2.9, and C.2.10 respectively while a sample listing is given in Table C.2.11.

C.3 Program execution

The execution of quantum mechanical calculations in the OLCAO suite is accomplished primarily with a Perl script called olcao. This script will, according to given command line options, execute an ordered sequence of Fortran 90 programs. Each Fortran 90 program requires its own command line parameters, management of its input and output files, and record keeping of its execution. Although it is possible to perform these tasks by hand with standard UNIX commands the olcao script will automatically take care of all these issues for each of the Fortran 90 programs such that a single use of the script is sufficient to complete a particular calculation (e.g., the command, "olcao -dos" is sufficient to compute the total and partial DOS of the given system). If the output file from any part of a previous computation exists, the script will detect it and refrain from duplicating that part. In this sense, the olcao program has a built in check pointing scheme that is crude but effective. The olcao script itself is shown schematically in Fig. C.3.1 to illustrate its command line options and what it produces as output. As with Fig. C.2.1, files are indicated by ellipses and command line parameters by cross boxes. The set of OLCAO input files and OLCAO output files depend on the specific computation requested by the command line options and are detailed later. The basis can be altered to minimal, full, or extended for both the SCF and PSCF parts independently. The computation can use an alternate set of Fortran 90 binaries for testing purposes. The computation can be defined as either spin-polarized or spin-degenerate, and the exact type of computation can be specified (e.g., density of states, band structure, optical properties, etc.). Table C.3.1 presents a collection of example calls to the olcao script to demonstrate each type of calculation request and a few other possible options.

The Fortran 90 programs used by the olcao script and their purposes are described in Table C.3.2. A critical aspect of the implementation of the OLCAO method is that the suite has been divided into two distinct segments called the self-consistent field (SCF) part and the Post-SCF (PSCF) part. While it is possible to compute many electronic structure properties from the SCF part, it is instead often used solely to obtain an SCF converged description of the electronic potential function. This potential function is then typically used in

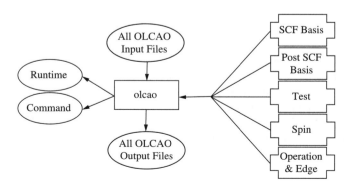

Fig. C.3.1.
Schematic of the OLCAO script.

Table C.3.1. Example "olcao" commands. The command line options for the olcao script typically take the form of a dash "-" followed by a short word. Some require parameters (e.g. -scf and -olcao) while other options can have optional parameters (e.g., -wave can be alone or given as -wave 1s to request calculation on a specific excited state from a XANES/ELNES type SCF calculation).

Command	Effect
olcao	Perform SCF calculation to obtain converged electronic potential function and total energy.
olcao -dos	Perform SCF calculation as above followed by a PSCF calculation of the total and partial DOS.
olcao -dos -scf MB -olcao MB	As above except with an explicit request to use a minimal basis set instead of the default full basis for SCF and DOS calculations.
olcao -sybd	Perform SCF calculation as above followed by a PSCF calculation of the so-called symmetric band structure. This is the band structure defined by a path among various high symmetry k points in the Brillouin zone.
olcao -bond	Perform SCF calculation as above followed by a PSCF calculation of the bond order and Mulliken effective charge.
olcao -optc	Perform SCF calculation as above followed by a PSCF calculation of the valence band optical properties in terms of the optical conductivity and imaginary part of the frequency dependent dielectric function.
olcao -sige	Perform SCF calculation as above followed by a PSCF calculation of the electrical conductivity.
olcao –pacs 1s	Perform a ground and excited state SCF calculation followed by a photo-absorption cross section XANES/ELNES calculation for the K edge.
olcao -wave	Perform SCF calculation as above followed by a PSCF calculation to evaluate the wave function or charge density on a numerical mesh.
olcao -test	Perform SCF calculation as above except that an alternate set of executables for testing purposes will be used.
olcao -spinpol -bond	Perform a spin polarized SCF calculation followed by a spin polarized PSCF calculation of the bond order and Mulliken effective charge.
olcao -help	Print a helpful set of instructions on how to use the olcao program.

the PSCF part with higher resolution by increasing the number of k points. This division provides considerable flexibility because the PSCF part can be repeated quite quickly for different purposes, with different levels of accuracy (by changing k points), and with different computational resource requirements for memory and intermediate disk space. As an example, the choice of basis, which plays an important role in the OLCAO method, can be altered between the SCF and PSCF parts. The SCF part is commonly performed using the so-call full basis (FB) (see Appendix) while bond order and Q* PSCF calculations, which use the Mulliken scheme, are always performed with a so-called minimal basis (MB). In the case of optical properties calculations the SCF part is by default performed with a FB but in the PSCF part the higher energy

Table C.3.2. OLCAO Fortran 90 programs and their purposes.

Fortran 90 program	Purpose
setup	Compute multi-center Gaussian integrals for the overlap, kinetic energy, nuclear Coulomb potential, and electronic Coulomb potential for as yet unknown electronic potential coefficients for all k points; Evaluate core charge and cast into Gaussians; Prepare the exchange-correlation mesh; Prepare Ewald summation terms for long range Coulomb interaction.
main	Construct the secular equation from a given set of electronic potential coefficients and evaluate it; Populate the computed electron levels; Fit the electron charge to Gaussians; Evaluate the exchange-correlation energy and potential; Construct a new set of electronic potential function coefficients for the next SCF iteration. Can also be used for properties evaluation such and bond order and density of states.
intg	Compute multi-center Gaussian integrals for the overlap, kinetic energy, nuclear Coulomb potential, electronic Coulomb potential, and optionally momentum matrix elements for a known set of electronic potential coefficients, but undefined set of k points.
band	Construct and evaluate the secular equation for a given set of k points. Can be used to evaluate the symmetric band structure where the k points are chosen as a path between high symmetry points within the Brillouin zone and where explicit evaluation of the wave function states is not necessary.
dos	Evaluate the density of states (DOS) and atom, orbital, and spin-resolved partial DOS (PDOS). Results may be in terms of species or individual atoms. Also obtain the localization index.
bond	Evaluate the bond order between all atomic pairs; Evaluate the effective charge for each atom. Both methods are based on the Mulliken scheme.
optc	Compute transition probabilities between select sets of occupied and unoccupied states. Used for valence band optical properties, XANES/ELNES spectroscopy, and electrical conductivity $\sigma(E)$ calculations.
wave	Evaluate the charge density, specific states of the wave function, or contributions of specific atoms or groups of atoms over a three-dimensional real-space mesh. Can produce data for easy plotting with OpenDX or other visualization programs or styles.

excited states are computed more accurately with a so-called extended basis (EB). The "setup" and "main" programs are the two members of the SCF part while all the other programs are members of the PSCF part.

The basis designations can be understood by the following simple rule of thumb. A minimal basis for an atom consists of all occupied orbitals. A full basis consists of the minimal basis plus one more shell of unoccupied orbitals. An extended basis consists of the full basis plus one further shell of unoccupied orbitals. This rule of thumb can be altered as necessary for different calculations to include more orbitals of different angular momentum character. For example, the minimal basis for Si contains 1s, 2s, 3s, 2p, and 3p orbitals. A full basis for Si will have the MB plus 3s, 4p, and 3d orbitals, but could contain only 3s and 4p with little loss in accuracy and some savings in the computational burden.

Each of the programs in the OLCAO suite has a specific relation to the other programs in terms of how the output of one program is used as input

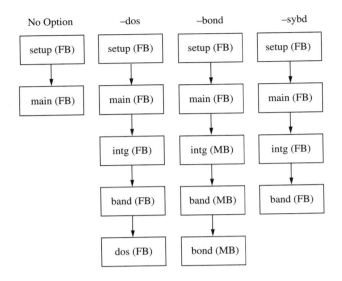

Fig. C.3.2.
Execution flow for no option and dos, bond, and sybd options.

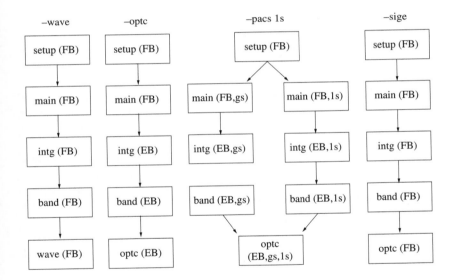

Fig. C.3.3.
Execution flow for wave, optc, pacs, and sige options.

for another. The exact sequence of programs executed depends on the type of calculation requested. Additionally, the basis set (MB, FB, or EB) used for each phase of a calculation is assigned by default but can be overridden by options given on the olcao script's command line. Figures C.3.2 and C.3.3 show the flow of Fortran 90 programs that are used for the various olcao script options. The default basis set is indicated in parenthesis where MB = minimal basis; FB = full basis; and EB = extended basis. A complicating factor can be seen in Fig. C.3.3 where an excited state calculation is performed for the photo-absorption cross-section (PACS) calculation to obtain XANES/ELNES

Table C.3.3. Dominant naming convention for computed OLCAO files.

Component	Description
xx_yyyy-zz.aaa.bbb	Complete Form
xx	Name prefix defining the core excitation state this file is associated with. Valid values proceed from gs (ground state) to 1s, 2s, 2p, 3s, 3p, 3d, ...
yyyy	Primary name for the nature of the computation that was done to obtain the contents of this file. Valid values are setup, main, scfV, intg, band, sybd, bond, dos, optc, pacs, sige, and wave. Some programs are responsible for multiple types of output (e.g. The Fortran 90 band program produces band and sybd output forms depending on the command line request).
zz	Name suffix defining the basis set used. Valid values are mb, fb, and eb.
aaa	Secondary name used to distinguish different types of computed data from one calculation. From the optc program: elf, eps1, eps2, eps1i, cond, and refl. From the dos program: t, p, and li for total and partial DOS and the localization index.
bbb	Suffix used to identify the type of data this file holds. Valid values are out, dat, plot, dx, hdf5, and raw.

spectra. The program flow for that case shows that a 1s core electron has been moved to the conduction band during a second SCF process to produce a new SCF potential that is different from the ground state potential. Note that in Fig. C.3.3 the PACS flow diagram is shown for the K edge case only, but, other edges would be similar with the notation that gs = ground state and 1s = K edge excited state. It is also important to see that the optc program is responsible for three different types of calculations (valence band optical, PACS, and σ(E) electric conductivity).

Because there are many input, output, and results files, there is naturally a need for a succinct and clear naming convention for all OLCAO files such that users know how each file was produced, what it contains, and what its purpose is. The general form for the name of an OLCAO file is xx_yyyy-zz.aaa.bbb where each component of the file name is used to help describe the nature of the file. The naming convention is detailed in Table C.3.3 and examples are given in Table C.3.4. These file names are the formal designations for files mentioned in the next discussion and set of figures. Naturally, there are a small number of exceptions to the naming convention that exist but they can be treated on a case by case basis when needed.

Each of the Fortran 90 programs that the olcao script executes requires specific input files and command line parameters. Each program then produces specific required output files and other sometimes optional output files depending on the command line parameters. Figures C.3.4 through C.3.11 show the input files, output files, optional output files, and command line parameters of each Fortran 90 program as managed by the olcao script. For each of the figures (C.3.4 through C.3.11) the executable programs are in rectangles, text files are in ellipses, command line options are in cross boxes, intermediate data files are in rhombi. Input is shown entering the program from the top, output is

Table C.3.4. Example names of files from OLCAO calculations.

File name	Description
gs_main-fb.out	Output file for a full basis ground state SCF calculation.
gs_scfV-fb.dat	Converged potential produced from a ground state SCF calculation.
1s_main-eb.out	Output file for a full basis 1s excited state SCF calculation.
gs_dos-fb.t.plot	Total density of states from the dos program. Can be plotted directly.
gs_dos-fb.p.raw	Partial density of states from the dos program that must be post-processed by the user before plotting.
gs_optc-eb.eps2.plot	Real part of the optical dielectric function containing the total and xx, yy, zz components from the optc program. Can be plotted directly.
gs_sybd-fb.plot	Symmetric band structure from the band program when the -sybd option is given to the olcao script. Can be plotted directly.

shown flowing to the bottom, and command line parameters enter from the right side. All optional components are drawn with dashed symbols. Optional output is shown exiting the executable to the left while optional input is shown entering the executable from the top along with required input. The command line parameters control what optional files are needed or produced and typically they each take the form of a single integer number. All of the computed results produced by the "setup" calculation are stored in a single compressed file using the hierarchical data format version 5 (HDF5) library.

A variety of electronic structure results can optionally be obtained by the "main" program as illustrated in Fig. C.3.5. The most useful result of main is the file that contains the SCF converged potential function. This file also contains the electron charge (total and valence) as cast into the same Gaussians that are used for the potential function. For spin-polarized calculations it contains the spin-density function also cast into the same Gaussians. The next most useful file is the iteration log because this provides the user with a simple way to observe the success of the SCF iterations and the converged total energy. The "BO + Q*" oval represents the computation of bond order and effective charge data. The "Excited QN_n,l" is a pair of integer numbers for the command line defining the "n" and "l" quantum numbers identifying which orbital of a target atom should be excited if a XANES/ELNES type calculation

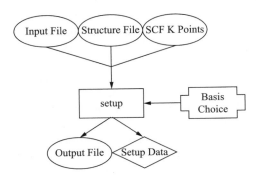

Fig. C.3.4.
Schematic of the "setup" program.

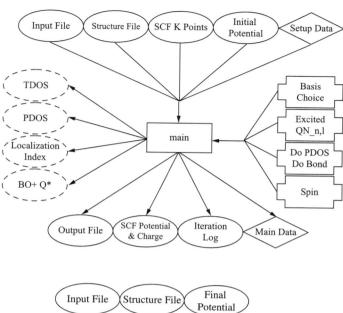

Fig. C.3.5.
Schematic of the "main" program.

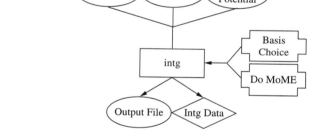

Fig. C.3.6.
Schematic of the "intg" program.

is being performed. The "Spin" command line parameter is an integer that can be 1 for non-spin-polarized or 2 for spin-polarized calculations.

Figure C.3.6 details the "intg" Fortran 90 program. The "Do MoME" option is a request to also include the momentum matrix element interaction integrals in the calculation. This will increase the amount of computed data by a factor of about 2.5 but it will not substantially increase the computation time because they are two center integrals that are performed in conjunction with the other integrals. A key aspect of this computed result is that it is k point independent.

The input, output, and command line parameters for the "band" Fortran 90 program are shown in Fig. C.3.7. Note that the primary result can take on one of two forms. Either the band structure can be produced as dispersion curves on specific high symmetry k point lines and then organized for plotting, or the band structure can be evaluated on the usual Monkhorst–Pack k point mesh and stored in HDF5 form for use by other later programs. The "Excited QN_n,l" and "Spin" shown here have the same meaning as in Fig. C.3.5.

The schematic for the input, output, and command line parameters for the "dos" Fortran 90 program are presented in Fig. C.3.8. Note the designation

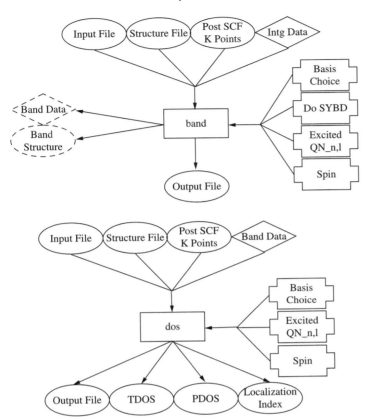

Fig. C.3.7.
Schematic of the "band" program.

Fig. C.3.8.
Schematic of the "dos" program.

"raw" for the PDOS file. This indicates that the data contained in this file is unsuitable for direct plotting and that it requires some level of post-processing. Depending on the input to the dos program the raw file may contain the PDOS of individual atoms resolved down to the l or m quantum numbers. As with the band calculation, the "Excited QN_n,l" and "Spin" shown here have the same meaning as in Fig. C.3.5.

Fig. C.3.9 shows the input, output, and command line parameters for the "bond" Fortran 90 program. Again, the "raw" designation indicates that the file contains information unsuitable for direct plotting and that it must be post-processed before it can be plotted in the desired way. The "Excited QN_n,l" and "Spin" shown here have the same meaning as in Fig. C.3.5.

Fig. C.3.10 presents a schematic of the input, output, and command line parameters for the "optc" Fortran 90 program. The most important command line option is "State Set Code." This will define the set of initial and final states that transition probabilities will be computed over and in effect will distinguish a standard optical properties calculation from a XANES/ELNES type calculation from other possible calculation types to be implemented in the future. The "Serial xyz" option is a toggle switch that will cause the computation to perform the xx, yy, and zz component calculations in serial

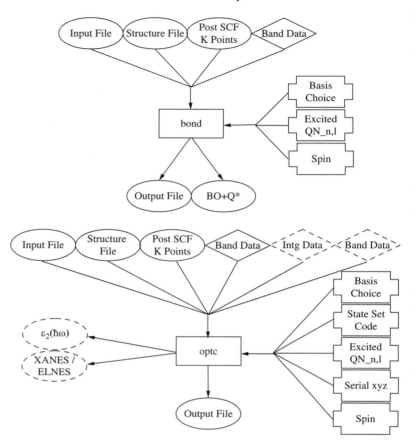

Fig. C.3.9.
Schematic of the "bond" program.

Fig. C.3.10.
Schematic of the "optc" program.

as opposed to simultaneously so as to save memory at a slight expense in computation time. The "Excited QN_n,l" and "Spin" shown here have the same meaning as in Fig. C.3.5.

The final Fortran 90 program is "wave" and this is presented in Fig. C.3.11 as a schematic showing the input, output, and command line parameters. The term "ODX" represents the program Open DX and so the files listed with ODX are in a format for use by Open DX. The "Excited QN_n,l" and "Spin" shown here have the same meaning as in Fig. C.3.5.

One distinguishing feature of the OLCAO suite is that its operation is divided between two separate directories in the file system. The primary directory, or so-called project directory, is typically located within the users $HOME part of the directory tree. This directory will contain all the program input files and key results files. However, the actual execution of the Fortran 90 programs takes place in a separate directory. This other directory is linked to by a soft link named "intermediate" which is created by the olcao script in the project directory. The intermediate directory is typically rooted to a different disk partition where scratch or temporary work can be done. Most high performance computing centers will separate a user's home directory from the place where

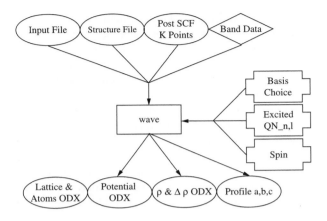

Fig. C.3.11.
Schematic of the "wave" program.

production calculations should be performed. The available space on the home directory is often small compared to the space available on a scratch disk. However, the home directory is often automatically backed up while the scratch work area is not. Further, in relation to the compute nodes where the calculation is done, the scratch disk may have much better performance characteristics than the home disk. The OLCAO suite is well designed to fit this paradigm because the results of interest are typically small data files that can be kept in a user's home partition while the large intensive computations are done on the high performance scratch disk. The naming convention for files inside the intermediate directory follows the naming convention discussed previously for files in the project directory with one additional characteristic. The Fortran 90 unit number that is used to access a particular file is appended to the end of the file name. For example, a program's standard output file will thus always have a ".20" appended to it and the raw PDOS output file with have a ".70" appended to it. The choice of number itself is arbitrary and has little meaning outside of the source code.

The one exception to this intermediate file naming convention concerns the set of files that store so-called intermediate results. These results are typically not of direct interest for analysis but rather are used by subsequent programs to produce results that are of interest. The most straightforward example is that of the computed results from the setup program being used by the main program to obtain a self-consistent potential function. All OLCAO programs that produce intermediate data files use version 5 of the hierarchical data format (HDF5) library for storing their computed results. The reasons for using this particular storage method are numerous but the most important ones are to conserve space on the scratch disk and to organize the data for easy understanding and access both within the program source code and by possible third party or supplementary programs. The HDF5 library provides the ability to store large amounts of disparate blocks of data in a compressed format and with an internal structure to the file. Each program in the OLCAO suite that produces intermediate data uses its own form of HDF5 file. Gaining access to this data from a program outside of the OLCAO suite can be useful for many reasons (e.g. third party add-on development) and to that end it is necessary to understand the structure of those files. Figures C.3.12 through C.3.15 show schematic diagrams of the structure of the HDF5 files produced by each of the OLCAO Fortran 90 programs that use them. The HDF5 files used in OCLAO are organized akin to a directory tree in the UNIX file system. There are groups (represented by circles) that are analogous to directories (with the "/" representing the root group) and data sets (represented by blocks) that are analogous to files. Depending on the type of calculation, some data sets or groups may not be present and this is represented by a dashed arrow from the group to the data set in question. An example of this would be the presence of dashed arrows to all the imaginary components of the interaction integral matrices. The dashes exist because a calculation with one k point at the Γ site will produce interaction matrices that are real and hence no imaginary valued data sets are necessary.

The "setup" HDF5 file (Fig. C.3.12) contains three main groups with some subgroups and data sets as leaves in the tree structure. The main groups are the inter-atomic interaction matrices that are the results of multi-center (2 or

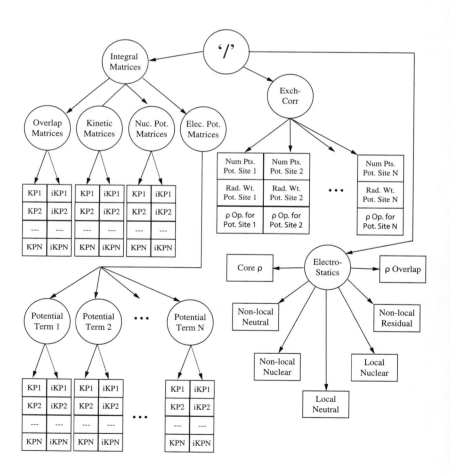

Fig. C.3.12.
Structure of the "setup" HDF5 file.

3 center) Gaussian integrals, the description of the exchange-correlation mesh, and the long range electrostatic interactions derived from Ewald summation. The integral matrices are subdivided into four groups for the overlap, kinetic energy, nuclear–electron potential, and electron–electron potential interactions. Within the electron–electron potential group are further subgroups, one for each term in the potential function. One dataset exists for each real and imaginary contribution for each k point. In the case of a Γ k point calculation, all the interaction matrices are real and so the imaginary data sets will not be created. The matrices are stored in packed format because they are Hermitian. The exchange-correlation mesh is defined using atom-centered spherical meshes. Each atomic site has a record of the number of points, radial weighting factor, and a matrix representing the contribution of all atomic sites to the mesh points associated with this site. The electrostatic interactions group consists of many different components.

The "main" HDF5 file (Fig. C.3.13) contains only the eigenvalues and eigenvector solutions of the secular equation for each k point. In the case of a Γ k point calculation the imaginary component is not necessary. Indeed, for any one k point calculation the eigenvectors are not saved until the computation has converged. The word "spin" used in some of the blocks indicates that these data sets are used only in the case of spin-polarized calculations. For spin polarized calculations the data sets without the word spin will hold spin-up data while the datasets labeled with the word spin here will hold spin-down data.

The "intg" HDF5 file (Fig. C.3.14) is the most complicated. The symbol definitions used here are the same as in Fig. C.3.12 with the addition of a triangle to represent a so-called data attribute and the modification that the dimensions of the boxes carry some meaning. The data attribute of the XX-Momentum group is used as a simple flag to indicate whether the momentum matrix elements have been computed or not. This is necessary because the momentum matrix elements are not typically computed unless some form of optical properties calculation is being done. The box height decreases for each successive atom because there is no double counting. The box width varies

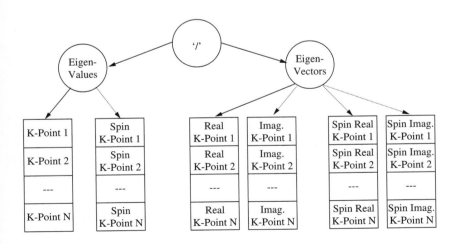

Fig. C.3.13.
Structure of the "main" HDF5 file.

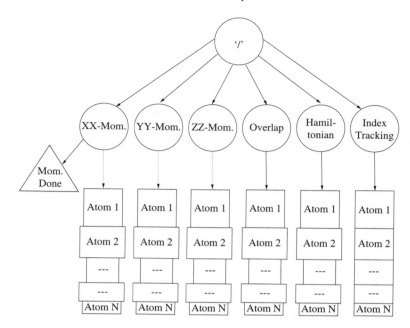

Fig. C.3.14.
Structure of the "intg" HDF5 file.

because different elements require a different number of Gaussian terms in their basis functions. The stored data sets are the interactions between specific atoms and all other atoms.

The "band" HDF5 file (Fig. C.3.15) contains the eigenvalues and eigenvector solutions of the secular equation for each k point. In addition, the overlap matrix for each k point and the orthogonalization coefficients for each k point are also stored. The overlap matrix is needed for partial charge analysis as

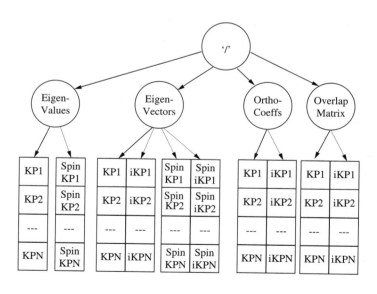

Fig. C.3.15.
Structure of the "band" HDF5 file.

part of Mulliken analysis while the orthogonalization coefficients are often needed for orthogonalization of the momentum matrix elements. In the case of a Γ k point calculation the imaginary components of these matrices are not necessary. Note that the overlap matrix is stored in packed format because it is Hermitian while the other large data sets are not stored this way.

C.4 Results analysis

One of the key strengths of the OLCAO method is its ability to allow for direct and intuitive analysis of a solid state system's wave function (and hence all computed properties) in terms of the system's atomic constituents down to the level of their n, l, m, and s quantum numbers. This ability, of course, derives from the use of an atomic orbital basis set but it also depends on many flexible and powerful auxiliary scripts and programs that are used to manage the OLCAO output data for plotting, visualization, and analysis. In this section of the appendix these scripts and programs will be explored with an eye toward providing a flavor of the types of analysis and processing that is possible in the OLCAO suite. The information presented here should be viewed as the unifying scheme of data analysis in OLCAO rather than as a comprehensive manual. Each of the scripts and programs in the suite has its own section in the OLCAO suite's manual that describes its purpose, functionality, and operation. Additionally, the documentation internal to the source code contains substantial instruction about particular methods that are used. Readers who are interested in detailed aspects of these programs are referred to those sources of information.

The OLCAO program suite produces a variety of output file formats depending on the type of calculation that was done and to a small extent on parameters in the olcao.dat input file. The results can be divided into three general categories: raw, plot, and dx. The terms "raw," "plot," and "dx" refer to the final component in the data file's name and are meant to indicate what can be done with the file. Files with the "dx" extension contain data that can be displayed in three dimensions using the third party Open DX program. This data is either from the evaluation of some function (e.g., charge density or electronic potential) on a three-dimensional mesh or it is used to display characteristics of the system's atomic structure through the use of icons or glyphs in three dimensions (e.g., spheres of various colors and sizes to represent atoms and their effective charge.) Files with the "plot" extension contain data that typically represent some numerical function that can be plotted on Cartesian axes (e.g., total DOS, optical spectra, band structure). The "plot" files contain data organized as columns with headings and they can be viewed immediately with a combination of the third party gnuplot program and a simplifying interface script called plotgraph that serves as an automating agent for gnuplot. In some cases, the immediate output results of the OLCAO program suite are not suitable for direct interpretation or plotting and are considered to be "raw" data. These data files typically have a complex internal structure that needs to be parsed by an auxiliary script before it can be plotted.

The auxiliary scripts have the cross-purpose goals of making extraction and organization of a typical subset of data painless while making extraction and organization of a highly complex subset of data possible. The three most commonly used scripts in this regard are "makePDOS," "makeBOND," and "makeSYBD." The makePDOS script is used to post-process the raw partial density of states data produced by OLCAO. It is important to note that the data produced by OLCAO may already have some level of organization to it depending on the option settings at the end of the olcao.dat input file. For example, the output data may contain the complete DOS of every single orbital of every atom or it may contain the DOS of only the uniquely identified atomic types resolved into orbitals of distinct n and l quantum number. The choice of what content the raw data file contains is defined by a flag in the olcao.dat input file. The makeBOND program can be used to analyze the bond order and effective charge data in a number of different ways including scatter plots, three-dimensional visualization in Open DX with glyphs, simple tables, and bond order averages over different axes. As with the raw PDOS file, this file is not designed to be easily read by humans. The last of the raw data analyzers is the makeSYBD script. Because of its limited options it is run automatically after any symmetric band structure calculation so that the band structure along a high symmetry path in the irreducible portion of the Brillouin zone can be plotted immediately with gnuplot. As such, the raw data file is typically not ever encountered by the user.

An important capability of the OLCAO method is the ability to visualize a number of functions of space evaluated on a three-dimensional mesh. This includes the wave function (resolved into particular states and spin directions), the valence charge density derived from the wave function, the fitted charge density divided into core and valence components, and the electronic potential function. These data can also be compared to data constructed from the neutral atom system. In this way changes such as charge transfer and bond formation due to the interacting nature of the atoms can be observed. This type of data is stored in the OpenDX file format and can be viewed with an OpenDX program included with the OLCAO suite. This same OpenDX program can be used to visualize data produced by the makeBOND post-analysis program so that the results can be overlaid with each other.

A key goal of the OLCAO suite is to be a computational scheme that provides tools for the user from importing or developing a model, through the actual calculation, into post-calculation analysis, and finally to data visualization and presentation. There exist a number of programs at each phase to make the transition between stages of research as smooth as possible with many interconnections to other popular third party programs for maximum efficiency and utility.

Appendix D
Examples of Computational Statistics

As with many programs, the efficiency of the OLCAO method is dependent on a large number of factors such as the number of atoms and their elemental makeup, the number of atoms with crystallographic equivalence or near-equivalence, the number of k points, the convergence criterion, and the size of the basis set used among others. Importantly, the rate at which the program is able to complete its assigned task is not the only factor that is critical to its overall performance. Consumption of other resources such as main memory and hard disk space will also play an important role in determining whether or not a certain calculation can be accomplished within the constraints of the hardware applied to the task. To provide a feeling for the capabilities of the OLCAO program and the rate at which it consumes resources we have compiled a few statistics for different types of systems and calculations. The five systems that were selected for this brief analysis are α-B_{12}, hydroxyapatite, a ten base-pair model of b-DNA with all cytosine–guanine pairs, a model of crystalline herapathite, and a model of an inter-granular glassy film between the prismatic planes of crystalline β-Si_3N_4. Certain structural and input data for these systems are presented in Table D.1.

There are three important parameters in Table D.1 that are particularly important for determining the computational cost of the system. The first is the valence dimension. This number is equal to the number of terms in the basis expansion of the states of the solid state wave function and its value will strongly affect the amount of disk space required and the amount of time needed for any calculation. The duration of the diagonalization procedure used in the solution of the secular equation depends on this value to the third power. All the interaction integrals are Hermitian matrices with dimension equal to the valence dimension. This leads to the second parameter which is the potential dimension. This value defines the number of terms in the function used to describe the electronic potential and the fitted charge. As this value increases, the number of three center interaction integral matrices also increases. Thus, if there are a large number of matrices and the matrices themselves are large, then the computation will take a longer time. The final term is the number of k points. When this is equal to 1 and the Γ point is chosen, then all the matrices are real instead of complex. This allows for much faster operations such as multiplication and diagonalization and it can thus greatly increase the

Table D.1. Summary of structural and input data for electronic structure calculations of five systems.

	$\alpha - B_{12}$	HAP	DNA	Herapathite	IGF
Lattice Vector Magnitudes	a = 5.0574Å	a = 9.4302Å	a = 30.000Å	a = 15.247Å	a = 14.533Å
	b = 5.0574Å	b = 9.4302Å	b = 30.000Å	b = 18.885Å	b = 15.225Å
	c = 5.0574Å	c = 6.8911Å	c = 38.800Å	c = 36.183Å	c = 47.420Å
Lattice Angles	α = 58.055°	α = 90.0°	α = 90.0°	α = 90.0°	α = 90.0°
	β = 58.055°	β = 90.0°	β = 90.0°	β = 90.0°	β = 90.0°
	γ = 58.055°	γ = 120.0°	γ = 90.0°	γ = 90.0°	γ = 90.0°
Cell Volume	87.39Å^3	530.71Å^3	30420.00Å^3	10418.69Å^3	10492.78Å^3
# of Atoms	12	44	650	998	907
# of Electrons	36	268	2220	2864	4288
Valence	48 (MB)	220 (MB)	1940 (MB)	2512 (MB)	3628 (MB)
Dimension	96 (FB)	476 (FB)	4740 (FB)	6644 (FB)	9111 (FB)
	144 (EB)	842 (EB)	8970 (EB)	14196 (EB)	15090 (EB)
Potential Dimension	32 Terms	134 Terms	788 Terms	1006 Terms	432 Terms
# of k points	60 SCF	23 SCF	1(Γ) SCF	1(Γ) SCF	1(Γ) SCF
	110 PostSCF	56 PostSCF	8 PostSCF	1(Γ) PostSCF	1(Γ) PostSCF

completion rate. For other quantities of k points it essentially becomes a linear multiplier on the expense of the calculation in terms of time and disk space. For the OLCAOsetup calculation, the number of k points is also a multiplier for the amount of main memory required.

The calculations for these material systems were performed on a computer managed by the University of Missouri Bioinformatics Consortium. This computer is an SGI Altix 3700 BX2 with 64 1.5 GHz Itanium2 processors in a symmetric multi-processor configuration. There are 128 gigabytes of shared memory and 4 terabytes worth of disk storage via an SGI TP9500 InfiniteStorage RAID system. The operating system is SuSE Linux Enterprise Server 11 with SGI ProPack 6.

Each of the material systems was used to perform a series of computations with the OLCAO package. The sequence of programs that was run is considered rather typical for electronic structure calculation. The first two programs (OLCAOsetup and OLCAOmain) are used to compute a converged SCF electronic potential and then the remaining programs are used for computing specific electronic structure results (post-SCF). A detailed explanation for this division and other aspects of the OLCAO program execution is provided in Appendix C Section 3. The post-SCF calculations can be run for different numbers of k points than the SCF calculation or with a different basis set (full, minimal, or extended). The results for full basis SCF calculations and post-SCF calculations in a number of basis sets for a few different electronic structure results are presented in Table D.2.

While this information does provide clear evidence of the degree of the increased cost of larger calculations, the primary purpose is not to form a rigorous set of scaling laws for changes in input parameters. Rather the goal is to establish a broad sense of the cost of a typical series of calculations for a set of typical systems of interest. For the OLCAOsetup portion, changes in

Table D.2. Performance in minutes for different calculation components of five material systems.

	α-B$_{12}$	HAP	DNA	Herapathite	IGF
Setup (FB)	2	17	1210	2067	4618
Main (FB)	1	35	1106	3972	5599
Intg (MB)	1	3	61	314	365
Intg (FB)	1	3	61	314	366
Intg (EB)	1	3	–	411	344
Band (MB)	<1	1	24	3	9
Band (FB)	1	3	340	47	139
Band (EB)	1	20	–	444	542
PDOS (FB)	<1	1	113	9	30
Bond (MB)	<1	1	197	73	90
Optc (EB)	1	14	–	214	178

the valence dimension from one system to another have some impact, but a stronger effect is likely the increased potential dimension. This is also demonstrated by the stable times for different basis sets for the integral calculation where the potential dimension parameter plays no role. Interestingly, the variations in times for different basis sets for the integral calculation are probably due to machine load. The effect of the $O(N^3)$ scaling for the diagonalization process that is used in OLCAOmain and OLCAOband is quite evident for all the calculations. For the most part, the actual electronic structure calculations (PDOS, Bond, and Optc) tend to represent a small portion of the overall cost of the calculation.

Index

Note: Greek letters are spelled out (alpha, beta, gamma, kappa)

Index

Index

Index